The Biology of Crustacea

VOLUME 8

ENVIRONMENTAL ADAPTATIONS

The Biology of Crustacea

Editor-in-Chief

Dorothy E. Bliss

Department of Invertebrates
The American Museum of Natural History
New York, New York*

*Present address: Brook Farm Road, RR5, Wakefield, Rhode Island 02879

The Biology of Crustacea

VOLUME 8

Environmental Adaptations

Edited by

F. JOHN VERNBERG

Belle W. Baruch Institute for Marine Biology
and Coastal Research
University of South Carolina
Columbia, South Carolina

Winona B. Vernberg

College of Health
University of South Carolina
Columbia, South Carolina

ACADEMIC PRESS 1983

A Subsidiary of Harcourt Brace Jovanovich, Publishers

New York London
Paris San Diego San Francisco São Paulo Sydney Tokyo Toronto

ACADEMIC PRESS, INC.
111 Fifth Avenue, New York, New York 10003

United Kingdom Edition published by
ACADEMIC PRESS, INC. (LONDON) LTD.
24/28 Oval Road, London NW1 7DX

Library of Congress Cataloging in Publication Data
Main entry under title:

The Biology of crustacea.

Includes bibliographies and indexes.
Contents: v. 8. Environmental adaptations.
Dorothy E. Bliss, editor-in-chief / edited by
F. John Vernberg and Winona B. Vernberg.
1. Crustacea. I. Bliss, Dorothy E. II. Vernberg, F. John
III. Vernberg, Winona B. IV. Series.
QL435.B48 595.3 82-4058
ISBN 0-12-106408-5 (v. 8)

PRINTED IN THE UNITED STATES OF AMERICA

83 84 85 86 9 8 7 6 5 4 3 2 1

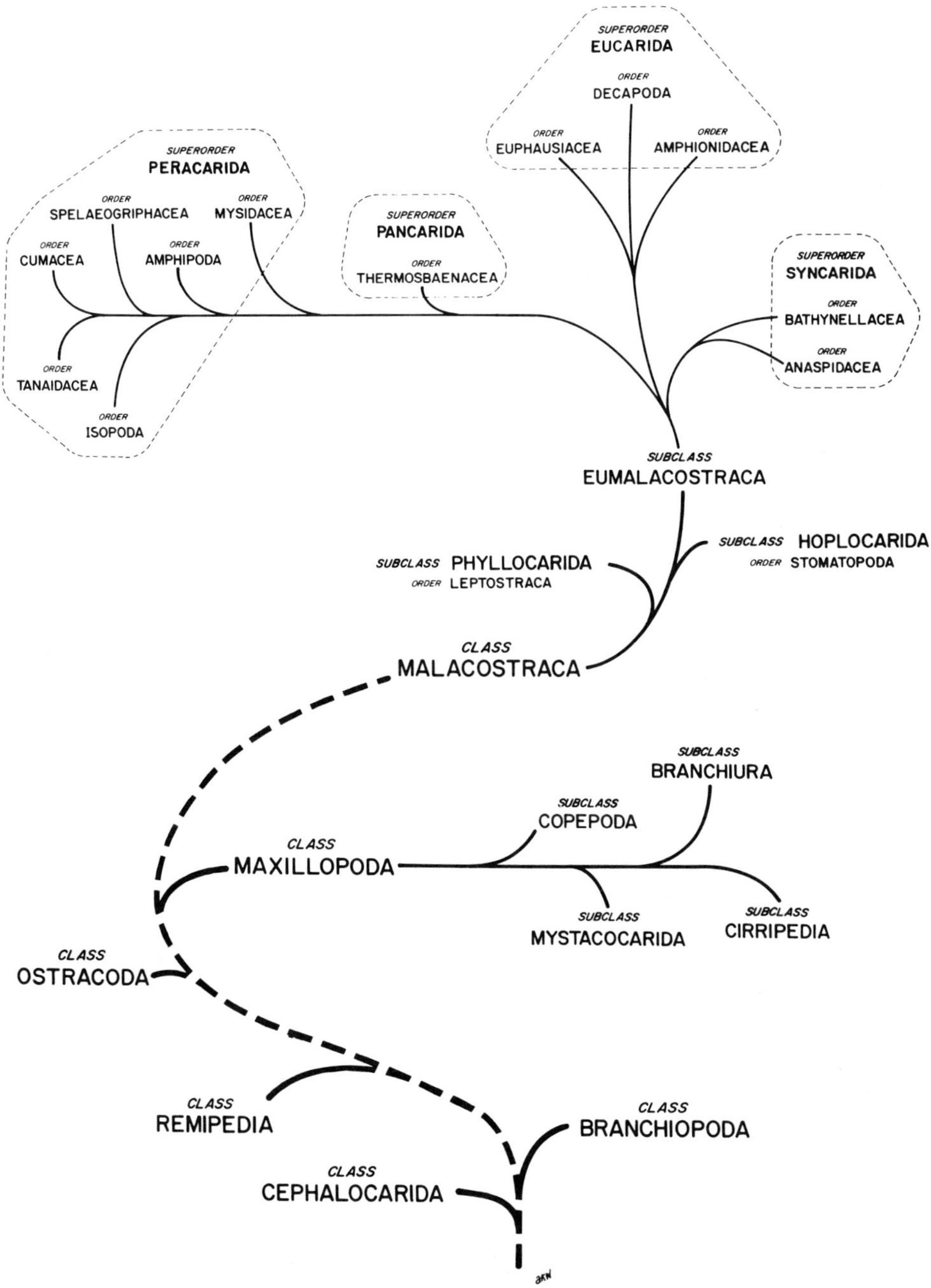

A visual representation of the Bowman and Abele classification of Crustacea (see Chapter 1, Volume 1). This is not intended to indicate phylogenetic relationships and should not be so interpreted. The dashed line at the base emphasizes uncertainties concerning the origin of the five classes and their relationship to one another.

Contents

6. Terrestrial Adaptations

LAWRENCE W. POWERS AND DOROTHY E. BLISS

7. Freshwater Adaptations

WINONA B. VERNBERG AND F. JOHN VERNBERG

List of Contributors

Numbers in parentheses indicate the pages on which the authors' contributions begin.

Dorothy E. Bliss[1] (271), Department of Invertebrates, The American Museum of Natural History, New York, New York 10024

James N. Cameron (43), The University of Texas at Austin, and Port Aransas Marine Laboratory, Port Aransas,Texas 78373

R. Gilles (109), Laboratory of Animal Physiology, University of Liege, B-4020 Liege, Belgium

John Grahame[2] (65), Wellcome Marine Laboratory, University of Leeds, Robin Hood's Bay, North Yorkshire, England

Charlotte P. Mangum (43), Department of Biology, College of William and Mary, Williamsburg, Virginia 23185

A. Pequeux (109), Laboratory of Animal Physiology, University of Liege, B-4020 Liege, Belgium

Lawrence W. Powers[3] (271), Department of Biology, City College of New York, New York, New York 10031

A. N. Sastry (179), Graduate School of Oceanography, University of Rhode Island, Kingston, Rhode Island 02881

F. John Vernberg (1, 335), Belle W. Baruch Institute for Marine Biology and Coastal Research, University of South Carolina, Columbia, South Carolina 29208

Winona B. Vernberg (335), College of Health, University of South Carolina, Columbia, South Carolina 29208

[1]Present address: Brook Farm Road, RR 5, Wakefield, Rhode Island 02879
[2]Present address: Pure and Applied Zoology, University of Leeds, Leeds LS2 9JT, England
[3]Present address: 109 South Snow Road, Semmes, Alabama 36575

General Preface

In 1960 and 1961, a two-volume work, "The Physiology of Crustacea," edited by Talbot H. Waterman, was published by Academic Press. Thirty-two biologists contributed to it. The appearance of these volumes constituted a milestone in the history of crustacean biology. It marked the first time that editor, contributors, and publisher had collaborated to bring forth in English a treatise on crustacean physiology. Today, research workers still regard this work as an important resource in comparative physiology.

By the latter part of the 1970s, need clearly existed for an up-to-date work on the whole range of crustacean studies. Major advances had occurred in crustacean systematics, phylogeny, biogeography, embryology, and genetics. Recent research in these fields and in those of ecology, behavior, physiology, pathobiology, comparative morphology, growth, and sex determination of crustaceans required critical evaluation and integration with earlier research. The same was true in areas of crustacean fisheries and culture.

Once more, a cooperative effort was initiated to meet the current need. This time its fulfillment required eight editors and almost 100 contributors. This new treatise, "The Biology of Crustacea," is for scientists doing basic or applied research on various aspects of crustacean biology. Containing vast background information and perspective, this treatise will be a valuable source for zoologists, paleontologists, ecologists, physiologists, endocrinologists, morphologists, pathologists, and fisheries biologists, and an essential reference work for institutional libraries.

In the preface to Volume 1, editor Lawrence G. Abele has commented on the excitement that currently pervades many areas of crustacean biology. One such area is that of systematics. The ferment in this field made it difficult for Bowman and Abele to prepare an arrangement of families of Recent Crustacea. Their compilation (Chapter 1, Volume 1) is, as they have stated, "a compromise and should be until more evidence is in." Their arrangement is likely to satisfy some crustacean biologists, undoubtedly not all. Indeed, Schram (Chapter 4, Volume 1) has offered a somewhat different arrangement. As generally used in this treatise, the classification of Crustacea follows that outlined by Bowman and Abele.

Selection and usage of terms have been somewhat of a problem. Ideally, in a treatise, the same terms should be used throughout. Yet biologists do not agree on certain terms. For example, the term *ostracode* is favored by systematists and paleontologists, *ostracod* by many experimentalists. A different situation exists with regard to the term *midgut gland,* which is more acceptable to many crustacean biologists than are the terms *hepatopancreas* and *digestive gland.* Accordingly, authors were encouraged to use *midgut gland.* In general, however, the choice of terms was left to the editors and authors of each volume.

In nomenclature, consistency is necessary if confusion as to the identity of an animal is to be avoided. In this treatise, we have sought to use only valid scientific names. Wherever possible, synonyms of valid names appear in the taxonomic indexes. Thomas E. Bowman and Lawrence G. Abele were referees for all taxonomic citations.

Every manuscript was reviewed by at least one person before being accepted for publication. All authors were encouraged to submit new or revised material up to a short time prior to typesetting. Thus, very few months elapse between receipt of final changes and appearance of a volume in print. By these measures, we ensure that the treatise is accurate, readable, and up-to-date.

Dorothy E. Bliss

General Acknowledgments

In the preparation of this treatise my indebtedness extends to many persons and has grown with each succeeding volume. First and foremost is the great debt owed to the authors. Due to their efforts to produce superior manuscripts, unique and exciting contributions lie within the covers of these volumes.

Deserving of special commendation are authors who also served as editors of individual volumes. These persons have conscientiously performed the demanding tasks associated with inviting and editing manuscripts and ensuring that the manuscripts were thoroughly reviewed. In addition, Dr. Linda H. Mantel has on innumerable occasions extended to me her advice and professional assistance well beyond the call of duty as volume editor. In large part because of the expertise and willing services of these persons, this treatise has become a reality. Also deserving of thanks and praise are scientists who gave freely of their time and professional experience to review manuscripts. In the separate volumes, many of these persons are mentioned by name.

Thanks are due to all members of the staff of Academic Press involved in the preparation of this treatise. Their professionalism and encouragement have been indispensable.

No acknowledgments by me would be complete without mention of the help provided by employees of the American Museum of Natural History, especially those in the Department of Invertebrates and in the Museum's incomparable library.

Lastly, I extend my appreciation and thanks to members of the staff of the Pell Marine Science Library of the University of Rhode Island for their thoughtfulness and generous help.

Dorothy E. Bliss

Preface to Volume 8

All the volumes in this series, The Biology of Crustacea, are interrelated to varying degrees. This volume, Environmental Adaptations, bridges the gap between those volumes which deal in detail with the functional machinery of crustaceans and those dealing with ecological, taxonomic, behavioral, and applied subjects. Our goal is to examine the dynamic interaction between an organism and its external environment in order to emphasize the adaptive significance of its responses. Although each chapter is a discrete entity dealing with a specific functional response, chapters in other volumes are cited where the reader may want to know more about anatomical and biochemical features. For example, the first two chapters in this volume highlight the range of respiratory responses exhibited by crustaceans when subjected to various environmental factors, either singly or in various combinations, whereas chapters of Volume 5 describe the respiratory system in detail. In Chapter 1 in this volume, F. J. Vernberg reviews the responses of the intact organism, as well as of some cellular systems, to biotic and abiotic factors. Cameron and Mangum (Chapter 2) deal more specifically with ventilation, circulation, and oxygen transport.

Energy is vital to biotic systems, and Grahame (Chapter 3) summarizes the mechanisms by which crustaceans take up energy (food) from the environment under fluctuating conditions. The assimilation efficiencies of crustaceans as well as the predator–prey interaction are reviewed. Grahame feels that this is a challenging field in that we have a general framework of knowledge but that the adaptive significance of feeding needs attention.

Gilles and Pequeux (Chapter 4) deal with the complex and voluminous subject of ionic and osmoregulation from the view of its interaction with the environment. They emphasize the adaptive nature of chemical regulation that enables various crustaceans to inhabit waters ranging from hypersaline ponds to fresh water.

The perpetuation of a species is a fundamental biological necessity. How the various species are reproductively adapted to generate offspring in the tremendous diversity of habitats occupied by crustaceans is discussed by Sastry (Chapter 5).

Powers and Bliss (Chapter 6) and W. B. Vernberg and F. J. Vernberg (Chapter 7) report on the specific adaptations of crustaceans to terrestrial and freshwater habitats.

This volume is organized to highlight some of the major current research areas, and we hope that it will stimulate further investigations.

We gratefully acknowledge the invaluable contributions made by our colleagues who critically reviewed all of the chapters. Ms. Nanette C. Baker assisted in many ways in the preparation of this volume, and Ms. Anne B. Miller aided in editing.

F. John Vernberg
Winona B. Vernberg

Classification of the Decapoda*

Order Decapoda Latreille, 1803
Suborder Dendrobranchiata Bate, 1888
Family Penaeidae Rafinesque, 1815, *Penaeus, Metapenaeus, Penaeopsis, Trachypenaeopsis*
Aristeidae Wood-Mason, 1891, *Gennadus, Aristeus*
Solenoceridae Wood-Mason and Alcock, 1891, *Solenocera, Hymenopenaeus*
Sicyoniidae Ortmann, 1898, *Sicyonia*
Sergestidae Dana, 1852, *Sergestes, Lucifer, Acetes*
Suborder Pleocyemata Burkenroad, 1963
Infraorder Stenopodidea Claus, 1872
Family Stenopodidae Claus, 1872, *Stenopus*
Infraorder Caridea Dana, 1852
Family Procarididae Chace and Manning, 1972, *Procaris*
Oplophoridae Dana, 1852, *Oplophorus, Acanthephyra, Systellaspis*
Atyidae De Haan, 1849, *Atya, Caridina*
Nematocarcinidae Smith, 1884, *Nematocarcinus*
Stylodactylidae Bate, 1888, *Stylodactylus*
Pasiphaeidae Dana, 1852, *Leptochela, Parapasiphae*
Bresiliidae Calman, 1896, *Bresilia*
Eugonatonotidae Chace, 1936, *Eugonatonotus*
Rhynchocinetidae Ortmann, 1890, *Rhynchocinetes*
Campylonotidae Sollaud, 1913, *Bathypalaemonella*
Palaemonidae Rafinesque, 1815, *Palaemon, Palaemonetes, Macrobrachium*
Gnathophyllidae Dana, 1852, *Gnathophyllum*
Psalidopodidae Wood-Mason and Alcock, 1892, *Psalidopus*
Alpheidae Rafinesque, 1815, *Alpheus, Synalpheus, Athanus*
Ogyrididae Hay and Shore, 1918, *Ogyrides*
Hippolytidae Dana, 1852, *Hippolyte, Thor, Latreutes, Thoralus, Lysmata*
Processidae Ortmann, 1896, *Processa*
Pandalidae Haworth, 1825, *Pandalus, Parapandalus, Heterocarpus*

**Prepared by Lawrence G. Abele*

Thalassocarididae Bate, 1888, *Thalassocaris*
Physetocarididae Chace, 1940, *Physetocaris*
Crangonidae Haworth, 1825, *Crangon, Pontophilus*
Glyphocrangonidae Smith, 1884, *Glyphocrangon*
Infraorder Astacidea Latreille, 1803
Family Nephropidae Dana, 1852, *Nephrops, Homarus*
Thaumastochelidae Bate, 1888, *Thaumastocheles*
Cambaridae Hobbs, 1942, *Cambarus, Orconectes, Procambarus, Cambarellus*
Astacidae Latreille, 1803, *Astacus, Pacifastacus, Austropotamobius*
Parastacidae Huxley, 1879, *Euastacus, Cherax, Astacopsis, Engaeus*
Infraorder Thalassinidea Latreille, 1831
Family Thalassinidae Latreille, 1831, *Thalassina*
Axiidae Huxley, 1879, *Axius, Calocaris*
Laomediidae Borradaile, 1903, *Naushonia, Jaxea, Laomedia*
Callianassidae Dana, 1852, *Callianassa*
Callianideidae Kossmann, 1880, *Callianidea*
Upogebiidae Borradaile, 1903, *Upogebia*
Axianassidae Schmitt, 1924, *Axianassa*
Infraorder Palinura Latreille, 1903
Family Glypheidae Zittel, 1885, *Neoglyphea*
Polychelidae Wood-Mason, 1874, *Polycheles*
Palinuridae Latreille, 1803, *Panulirus, Palinurus, Jasus*
Scyllaridae Latreille, 1825, *Scyllarus, Scyllarides*
Synaxidae Bate, 1881, *Palinurellus*
Infraorder Anomura H. Milne Edwards, 1832
Family Pomatochelidae Miers, 1879, *Pomatocheles*
Diogenidae Ortmann, 1892, *Paguristes, Dardanus, Diogenes, Clibanarius*
Coenobitidae Dana, 1851, *Coenobita, Birgus*
Lomisidae Bouvier, 1895, *Lomis*
Paguridae Latreille, 1803, *Pagurus*
Lithodidae Samouelle, 1819, *Lithodes, Paralithodes*
Parapaguridae Smith, 1882, *Parapagurus*
Galatheidae Samouelle, 1819, *Galathea, Munida, Pleuroncodes*
Aeglidae Dana, 1852, *Aegla*
Chirostylidae Ortmann, 1892, *Chirostylus*
Porcellanidae Haworth, 1825, *Porcellana, Petrolisthes*
Albuneidae Stimpson, 1858, *Albunea*
Hippidae Latreille, 1825, *Hippa, Emerita*
Infraorder Brachyura Latreille, 1803
Section Dromiacea De Haan, 1833
Family Homolodromiidae Alcock, 1899, *Homolodromia*
Dromiidae De Haan, 1833, *Dromia*
Dynomenidae Ortmann, 1892, *Dynomene*
Section Archaeobrachyura Guinot, 1977
Family Cymonomidae Bouvier, 1897, *Cymonomus*
Tymolidae Alcock, 1896, *Tymolus*
Homolidae De Haan, 1839, *Homola*
Latreilliidae Stimpson, 1858, *Latreillia*
Raninidae De Haan, 1839, *Ranilia, Ranina*

Section Oxystomata H. Milne Edwards, 1834
  Family Dorippidae MacLeay, 1838, *Ethusina, Dorippe*
    Calappidae De Haan, 1833, *Calappa*
    Leucosiidae Samouelle, 1819, *Persephona, Randallia*
Section Oxyrhyncha Latreille, 1803
  Family Majidae Samouelle, 1819, *Maja, Hyas*
    Hymenosomatidae MacLeay, 1838, *Hymenosoma*
    Mimilambridae Williams, 1979, *Mimilambrus*
    Parthenopidae MacLeay, 1838, *Parthenope*
Section Cancridea Latreille, 1803
  Family Corystidae Samouelle, 1819, *Corystes*
    Atelecyclidae Ortmann, 1893, *Atelecyclus*
    Pirimelidae Alcock, 1899, *Pirimela*
    Thiidae Dana, 1852, *Thia*
    Cancridae Latreille, 1803, *Cancer*
Section Brachyrhyncha Borradaile, 1907
  Family Geryonidae Colosi, 1923, *Geryon*
    Portunidae Rafinesque, 1815, *Portunus, Carcinus, Callinectes, Scylla*
    Bythograeidae Williams, 1980, *Bythogrea*
    Xanthidae MacLeay, 1838, *Rhithropanopeus, Panopeus, Xantho, Eriphia, Menippe*
    Platyxanthidae Guinot, 1977, *Platyxanthus*
    Goneplacidae MacLeay, 1838, *Frevillea*
    Hexapodidae Miers, 1886, *Hexapodus*
    Belliidae, 1852, *Bellia*
    Grapsidae MacLeay, 1838, *Grapsus, Eriocheir, Pachygrapsus, Sesarma*
    Gecarcinidae MacLeay, 1838, *Gecarcinus, Cardisoma*
    Mictyridae Dana, 1851, *Mictyris*
    Pinnotheridae De Haan, 1833, *Pinnotheres, Pinnixa, Dissodactylus*
    Potamidae Ortmann, 1896, *Potamon*
    Deckeniidae Bott, 1970, *Deckenia*
    Isolapotamidae Bott, 1970, *Isolapotamon*
    Potamonautidae Bott, 1970, *Potamonautes*
    Sinopotamidae Bott, 1970, *Sinopotamon*
    Trichodactylidae H. Milne Edwards, 1853, *Trichodactylus, Valdivia*
    Pseudothelphusidae Ortmann, 1893, *Pseudothelphusa*
    Potamocarcinidae Ortmann, 1899, *Potamocarcinus*
    Gecarcinucidae Rathbun, 1904, *Gecarcinucus*
    Sundathelphusidae Bott, 1969, *Sundathelphusa*
    Parathelphusidae Alcock, 1910, *Parathelphusa*
    Ocypodidae Rafinesque, 1815, *Ocypode, Uca*
    Retroplumidae Gill, 1894, *Retropluma*
    Palicidae Rathbun, 1898, *Palicus*
    Hapalocarcinidae Calman, 1900, *Hapalocarcinus*

Contents of Previous Volumes

Volume 3: Neurobiology: Structure and Function

Edited by Harold L. Atwood and David C. Sandeman

Volume 4: Neural Integration and Behavior

Edited by David C. Sandeman and Harold L. Atwood

Volume 5: Internal Anatomy and Physiological Regulation
Edited by Linda H. Mantel

Volume 6: Pathobiology

Edited by Anthony J. Provenzano, Jr.

Volume 7: Behavior and Ecology

Edited by F. John Vernberg and Winona B. Vernberg

1

Respiratory Adaptations

F. JOHN VERNBERG

I. INTRODUCTION

A dynamic interaction exists between the functioning of the internal machinery of a crustacean and its external environment. In order for an organism to survive, the various factors constituting the ambient environmental complex must not exceed genetically determined tolerance limits charac-

THE BIOLOGY OF CRUSTACEA, VOL. 8

ISBN 0-12-106408-5

teristic of a given organism. To understand the ecological potential of crustaceans, a basic understanding of this dynamic interaction between environment and metabolic capability is vital. Crustacea represent a unique group of organisms for study because during the evolutionary process of adaptive radiation, various species have occupied extremely diverse habitats ranging from the deep sea to terrestrial mountains and from the intestines of parasitized hosts to free-living stages swimming freely in the open ocean.

The metabolic response of that intact organism to an environmental change is an overall indicator of adaptive capacity at the organismic level, but it does not explain in detail what is happening to the internal machinery of the organism. Success of a species or a community, as measured by its ability to survive and maintain itself, is the sum of a complex interacting continuum of factors ranging from molecules to organ systems to species to communities. One of the greatest challenges to future adaptation biologists is to synthesize the numerous suborganismic responses into a meaningful explanation of how an organism functions internally to meet external stresses. Concomitantly, the role of the species in the functioning of an ecosystem must be better understood. As an organism must be able to functionally respond to a changing environment, so must an ecosystem. Although detailed studies of adaptive phenomena must and will continue at all levels of biological organization, each must not exist in a vacuum, but must be integrated into a unified concept of environmental adaptation.

Numerous studies at the tissue, cellular, and molecular levels of biological organization have attempted to describe the underlying functional mechanisms to explain how crustaceans metabolically adapt to changes in the ambient environment. Since many of these studies are discussed in other volumes (e.g., Volumes 2, 5, and 7) and in Chapters 3 and 4 in this volume, a comprehensive discussion of this subject is not the intent of this chapter; however, the subject is of fundamental importance in understanding the broader picture of environmental adaptation. This chapter deals with one aspect of metabolism, respiration, and how it is influenced by various ecological factors.

A physiological–ecological approach provides a comparative basis for describing and understanding the scope of evolutionary diversity exhibited by crustaceans in adapting both to different modes of existence and to different habitats. Unfortunately, a voluminous literature exists on certain phases of this general topic, and I am forced by limitations of space and objective of this volume to select a sampling of papers that illustrates important principles. However, a liberal listing of review papers and textbooks is included.

The following abiotic and biotic environmental factors have been shown to influence the respiratory response of organisms and to have ecological significance.

II. ABIOTIC FACTORS

A. Temperature

1. INTRODUCTION

The inability of crustaceans to control their body temperature to a significant degree suggests that their metabolic rate would vary directly with a change in ambient temperature (the graphic plotting of respiration rate versus temperature results in an M–T curve). In general this response is typical for crustaceans (see Wolvekamp and Waterman, 1960; Bliss and Skinner, 1963; Prosser, 1973; Vernberg and Vernberg, 1972; Lockwood, 1967; Kinne, 1970); however, variations on this theme that have ecological significance have been noted. Animals from cold environments have M–T curves displaced to the left of those of animals from warm environments, whereas animals from the temperate zone experience marked seasonal thermal variations and may have M–T curves that may shift to the right or left depending on field temperatures and overlap curves of animals in both cold and warm environments (for review, see Vernberg, 1962). Both Prosser (1958) and Precht (1958) proposed schemes for graphically representing M–T curves. The slope and shape of the M–T curve may vary depending on many factors, including season, starvation, sex, and previous thermal exposure (Vernberg and Vernberg, 1972).

Although the respiration rate generally decreases with decreasing temperature and increases with increasing temperature, some crustaceans are thermally insensitive (low Q_{10} values) over a wide temperature range (Vernberg and Vernberg, 1972; Newell, 1976). For example, *Uca* (fiddler crabs) have a range of insensitivity (20°–30°C) that coincides with the thermal range in which the animal is typically active in the intertidal zone (Vernberg, 1969). This response could result in the conservation of energy. Larval fiddler crabs also showed a similar response (Vernberg and Costlow, 1966). In contrast, no evidence of a "plateau" of temperature independence in the shore crab, *Carcinus maenas,* was reported for the adult (Wallace, 1972) or the juvenile stage (Breteler, 1975).

When measuring the oxygen consumption rate of the sand-living shrimp *Crangon crangon* over a range of temperatures from 0° to 20°C, Hagerman (1970) did not detect any marked adaptation to temperature in that the Q_{10} values were between 2.5 and 3.6. Leffler (1972) reported a similar response for juvenile blue crabs (*Callinectes sapidus*).

Some freshwater species showed a similar trend. For example, over the temperature range of 16°–35°C, two species of crayfish (*Orconectes immunis* and *O. nais*) consumed oxygen at an increasing rate with Q_{10} values of about 1.5. In contrast, other species showed metabolic compensation. Over the thermal range of 2°–7°C, the respiration of *Mysis relicta,* a freshwater malacostracan, is stable (Ranta and Hakala, 1978). This range corresponds to that which this species normally encounters. In addition, when the oxygen consumption rates of two freshwater copepods (*Diaptomus dorsalis* and *Mesocyclops edax*) were estimated by Smith *et al.*, (1978) at 10°, 15°, 25°, and 30°C, low Q_{10} values were reported, especially between 25° and 30°C (1.26 and 1.66, respectively). The previous thermal regime to which an organism is subjected may influence its metabolic response (see reviews of Bullock, 1955; Prosser, 1955; Kinne, 1970; Vernberg and Vernberg, 1972).

Temperature–metabolic interaction may also vary on a temporal basis, as illustrated in a study on the estuarine amphipods *Bathyporeia pilosa* and *B. pelagica,* (Fish and Preece, 1970). *Bathyporeia pilosa,* which occurs intertidally, had higher metabolic rates over the thermal range of 15°–25°C during the winter than in the summer. In contrast, metabolic peaks in the subtidal *B. pelagica* occurred in May and September. These differences were correlated with reproductive behavior since *B. pilosa* reproduced throughout the winter, but *B. pelagica* did not. However, since there was no observed metabolic adaptation at low temperature, the significance of this response is not clear. Of more demonstrable biological meaning is the marked reduction in metabolic rate observed at higher temperatures during the summer in *B. pilosa.* This response would tend to conserve energy, and it indicates metabolic regulation. Other examples of seasonal differences are discussed elsewhere in this chapter.

The respiration of developmental stages of crustaceans is affected by temperature, but not always in a predictable manner. For example, Epp and Lewis (1979) reported that the oxygen uptake rates through naupliar development showed a sudden jump between nauplii and copepodite. Also, adults and copepodites, but not nauplii, showed a metabolic plateau at higher temperatures. Larval fiddler crabs have a different metabolic response than adults (Vernberg and Vernberg, 1972). Sastry (1979) found that the metabolic response of larval *Cancer irroratus* subjected to a cyclic thermal range was different from that of animals maintained at a constant temperature. In addition, the larvae of two decapod crustaceans (*Emerita talpoida* and *Libinia emarginata*) showed different metabolic responses to thermal change.

Temperature may influence the standard and active metabolism of crustaceans as shown by Halcrow and Boyd (1967). They reported that for *Gammarus oceanicus* the rate for active metabolism was more affected than was the rate of standard metabolism. Other examples are included in Section III,A.

In many temperature–metabolic studies, the animals are subjected to constant temperature for various periods of time before oxygen consumption rates are determined. Although data from these studies are invaluable, the question can be raised about the physiological–ecological significance of these results, since many crustaceans are subjected to fluctuating habitat temperatures.

The metabolic rates were determined for two intertidal crabs (*Panopeus herbstii* and *Uca pugilator*) subjected to three photoperiods corresponding to those typical of different seasons of the year and to four cyclic thermal regimes (Dame and Vernberg, 1978). Although interspecific differences were observed, cyclic acclimation temperatures of both species significantly depressed oxygen consumption in the 15°–25°C temperature range in comparison with rates of animals subjected to constant acclimation temperatures. Since this depression of metabolic rates occurs over that portion of the yearly temperature range within which the animals are most active, it is suggested that these organisms utilize energy more efficiently when subjected to natural cyclic temperature conditions than when subjected to constant temperature environments. This difference in metabolic data would be significant in analyzing the role of the yearly energy budgets of crabs in ecosystem energetics.

Fluctuating temperatures markedly influenced the metabolic response of the terrestrial wood louse (*Oniscus asellus*) (Nash, 1979). Temperature compensation was not apparent in animals kept in fluctuating temperature conditions; however, partial metabolic acclimation was observed in animals acclimated to constant temperature.

Similar studies were done on developmental stages of subtidal crab (*Cancer irroratus*) by Sastry (1979). The M–T curves of larval *C. irroratus* cultured under a cyclic regime differed from those held at constant temperatures by showing (1) increased respiration, (2) metabolic rate compensation between 20° and 25°C, and by (3) extension of the metabolically active range toward higher temperatures.

Salinity and temperature may interact to influence the metabolic response. For example, Laird and Haefner (1976) reported that blue crabs (*Callinectes sapidus*) acclimated to cold (10°C) consumed more oxygen at low salinity than at high salinity. Cold-acclimated crabs had higher metabolic rates than warm-acclimated animals.

2. LATITUDINAL VARIATION

Earlier studies of Fox, Sparck, and Thorson (see review by Vernberg, 1962) demonstrated the correlation of metabolism and the biogeographical distribution of a species. Benthic polar animals exhibited little or no metabolic control at elevated but sublethal temperatures, with the result that they could not eat enough to supply their energy demands. In contrast, animals from warmer environments could not maintain high metabolic rates to be active and feed at lower temperatures. Therefore, the equatorial spread of polar animals is limited by increasing metabolic demands at higher temperatures, and the polar migration of warm water animals is restricted by metabolic insufficiency at low temperatures. In general, tropical animals consume oxygen at similar rates at their typical ambient temperatures as do related cold-water species at their ambient temperatures. Animals from both polar and tropical regions are not greatly influenced by thermal acclimation, whereas animals from intermediate latitudes experience wider temperature changes, and many species have great metabolic acclimatory ability to compensate for changes in temperature.

In a series of ten papers, the Vernbergs (1972) demonstrated marked physiological differences between temperate and tropical zone fiddler crabs. Differences were observed in the respiration of whole organisms and tissues and enzyme activity as influenced by the thermal acclimation.

Polar and temperate zone animals also show marked differences. For example, the metabolic rate of an Antarctic copepod, *Rhincalanus gigas,* was markedly lower than that of a related species from lower latitudes (Rakusa-Suszczewski *et al.,* 1976). Because of a low metabolic rate and high lipid content, these authors suggested that energy was derived from lipid oxidation.

Crustacea living at different latitudes may experience dissimilar thermal ranges that may be reflected in their metabolic response to temperature. It is not always certain, however, whether observed differences in metabolic response are genetically or environmentally induced. Consider, for example, the study on metabolic adaptation in the widely distributed xanthid crab *Rhithropanopeus harrisii.* This crab is distributed along the east coast of the Americas from New Brunswick, Canada, to northeast Brazil. Field-collected populations from Maine, North Carolina, and Florida all had distinctive metabolic–temperature curves. When metabolic–temperature patterns were determined on laboratory-reared animals from the three populations, however, most of the interpopulation differences disappeared. Therefore, environmentally-induced phenotypic variation, rather than genetic factors, accounted for differences in metabolic–temperature patterns between populations (Schneider, 1967).

Diversity in metabolic–temperature response has been reported for latitudinally separated populations of the coastal hermit crab *Pagurus longicarpus* (Vernberg, 1972a). Populations were from Massachusetts, from north and south of Cape Hatteras, North Carolina, and from the east and west coasts of Florida. In general the metabolic–temperature curves for the three northern populations demonstrated a common adaptational response in that cold-acclimated animals consumed oxygen faster than warm-acclirnated animals. In contrast, the metabolic–temperature response of the two Florida populations was different. Comparison of the absolute metabolic rates of these geographically separated populations reveals certain generalities. Warm-acclimated hermit crabs from Florida exhibit higher metabolic rates at elevated temperatures than do northern populations, but at lower temperatures the cold-water populations tended to have the higher rates. Since the northern animals normally experience lower temperatures, their relatively higher metabolic rate at lower temperatures would have adaptive significance.

Geographical differences were noted in that metabolic rates for congeners of *Diaptomus dorsalis* from northern populations were higher at similar temperatures than those of animals from warmer waters (Smith *et al.*, 1978).

After acclimating adult blue crabs from Maryland and Florida to 10° and 20°C, Robert and Gray (1972) reported that the Florida crabs were less metabolically adaptable to low temperature stress than was the more northern population. They also found in midgut gland a significantly increased specific activation of glucose-6-phosphate dehydrogenase and 6-phosphogluconate dehydrogenase, but only in the Maryland crabs, a fact that might account for the observed metabolic compensation of the whole animal.

Also, temperature adaptation at the enzyme level has been demonstrated for fiddler crabs. Cytochrome *c* oxidase activity was measured in muscle tissue and the supraesophageal ganglion (SEG) from cold- and warm-acclimated crabs from Florida and Puerto Rico (*Uca rapax*) and North Carolina (*Uca pugnax*) (Vernberg and Vernberg, 1968). After a period of exposure to 15°C, cytochrome *c* activity in SEG from *U. pugnax* was enhanced over that in preparations from animals acclimated to 25°C. Cytochrome *c* activity for SEG of *U. rapax* from Florida was the same for cold- and warm-acclimated crabs, whereas cold acclimation reduced this enzyme activity in animals from Puerto Rico. For muscle tissue, enzyme activity was higher for cold-acclimated animals from North Carolina and Florida, but it was relatively unchanged in Puerto Rican fiddler crabs. These results are consistent with the general theory of metabolic-temperature adaptation, which states that the activity rate of a physiological function is higher in a cold-acclimated animal than in a warm-acclimated animal. Cytochrome *c* activity of the temperate zone *U. pugnax* demonstrated this response, whereas the activity of tropical crabs from Puerto Rico either was not influenced or was inhibited.

The metabolic–temperature response of some crustaceans has been shown to be influenced by eyestalk extracts. Silverthorn (1973, 1975a,b) suggested that hormone-like substances from the eyestalks of cold-acclimated fiddler crabs (*Uca pugilator*) increased the metabolic rate of warm-acclimated animals. Pruitt and Dimock (1979) reported that eyestalk extracts from warm-acclimated crayfish (*Cambarus acuminatus*) significantly decreased the oxygen consumption rate of eyestalkless cold-acclimated animals. The model of action of neurohormonal factors is unknown.

3. OCEANIC HABITAT

Relatively few definitive studies exist on the effects of temperature on the metabolic rates of oceanic invertebrates. For pelagic crustaceans living in midwater environments, the available data suggest that oxygen consumption rates increase with temperature, with Q_{10} values of 1.5 to 3.0 (Childress, 1976, 1977; Marshall, 1973; Mauchline and Fisher, 1969; Teal, 1971). However, many factors interact with temperature in determining particular metabolic rates. For pelagic invertebrates these factors include ambient levels of dissolved oxygen and hydrostatic pressure (see other sections in this chapter for further discussion).

The regression equation relating metabolic rate to body weight and temperature of three species of copepods from New Zealand (Green, 1975) is similar to that of six species of *Diaptomus* from the northern hemisphere (Comita, 1968).

There have been few studies on the influence of temperature acclimation on pelagic species. In one such study, Halcrow (1963) indicated a considerable capacity in the copepod *Calanus finmarchicus* to acclimate to different temperatures, a characteristic that has interesting implications for defining their ecological energetics.

The ability to acclimate metabolically to changing temperatures is influenced by many factors. For example, Conover (1962) reported that Stage 5 of *Calanus hyperboreus* almost perfectly acclimated its oxygen uptake over normal temperature ranges (2°–8°C). Spring populations of *C. finmarchicus* did not acclimate over a temperature range of 4°–10°C, but they did at temperatures of 10°–20°C. During the summer, *Calanus* showed no acclimation over the entire range from 4° to 20°C (Halcrow, 1963). Small and co-workers (1966) reported no acclimation in *Euphausia pacifica* at temperatures between 5° and 10°C, but they did find acclimation at 15°C.

Differences have been reported in respiration rates of populations of the copepod *Acartia clausi* from England and Long Island, New York. The forms from the English waters had a statistically higher metabolic rate during the colder months of the year, suggesting that the American form had the greater capacity for control of metabolism under varying thermal conditions (Conover, 1959).

McWhinnie (1964) measured the metabolic responses of the antarctic plantonic crustacean *Euphausia superba,* collected at −1.27° to 2°C from surface waters in the Bransfield Strait and Weddell Sea, and *Munida gregaria* (grimothea, or adolescent, stage), collected at 8.8°C from subantarctic waters in the Straits of Magellan. Oxygen consumption of *E. superba* at temperatures between 0° and 15°C was measured within 30 min of collection because of their high mortality rate in the laboratory. Data obtained from two groups collected at 62°S, 57°W and 62°S, 63°W showed no significant difference; the general response to increasing temperature was a slight increase in O_2 consumption between 0° and 5°C with no further increase above the 5°C level. The normal temperature range of *E. superba* at the sea surface seems to be 0° to 2°C; the responses illustrated indicate a considerable degree of metabolic independence within this range. Grimothea stages of the anomuran decapod *Munida gregaria* were transferred to 2° and 13°C. Oxygen consumption was measured 4 days after introduction to the test temperatures. The differences in O_2 consumption between 2°C individuals and the 13°C individuals were not significant and indicated no marked increase in O_2 uptake by specimens held at 2°C relative to those kept at 13°C. However, extension of the stabilization period resulted in metabolic acclimation.

Temperature influenced the oxygen consumption rate of *Phronima sedentaria,* an amphipod from the western Mediterranean. The Q_{10} value over the range of 5° to 15°C is 2, but this value is only 1.3 over the range of 10.5° to 19.5°C, which suggests compensation at higher temperatures.

Distinctive differences in thermal-metabolic acclimation patterns have been reported within one species. In the copepod *Euterpina acutifrons,* dimorphic male forms occur. One is distinctly smaller than the other, and there are also morphological differences in the antennules, antennae, and second pair of legs. Haq (1965) reported that the dimorphism was genetically controlled and independent of environmental temperature. Moreira and Vernberg (1968) found that enhanced metabolic activity occurred in small males at low temperatures (15°C), but not at high temperatures (25°C). In large males the reverse was observed; cold-acclimation enhanced metabolic activity at high temperatures, but not at low temperatures. Female *E. acutifrons* show metabolic adaptation to both high and low temperatures. Thus, the female would appear to be more labile metabolically than either large or small males (Moreira and Vernberg, 1968). These differences in metabolic responses in the female and dimorphic males could be of value to the species in invading new areas with different thermal regimes.

4. FRESHWATER HABITAT

Although freshwater ecosystems in middle or high latitudes experience long periods of low temperature, little information was available on the influence of temperature on the energetics of plankton. To help fill this void, Ostapenya *et al.* (1969) investigated the metabolic–thermal interaction for one of the dominant members of the winter plankton, the copepod *Diaptomus graciloides.* A large increase in metabolism (Q_{10} of 5.1) was reported over the short range of 0.5° to 2.5°C, but a thermal increase from 2.5° to 3.9°C resulted in a Q_{10} value of 1.8. These data are similar to those of other planktonic crustaceans living in an ice-covered lake (Scherbakov, 1935). The winter population of *Diaptomus* has a higher calorific content (7.17 cal · mg^{-1} dry weight) than the summer population (5.5 cal · mg^{-1} dry weight), due to increased lipid content. *Limnocalanus macrurus,* a copepod inhabiting arctic lakes, showed no metabolic–temperature adaptation in that the Q_{10} values (1.93 for temperature of 0°–10°C, and 2.2 for 5°–15°C) were typical of chemical reactions (Roff, 1973). No thermal–metabolic plateau was observed.

As temperature was increased from 15° to 27°C, the respiration rate of *Ceriodaphnia reticulata* living in Lake Kinneret, Israel, increased, especially at the higher temperatures (Gophen, 1976). This response at high temperature resulted in an energy deficit causing a reduction in body size and egg production, which led to a progressive decline of the population size, a response correlating with observations on natural populations in the field.

Epp and Lewis (1979) reported on the influence of temperature on various stages of a freshwater copepod (*Mesocyclops brasiliensis*) from Venezuela. A Q_{10} of 1 was found for copepodites and adults over the range of temperatures normally encountered during the daily vertical migration when the lake they inhabit is thermally stratified; a very high Q_{10} value between a lower temperature and the normal temperature range was found.

Temperature also similarly influenced the oxygen consumption of another freshwater zooplankter *Daphnia ambigua* (Armitage and Lei, 1979). The M–T curve was shifted to the left in cold-acclimatized animals and the authors suggested that a temperature near 12°C separates warm and cold acclimatization.

Cold-acclimated (20°C) *Parathelphusa hydrodromus,* a tropical freshwater crab, consumed oxygen more rapidly than warm-acclimated (35°C) crabs (Kotaiah and Rajabai, 1972). In both groups of animals, the metabolic rate increased with temperature.

B. Salinity

1. INTRODUCTION

Certain crustaceans occupy only selective segments of a salinity gradient ranging from hypersaline waters to fresh water, whereas others are able to cope with extremely wide variations in salinity. For example, the blue crab (*Callinectes sapidus*), the fiddler crabs (genus *Uca*), and the Chinese wool-handed crab (*Eriocheir sinensis*) are found in habitats ranging from fresh water to hypersaline waters, whereas the spider crabs are stenohalinic, oceanic species. Along with this diversity in selection of habitats based on salinity, four major types of respiratory responses to salinity change have been described by Kinne (1964) along with some examples.

Type 1: The rate increases in subnormal salinities and/or decreases in supranormal salinities. Examples are *Carcinus maenas, Eriphia verrucosa, Gammarus locusta, Potamon edulis, Gammarus duebeni, Uca* spp., *Hemigrapsus oregonensis,* and *Semibalanus balanoides.*

Type 2: The rate increases in sub- and supranormal salinities. Examples are *Ocypode quadrata, Palaemonetes varians,* and *Metapenaeus monoceros.*

Type 3: The rate decreases in sub- and supranormal salinities. No crustaceans were cited.

Type 4: The rate remains essentially unaffected. Examples are *Eriocheir sinensis, Artemia salina, Cyathura polita,* and *Corophium volutator.*

These changes in respiratory rates may reflect a number of physiological alterations, including ionic and osmoregulation, feeding, reproduction, and locomotion. A number of studies have further described the complex interaction between salinity and respiration with special emphasis on modifying factors. A few examples follow:

A change in salinity may alter the locomotor activity of an organism. Thus, observed changes in metabolic rate may reflect behavioral changes rather than the effect of salinity on basic metabolic processes. For example, the increased metabolic rate of the brackish water snail *Potamopyrgus jenkinsi,* when subjected to a new salinity regime, was reported to be due to increased locomotor activity (Duncan, 1966). In contrast, Dehnel (1960) suggested that the greater metabolic rate at low salinity of the crabs he studied did not result from muscular activity.

As mentioned earlier, the crab *Carcinus maenas* shows an increase in the rate of respiration when exposed to low salinity (Schlieper, 1929). Taylor (1977) determined the response of *Carcinus* when subjected to sudden changes in salinity; the earlier work was done on animals acclimated to low salinities. In nature, this crab is frequently subjected to marked fluctuation in salinity, since it lives in estuaries and in rock pools. After exposure to reduced salinity, the oxygen consumption rates were highest during the ensuing 2–3 hr, after which they declined. However, the rate was always higher than that reported during the periods before exposure to low salinity. This increased oxygen consumption rate lasted even after 3–4 days in water of reduced salinity. This response indicated to Taylor that the respiratory rate of *Carcinus* does not acclimate rapidly to reduced salinity. Associated with the increased oxygen consumption is an increase in ventilation volume resulting from a more rapid rate of beating of the scaphognathites. Little change was observed in the oxygen transporting properties of the blood under the conditions of reduced salinity. However, Taylor suggested that the increased oxygen demand of the tissues was met by an increase in heart rate, resulting in a rise in cardiac output.

The mystacocarid, *Derocheilocaris remanei,* which is a member of the meiofaunal community, exhibited peak respiratory rates at both high and low salinities (Lasserre and Renaud-Mornant, 1973). This response was suggested to be an energetic relationship between ionic regulation and metabolic rate.

The size of an animal may influence metabolic response to salinity stress as in the crab *Sesarma plicatum*. Juvenile crabs consume 50% less oxygen in 50% seawater than when they are in tap water, but in contrast, adult crabs have similar metabolic rates in both of these media (Madanmohanrao and Rao, 1962). The oxygen consumption of adults and nauplii of *Eurytemora hirundoides* increased when subjected to hypo- and hyperosmotic seawater (Gyllenberg and Lundqvist, 1978).

The metabolic rate of the copepod *Acartia tonsa* from coastal waters increased in low salinity waters and remained elevated for more than 24 hr (Lance, 1965). This increased rate of oxygen uptake at reduced salinity is accompanied by decreased grazing activity, which would suggest that this species would not do well in brackish water. However, since *A. tonsa* is one of the most common copepods in estuarine systems, Lance argued that acclimation to low salinity waters must occur in the field.

2. CHANGING SALINITY

A series of euryhaline amphipods were studied in reference to their ability to adapt to salinity change (Bulnheim, 1972). These species were subjected to abrupt changes in salinity (10–30‰, 30–40‰, 30–10‰, and 10–30‰)[1] and then their metabolic rate determined. In general, changes from more concentrated to more dilute media induce less extensive alterations in respiratory rates and lead to faster adaptation to new salinity levels than do transfers in the opposite direction. A new steady state of metabolic rate following salinity stress may be reached after 3 to about 30 hr. Species differences were observed that seem to be correlated with habitat differences. For example, *Gammarus duebeni* exhibits a higher capacity for compensatory adjustment than the other species do: this species is found in habitats with widely fluctuating salinity ranges. On the other hand, *Gammarus locusta* demonstrated the least ability to compensate for salinity change. This species typ-

[1]The symbol ‰ means parts per thousand (ppt).

ically is found in low salinity waters. Bulnheim (1972) found that other factors were involved in the distribution of these five species of amphipods, but salinity adaptation and osmotic and ionic regulation are important functional features.

In estuaries the salinity may fluctuate rapidly with the changing tides. Findley *et al.* (1978) examined the effects of a simulated tidal cycle on the respiration of two common estuarine species, the blue crab (*Callinectes sapidus*) and the oyster drill (*Thais haemastoma*). Respiration rates of animals acclimated to 10, 20, and 30‰ salinity were measured at one temperature (20°C) after exposure to various semidiurnal salinity regimes (10–5–10‰, 20–10–20‰, 30–10–30‰, and 10–30–10‰). Blue crab respired more rapidly at 10 and 20‰ S than at 30‰ S. With fluctuating salinity, their metabolic rate varied inversely, although relatively minor changes occurred. Typically the respiration rate dropped during the initial phase of declining salinity at a rate directly proportional to the rate of salinity decrease. This transitory response could represent a metabolic adjustment period because this species is capable of regulating the osmotic and ionic composition of extracellular fluid. In contrast, the oyster drill exhibits incomplete volume regulation and its metabolic rate was thus more directly influenced by salinity change.

The metabolic response to salinity may be an indicator of the distribution of species along a salinity gradient, as illustrated in a study of two brackish water isopods by Frier (1976). *Sphaeroma hookeri* is most abundant in Danish waters in salinities below 15‰ and may occupy waters with a salinity as low as 2‰, while *S. rugicauda* typically is not found in salinities lower than 10‰. The rate of respiration is higher in *S. hookeri* and increases more rapidly in low salinities than that of *S. rugicauda*. However, Frier reported no direct correlation between the degree of osmoregulation and the rate of respiration, since *S. rugicauda* maintained blood more concentrated at low salinities than did *S. hookeri*.

When confronted with changes in salinity, crustaceans have shown a number of adaptative mechanisms. One unusual method of surviving in changing salinity was demonstrated in the hermit crab *Pagurus bernhardus*, which lives in rocky shore tide pools and sand flats. Shumway (1978) exposed specimens of this species with and without their shells to gradual and abrupt changes in salinity. She measured the changes in hemolymph osmolality, tissue water content, and oxygen consumption. Although animals with shells showed significant changes in hemolymph osmolarity and total tissue water content during changing salinities, animals without shells showed far greater changes and were seen to become extremely swollen, thus impairing mobility in some cases. One example from her data to illustrate that the shell provided protection for the organism follows: When subjected to an abrupt salinity change, the blood osmolarity of crabs with shells fell to approximately 650 mosmoles after 6 hr in 50% seawater, while that of crabs without shells was isomotic with the external medium after only 2–3 hr. Crabs with shells showed a marked temporary increase in oxygen consumption when the external medium declined to approximately 75% seawater. In contrast, crabs without shells showed no such response.

3. TEMPERATURE AND SALINITY INTERACTION

Temperature and salinity do not always influence a species in a predictable manner. For example, the oxygen consumption rate of juvenile prawns (*Penaeus indicus*) increases with increasing temperature (19°–38°C), but changes in salinity (7–35‰) have no significant effect after long-term acclimation (Kutty *et al.*, 1971).

Hagerman (1970) measured the oxygen consumption rate of the shrimp *Crangon crangon* under different conditions of temperature and salinity. Over the temperature range of 0°–20°C, the oxygen consumption rate was found to increase at salinities lower than about 20‰, and at 6° and 13°C it also increased at higher salinities, while no significant increase at higher salinities was found at 20°C. Hagerman suggested that this metabolic pattern reflects the osmoregulatory work since *Crangon* is a homoiosmotic animal with isotonicity being in the range of 21–23‰.

A man-made lake (Lake Veere) was formed in 1961 when an area was closed off from the North Sea and the Oosterschelde estuary. A brackish water community (10–25‰ salinity) developed in which the isopod *Idotea chelipes* is abundant. Vlasblom *et al.* (1977) found that this species did not survive as well with increasing temperature or decreasing salinity. Based on respiration studies, these authors suggested that the metabolic cost of osmoregulation is detrimental to other necessary physiological processes when the temperature is 10°C and higher. At temperatures higher than 15°C and salinities lower than 8‰, the population of this species would be adversely affected.

4. CELLULAR RESPONSES

In addition to studies on the metabolic response of the blue crab to changes in salinity, Engel and Eggert (1974) and Engel *et al.* (1975) reported on the effects of salinity on respiration of blue crab gills as well as on ATP concentrations in excised gills. They reported seasonal differences in the metabolic rate of the gill tissue. During the winter, respiration rates decreased with increasing salinity with no difference between the response of males and females. However, during the summer months there was a distinct sexual difference in response. For example, at low salinity (5‰) gill tissue from females had a lower rate of respiration than that from males. It is interesting to note that the female crabs tend to be less abundant at low salinities during the summer than in the winter, a response coincidental with the breeding behavior of this species.

Engel *et al.* (1975) demonstrated that the ATP content of the blue crab gill tissue was shown to be directly dependent upon the salinity of the medium, and that the ATP content inversely related to the respiration rate of the tissue. They suggested that when gill tissue is subjected to osmotic stress, the process of cellular accommodation is an energetic one with the ultimate source of energy being ATP. Earlier King (1965) demonstrated with isolated gill mitochondria that the increased mitochondrial respiration rates were directly correlated with increased concentrations of citric acid cycle enzymes.

C. Oxygen

Crustaceans show a wide diversity in response to ambient oxygen concentrations. Some species are oxyconformers, while others are oxyregulators with Pc values (the critical level of oxygen tension below which the organism becomes an oxyconformer) varying with different factors (see review of Vernberg, 1972b). For example, small-sized prawns (*Penaeus indicus*) were less metabolically dependent on ambient oxygen tension below 100 mm Hg than large-sized animals (Subrahmanyan, 1962); Pc values are different in adult and larval stages of *Calanus finmarchicus* (Marshall *et al.*, 1935); and the Pc is higher in semiterrestrial amphipods than brackish and freshwater species (Walshe-Maetz, 1956). Further, other variables, such as food, molting, locomotor activity, and acclimation to various oxygen tensions influence metabolism. Some recent examples are discussed below.

In an attempt to present an empirical model to provide a quantitative description of the relationship between aerobic metabolism and decreasing environmental oxygen levels, Mangum and Van Winkle (1973) described a quadratic binomial that best fits data from 31 species of aquatic invertebrates. Based on this equation, Mangum and Van Winkle found that there appeared to be a phylogenetic trend of increasing regulation of aerobic metabolism in response to declining environmental levels. This trend could be correlated with the acquisition of structures by organisms that would effectively insulate their respiring tissue from the surrounding habitat. In many of these species the removal of oxygen would cease long before the available supply of oxygen was exhausted. These species seem to be characterized by having long-lasting internal oxygen reservoirs, such as gas bubbles pools of respiratory pigments with high oxygen affinity. Those species without substantial oxygen storage capabilities would switch to anaerobic pathways.

The metabolic rate of two crustaceans (*Derocheilocaris remanei* and *Paraleptastacus spinicauda*), which are meiofaunal species, was determined under different oxygen levels (Lasserre and Renaud-Mornant, 1973). The Pc value for *D. remanei* was 3 ml of oxygen per liter and 5 ml of oxygen per liter for *P. spinicauda*. Neither of these species show outstanding ability to withstand anaerobic conditions. Their metabolic responses were correlated with their habitat, in that both species are found in high energy beaches where there is apparently a high level of oxygen available to them.

Fiddler crabs *Uca pugilator* and *U. pugnax,* living in burrows in sandy-muddy substrates, may experience low oxygen tensions when the tide is in, since they apparently do not pump water through their burrows. These species are not only relatively resistant to anoxia, but also the critical oxygen tension is low: 1–3% of an atmosphere for inactive and 3–6% for active crabs. These crabs continue to consume oxygen down to a level of 0.4% of an atmosphere; in contrast, the nonburrowing wharf crab *Sesarma cinereum* stops respiring at a somewhat higher value (Teal and Carey, 1967a).

When crustaceans invaded land, a number of marked functional changes took place (see Chapter 6 of this volume); however, the present chapter emphasizes only respiratory adaptations. Earlier Pearse (1929) reported that the number and the total volume of gills of terrestrial crabs are less than in aquatic species. Later in a more comprehensive study, Gray (1957) found that the gill area per unit body weight tended to be less in terrestrial crabs, a trend correlated with the increased availability of oxygen in an aerial habitat. Along with this decrease in gill area, a tendency for an increase in oxygen consumption of crabs from more terrestrial habitats was reported by Vernberg (1956). This trend was not due solely to increased locomotor activity, because the respiration of isolated tissues and cytochrome *c* oxidase activity showed a similar response (Vernberg and Vernberg, 1968).

Recently Babula *et al.* (1978) reported a very high coefficient of correlation (0.9993) between oxygen consumption of the intact organism and the gill surface area in the marine isopod *Saduria entomon*. Working with a freshwater crab, *Barytelphusa guerini,* Ambore and Venkatachari (1978) found that the total gill plate number and total gill area increased with size, except when based on unit of total animal weight. Oxygen consumption was proportional to the amount of gill surface. Harding (1977) reported a similar relationship between total body surface and respiration in the euphausiid, *Thysanoessa raschii,* and described a method for estimating the surface of marine crustaceans.

When animals are exposed to air, the amount of oxygen is high and relatively stable, in contrast to the aquatic environments, where the oxygen content may vary sharply. Thus the respiration of aquatic animals is generally more sensitive to oxygen levels, and terrestrial animals are sensitized to CO_2 levels.

Variations in the respiratory adaptations of terrestrial crustaceans were reported by McMahon and Burggren (1979). *Birgus, Gecarcinus,* and *Cardisoma* are less influenced by hypoxia than hypercapnia, a response to be expected of animals living in a habitat rich in oxygen. However, the land hermit crab, *Coenobita clypeatus,* was more sensitive to hypoxia than hypercapnia. Apparently this difference in response can be correlated with the continued association of *C. clypeatus* with a fluid-filled molluscan shell, which permitted the crab to retain some aquatic tendencies.

Burke (1979) reported that when exposed to air, *Pachgrapsus crassipes,* the lined shore crab, decreased its oxygen consumption rate by 35% over that of animals during submersion. Perhaps this depressed metabolic rate influenced the activity of this species, in that submerged crabs maintained vigorous activity longer than those exposed to air.

The crab *Scylla serrata* consumed less than a tenth as much oxygen when exposed to air as when submerged (Veerannan, 1972). Wallace (1972) reported a similar response in the shore crab *Carcinus maenas;* however, the respiration rate was only 30% higher when the crab immersed.

Van Weel *et al.* (1954) studied seven species of shallow water crabs. They demonstrated that the rate of oxygen uptake of a mud-dwelling crab was independent of ambient oxygen tensions until a low level was reached; at this point the respiration pattern became oxygen dependent. Species from well-aerated waters, however, exhibited an oxygen-dependent response over a wide range of oxygen tensions. All species showed a steep increase in percentage oxygen utilization at oxygen concentrations of less than 2.5 ml oxygen/liter. This response was interpreted as an attempt by the organisms to ensure a minimum amount of oxidation at low tensions.

The oxygen uptake rate of two xanthid crabs, *Panopeus herbstii* and *Menippe mercenaria,* decreased in proportion to ambient oxygen levels (Leffler, 1973). However, the rate of decrease is less than that of the blue crab (*Callinectes sapidus*), a response apparently correlated with habitat differences in that the xanthid crabs often live in mud while the blue crab does so less frequently.

Two species of burrowing crabs, *Callianassa californiensis* and *Upogebia pugettensis,* regulate their metabolic rates over a wide range of oxygen concentrations. However, Thompson and Pritchard (1969) found species differences that correlated with habitat differences; *Callianassa* lives under more hypoxic conditions than *Upogebia* and has the lower metabolic rate, the lower Pc value, and is more resistant to anoxia. Recently, Torres *et al.* (1977) found that *C. californiensis,* when placed in simulated burrow conditions, regulates oxygen levels in its immediate microhabitat by using its pleopods. Body movement plays a large role in water exchange between the surface and the burrow. After oxygen deprivation, the presence of oxygen may stimulate feeding and activity at the surface of the burrow.

A related species, *C. jamaicense*, which lives in estuarine mud flats in the northern Gulf of Mexico where hypoxic conditions are common, was reported by Felder (1979) to be very tolerant of anoxia. It exhibited metabolic regulation to a low Pc, had a low metabolic rate, and showed toxic responses to altered oxygen tension: all adaptations to a hypoxic habitat. Another excellent example of physiological adaptation to habitat variation is seen in the isopod, *Cirolana borealis*, a scavenger which burrows into the flesh of dead fish. During feeding this isopod becomes anaerobic when it enters the deeper anoxic regions of the dead fish (deZwann and Skjoldal, 1979).

In 1970, Gamble compared the responses of free-living and tubicolous crustaceans. At low tide, very low oxygen tensions are found in the tubes of intertidal burrowing animals. After exposure to low oxygen levels, the ventilation rate increased for the free-swimming species, remaining unchanged for the tubicolous amphipods. Moreover, a marked change in the ventilation rhythm was found; at air-saturated oxygen tensions, only the tubicolous amphipods exhibited an intermittent rhythm of ventilation, but when the oxygen levels dropped, a continuous rhythm resulted. Coupled with Gamble's work on anaerobic resistance, these findings suggest that tubicolous amphipods are more resistant to anoxia than are other crustaceans of similar size, and that they do not attempt to hyperventilate at reduced oxygen tension. Because of the viscous drag resistance of the tube, water movement costs the organism energy, and hyperventilation would be energetically expensive. However, a minimal ventilation rate apparently is maintained, since the organisms show a continuous ventilation rhythm.

Unlike many other crustaceans, the subtidal, small-sized crab *Ebalia tuberosa* shows a very limited ability to regulate its oxygen consumption when the oxygen content decreases: the critical P_{O_2} is only about 130 mm Hg during activity, but the crab is a perfect conformer when active (Schembri, 1979). This species spends most of its time buried in a muddy-gravel substratum.

When experiencing external environmental changes, crustaceans may switch from aerobic to anaerobic metabolism (Theede, 1973; Breteler, 1975; Spaargaren, 1977). As an example, the responses of *Carcinus maenas* are discussed. Down to oxygen pressures of about 20 mm Hg, this species regulates oxygen uptake when measured at both 7° and 20°C (Spaargaren, 1977). The gill ventilation rate is unchanged, and the heart rate is more or less constant until the oxygen tension reaches 60–80 mm Hg (Taylor, 1976), and then it decreases (Uglow, 1973). The *Carcinus* survived anoxia for approximately 2 days. Irreversible damage resulted when animals were deprived of oxygen for too long a period of time. Apparently profound metabolic changes occurred even during the reversible phase of arrested activity because the organism took a long time to recover when subjected to normal oxygen concentrations. Although there is a reduction in energy production during anoxia, sufficient energy for ionic regulation results from anaerobic pathways to ensure internal constancy. Thus, under anoxic conditions the loss of locomotor ability makes species vulnerable to predation, but it can withstand changing salinities. These results demonstrate that biotic interaction is potentially more limiting under anoxic conditions than are abiotic factors.

Burke (1979) compared the respiratory response of *Carcinus maenas,* which encounters hypoxic conditions, and *Pachygrapsus crassipes,* an inhabitant of rocky zones where oxygen levels are high. Under hypoxic conditions, *C. maenas* is able to regulate metabolism to a lower Pc value and can shift to anaerobic metabolism as evidenced by increased lactate production. In sharp contrast, *P. crassipes* does neither.

The respiration of barnacles living subtidally or at different levels of the intertidal zone decreased with greater height of the habitat (Barnes and Barnes, 1964; Hammen, 1972). One upper-shore species, *Euraphia depressa,* regularly decreased its respiration rate with time, and it sometimes ceased respiring entirely, even when submerged. Hammen (1972) found, using data from Barnes and Barnes (1964) and from his own studies (1972), that *E. depressa* catalyzed the oxidation of L-lactate only and that this oxidation accounts for only 2% of the oxygen consumed. He suggested the need for further studies of anaerobic metabolism of barnacles.

Although direct correlation between oxygen uptake and heat production, as measured by calorimetric techniques, has been reported for many animals, Pamatmat (1978) found the two processes to be more or less independent of each other in the fiddler crab *Uca pugnax*. Hence, in this species and presumably in other species experiencing either hypoxic or anaerobic conditions, ecological energetics should be determined by measuring heat production, since "aerobic shutdown" and a switch to anaerobic metabolism may occur when relatively high levels of oxygen are present.

The oxygen-minimum layer, which occurs at intermediate depths in most of the world's oceans, is a special habitat. Although the oxygen content may be only 0.1 ml/liter, an abundance of animals has been reported there (Marshall, 1954). In addition to resident biota, the deep scattering layer has been recorded at the same depths. Some data on the metabolic capability of animals living in this low oxygen environment are available. Longhurst (1967) stated that resting stocks of zooplankton could reside for long periods of time in this layer at oxygen tensions as low as 0.2 ml/liter, or vertically migrating animals could survive here during their daytime sojourn at greater depths.

In one species of pelagic crustacean living continuously in this layer, the lophogastrid mysid *Gnathophausia ingens,* the physiological mechanisms that permit aerobic metabolism under extremely low oxygen conditions have been examined in detail. Childress (1968) showed that this species lives largely aerobically at oxygen concentrations as low as 0.20 ml/liter, with a relatively constant rate of oxygen consumption of 0.8 ml O_2/kg/min at 5°C. The ability of this species to maintain aerobic metabolism appears to result from a combination of several physiological and morphological factors. These include (1) large gill surface areas (between 5 and 15 cm^3/g); (2) low diffusion distances across the gills (1.5–2.5 μm); (3) high ventilation volumes (up to 8 body-volumes/min); (4) high utilization of the available oxygen in the respiratory streams (50–80%); (5) rapid blood flow through the gills (turnover time, 2–3 min); and (6) relatively low routine metabolic rates (Belman and Childress, 1976; Childress, 1968, 1971).

Studies of the relationships existing between low oxygen concentrations and the rates of oxygen consumption in a variety of midwater crustaceans have shown that most species can regulate their consumption down to the lowest oxygen concentrations at which they are found (as little as 0.1 ml O_2/liter); they appear to have generally limited or no anaerobic metabolic abilities (Childress, 1975). It has been concluded by Childress (1976) that the ability to regulate oxygen consumption rates at low oxygen tensions depends strongly on the maintenance of extremely low activity levels. Midwater species that cannot regulate their oxygen rates at low ambient oxygen levels generally have considerable anaerobic metabolic abilities; species relying on anaerobiosis face severe problems in the nutrient-poor deep sea. Childress (1975) noted that only three of twenty-eight species he examined exhibited substantial anaerobic metabolic abilities. Two of these are reported to be parasites on jellyfish and would therefore seem to have an abundant food supply. The third, a large (to 1 cm) copepod, *Gaussia princeps,* is a strong vertical migrator and presumably could metabolize anaerobic end products following ascent into more oxygen-rich water above the oxygen minimum layer.

Artemia salina lives in salt ponds, which have variable concentrations of oxygen due to changes in temperature and salinity. Vos *et al.* (1979) reported that adult males oxyregulate over a wide range of oxygen tensions. The Pc value is low and decreases when animals are acclimated to low oxygen concentrations. In addition to this shift in Pc values, hypoxic conditions induce the production of Hemoglobin III, resulting in a decrease in the percent ratio of HbI to HbIII; the percent of HbII present is relatively unchanged. A shift in metabolism occurs when animals are acclimated to various oxygen concentrations. For example, animals acclimated to high partial oxygen pressures and then exposed to a low oxygen environment produce lactic acid levels that may be lethal within a few hours. However, animals acclimated to low oxygen levels produce much less lactic acid when subjected to hypoxic conditions.

One of the decapod crustaceans living in Lake Tahoe, California– Nevada, is the crayfish *Pacifastacus leniusculus*. It is found in depths ranging from the surface to 60 m and under different conditions of low temperature and high oxygen concentration. Moshiri *et al*. (1970) demonstrated that this species is a oxycomformer over the range of 2–7 ml of oxygen per liter of water. The ventilation rate remained constant with a decrease in oxygen level down to 2.5 ml/liter, at which point it started to decline. The mean value for the efficiency of oxygen utilization was 22.1% down to a concentration of 3.88 ml/liter and then it began to increase. These authors suggest that in this species, there are no district adaptations for regulation of oxygen uptake at low oxygen tensions.

Dickson and Franz (1980) compared the respiratory response of gill tissue from three closely related species of crayfish living either in surface waters or in caves. They reported significantly lower gill tissue respiration rates and longer ATP turnover times in the subterranean crayfish when compared with the surface species. Differences in respiration rates could not be correlated with experimental temperature of oxygen tension of the original habitats. These authors felt that the respiration rates of tissues appeared to be adaptations associated with habitat conditions and food availability.

D. Hydrostatic Pressure

With every 10-m increase in water depth, the hydrostatic pressure exerted on aquatic species increases by 1 atm. Although crustaceans are known to inhabit various zonal regions of deep lakes or the world's ocean, their response to hydrostatic pressure is not extremely well known. Earlier, Fontaine (1929a,b) reported that the respiration rate of *Crangon crangon* and *Palaemon serratus* increased after exposure to 100 atm. Naroska (1968) also found a similar response by *Carcinus maenas;* however, after 14 hr of exposure, the rate had returned almost to its original rate. More recent studies generally show that in relatively shallow-water species that undergo vertical migration, temperature influences metabolic rate more markedly than hydrostatic pressure (Teal and Carey, 1967b; Pearcy and Small, 1968; Smith and Teal, 1973; Quetin and Childress, 1976; Childress, 1977). Further, the metabolism of deep-water crustaceans increases both with increased hydrostatic pressure and with temperature (Teal, 1971). Hence, a relatively constant rate of oxygen consumption results as these animals migrate to new depths. The respiratory rate of the giant ostracode, *Gigantocypris,* also exhibited variation at different temperatures, but the rate was relatively constant between 0 and 500 atm (MacDonald *et al.,* 1972).

The study of Childress (1977) on *Gaussia princeps,* a copepod inhabiting midwater depths, further illustrates the ecological significance of temperature–pressure interactions. This species migrates diurnally, spending the daylight hours below 400 m in the oxygen minimum layer and migrating to 100–300 m at night. After subjecting animals to various combinations of temperature, pressure, and oxygen, he found that at shallow nighttime depths, pressure was the dominant factor influencing respiratory rates, whereas temperature is more dominant at the deeper oxygen-poor depths. A lowered metabolic rate at greater depths would help in energy conservation, since this species is unable to regulate its oxygen consumption rate at the low oxygen levels in the oxygen minimum layer and appears to depend on anaerobic metabolic pathways.

George (1979a) further studied the interactions of temperature and pressure on metabolism of the eurythermal arctic amphipod, *Anonyx nugax*. The pressure effect is less pronounced in animals acclimated to low temperature. The maximum rate of O_2 uptake in three experimental temperatures occurs at about 140 atm, when amphipods are found to be in the hyperactive phase. Pressure effects on respiration are far more pronounced at 6° and 14°C than at −1.8°C, the temperature at which the experimental populations were captured in the field and acclimated in the laboratory. Although *A. nugax* is readily acclimated to higher thermal levels up to 14°C, sharp sensitivity to pressure is seen at elevated temperature. At all experimental temperatures, the biphasic mode of the curve is apparent, with the onset of an inhibitory influence of hydrostatic pressure on respiration of the amphipods beyond 150 atm. Based on various studies, George (1979b) proposed a hypothetical model of respiration as influenced by pressure and temperature (Fig. 1).

Recently George (1979a) presented data on the respiration of an arctic abyssal amphipod, *Eurythenes gryllus,* which lives at depths between 1000 and 6500 m. Little data are available on living animals inhabiting depths beyond 3000 m. Living individuals were maintained at five simulated hydrostatic pressures corresponding to depths between 80 and 340 atm in hyperbaric metabolic chambers that monitored their respiration. Respiration appears to be maintained at a fairly constant level within this range of more than 250 atm. Apparently this species is adapted to regulate respiratory rate to one constant level at the various depths of its occurrence. Furthermore, the pleopod activity is also sustained within a constant level throughout this pressure range. However, *E. gryllus* is extremely sensitive to subtle changes in ambient temperature. Respiration, as well as pleopod activity, is dras-

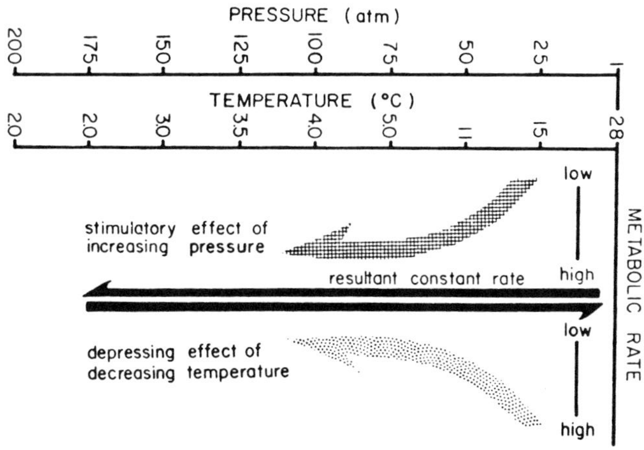

Fig. 1. A hypothetical model of respiratory adaptation of vertical migrators with constant metabolic rate as a result of increasing pressure and decreasing temperature in their normal vertical range of occurrence in the water column. (After George, 1979b.)

tically reduced beyond 3°C, whereas the inhibition of respiration is evident at 4° and 6°C. This information confirms that there is a very narrow thermal range within which normal activity persists. Such a stenothermal feature is possibly a "norm" for some deep-sea species that inhabit the cold depths below the thermocline in the temperate–tropical marine environment.

Excellent reviews of pressure effects on various functional attributes of many animal groups are those of Brauer (1972), Flugel (1972), MacDonald (1975), and George (1981).

E. Light

Earlier examples are available on the effect of photoperiod on metabolic rate in marine invertebrates from a report by Dehnel (1958). The decapod crustacean *Hemigrapsus oregonensis*—collected in the summer and maintained on a cycle of 8 hr light and 16 hr dark (LD 8 : 16)—showed a higher oxygen consumption than crabs kept under the same temperature and salinity conditions with LD 16 : 8 or constant darkness. Summer population individuals of the crab *H. nudus* exhibit the same increase in oxygen consumption as a result of maintenance under a short-day photoperiod (LD 8 : 16) as compared with a long-day photoperiod (LD 16 : 8) and constant darkness, but the differences are not as pronounced as in *H. oregonensis*. The response is evident after 1 week and persists for subsequent weeks. According to Dehnel (1958), the increase in weight-specific O_2 consumption in summer crabs exposed to short-day photoperiod may signify an adjustment of their metabolic rate to winter light conditions. If so, then the adaptive value of such a mechanism would be to prepare the animals to operate under low temperatures.

Daylight does not affect the level of metabolism of the crayfish *Orconectes nais* (Rice and Armitage, 1974). In contrast, a planktonic copepod (*Calanus finmarchicus*) responded to light by sharply increasing its respiratory rate (Marshall *et al.*, 1935).

The effect of light on the metabolism of animals that vertically migrate in the water column has been in question. For example, Buikema (1972, 1975) correlated the effects of light (nine light intensities, polarized light, and four wavelength ranges) on the energetics of a freshwater cladoceran (*Daphnia pulex*) which is known to vertically migrate. The net efficiency of respiration was lowest at light intensities less than 28 ft-candles. At 1.7 ft-candles this value was 9.8%, but at 110 ft-cancles it was 50.0%; however, values between 20 and 30% were typical at intermediate light intensities. Generally, the effect of polarized light was inseparable from the effect of light intensity. Also he reported that the metabolic rate was reduced after acclimation to different wavelengths. Working with marine species, Small and Hebard (1967), Bishop (1968), and Teal (1971) found that light had negligible effect on respiration of vertically-migrating crustaceans.

F. Drought

At certain times of the year, freshwater crabs may experience drought conditions. Diwan (1975) reported that the freshwater crab *Barytelphusa cunicularis* is able to carry on respiration under drought conditions by recycling over the gills water stored in the branchial chambers. The morphology of the cephalothoracic respiratory apparatus of this crab is well suited for recycling of stored water.

MacMillen and Greenaway (1978) made further observations on adaptive mechanisms of crabs exposed to drought conditions. They reported that the crab *Holthuisana transversa*, when transferred from hydrating to dehydrating conditions, reduced both the rate of O_2 consumption and evaporative water loss by one-half. When these crabs were placed in artificial burrows that then were subjected to progressive drying, the oxygen consumption rate was further reduced by 30–40% of the initial level, prior to the onset of conditions resulting in weight loss. These authors interpreted their data to mean that these crabs, which live in the Australian arid zone, exhibit a state of physiological dormancy as an energy-conserving device, permitting them to withstand long periods of drought when they are in humid burrows without immediate access to food.

Some crustaceans are well adapted to survive adverse conditions by forming cysts. The question has been raised about the respiratory metabolism of these cysts. Clegg (1976), working with cysts of the brine shrimp (*Artemia salina*), reported that no discernible respiration occurred until the level of hydration of the cyst reached a point of 0.45 g H_2O/g of cysts dried by $CaSO_4$, whereas above a value of 0.65 g, active respiration took place. Although these animals live in saline waters, Boulton and Huggins (1977) demonstrated that hydration, rather than salinity, determines the initiation of respiration in *Artemia* cysts.

G. Pollutants

With an increasing awareness of man's impact on the environment, attention has focused on the effects of various pollutants on the physiological ecology of crustaceans, especially since 1970. A few examples of pollutant effects are cited below; more detailed discussions have also been published (Khan and Bederka, 1974; Vernberg and Vernberg, 1974, 1981; Goldberg, 1976; W. B. Vernberg *et al.*, 1977, 1979; Cairns, 1980).

The oxygen consumption and ammonia production of the zooplankton *Calanus plumchrus* were not affected by exposure to copper until concentrations close to lethal dosage were reached (Reeve *et al.*, 1976). After exposure of adult crabs (*Cancer irroratus* and *Carcinus maenas*) to copper and cadmium, Thurberg *et al.* (1973) found that the oxygen consumption rates of isolated gill tissue were differentially influenced: cadmium reduced the metabolic rate of both species, while copper had no effect. Mercury and cadmium influences the respiration of larval and adult fiddler crabs (Vernberg *et al.*, 1974).

Freshwater crustaceans may also be influenced by heavy metals. For example, Chinnayya (1971) reported a considerable decrease in metabolic rates of the freshwater shrimp *Caridina rejadhari* after exposure to copper sulfate, lead nitrate, mercuric chloride, and zinc sulfate.

Pollutants other than heavy metals may affect metabolism. The respiratory rate of juvenile shrimp, *Crangon crangon,* was reduced after chronic exposure to water-soluble crude oil fractions at three temperatures; 10°, 15°, and 20°C (Edwards, 1978). In addition, the respiration of the Black Sea shrimp *Palaemon squilla* was significantly affected by sublethal concentrations of DDT (Andryushchenko, 1972).

Polychlorinated biphenyls (PCBs) are found in marine habitats throughout the world. The influence of two forms of PCB, Aroclors 1016 and 1254, on respiration of a fiddler carb (*Uca pugilator*) subjected to various temperatures was complex (Vernberg *et al.*, 1978). At certain temperatures respiratory inhibition was observed, but at others, either no effect was noted or respiration increased.

The effects of sodium pentachlorophenol on respiration of a grass shrimp

(*Palaemonetes pugio*) depends on the stage of the molt cycle, concentration, and extent of preexposure to Na-PcP (Cantelmo *et al.*, 1978; Rao *et al.*, 1979).

H. Other Factors

Alkalinity influences the respiration rate of *Arctodiaptomus spinosus*, a freshwater pond inhabitant: at an alkalinity of 100 mEq/liter the rate was lower than at either lower (10 mEq/liter) or higher (400 mEq/liter) levels (Newrkla, 1978).

Gradients of pH values have been reported in the ocean; at the surface the pH is about 8.3 and 7.5 or less in the oxygen minimum layer. However, Mickel and Childress (1978) found that the oxygen consumption rate of the mysid *Gnathophausia ingens*, an inhabitant of the oxygen minimum layer, was unaffected by pH in the range of pH 7.1–8.7.

Many planktonic crustaceans are limited to a specific water mass. Various suggestions have been proposed to explain this distributional pattern. Gilfillan (1975), determining the respiration rates of *Euphausia pacifica*, felt that temperature and salinity did not explain the distribution of subpopulations of this species endemic to specific fjords, but that differences in concentration of trace elements were important. In the section on pollution effects, the respiratory influence of various substances, including man-induced products, is discussed.

Multiple factor interaction may influence respiration. Two examples will illustrate this point. Roux (1975) compared the differences in metabolic temperature curves of two species of amphipods, *Gammarus pulex* and *G. fossarum*,as determined under three environmental conditions. The metabolism of *G. fossarum* is not influenced by extreme water velocity and is only slightly influenced by temperature and substratum. In contrast, the metabolism of *G. pulex* is influenced by all three factors. Roux expressed the opinion that *G. fossarum* is better adapted to specialized biotopes with running water and low temperature and that it has a competitive advantage over *G. pulex*. On the other hand, *G. pulex* is more adaptable to variable biotopes, especially at temperatures above 15°C. At these temperatures, it has greater energetic reserves then does *G. fossarum*.

III. BIOTIC FACTORS

Activity

The level of body activity is important in determining metabolic rates. However, with crustaceans it is difficult to quantify activity that ranges from maximum to basal metabolic rates. Quetin *et al.* (1978) described a method for simultaneously measuring the activity and oxygen consumption of pelagic crustaceans, and Wycliffe and Job (1977) discussed methods for determining standard, routine, and active oxygen consumption of freshwater shrimp.

Earlier, Halcrow and Boyd (1967) demonstrated a linear increase in metabolism of an amphipod (*Gammarus oceanicus*) with an increase in activity. The oxygen consumption of a mysid (*Gnathophausia ingens*) at maximum sustainable activity is about 9.5 times that at minimum activity (Childress, 1971).

Within the estuary, both activity and habitat influence metabolic patterns. In three genera of harpacticoid copepods, the lethargic *Enhydrosoma* had the lowest rate of oxygen uptake, whereas the active *Longipedia* had the highest metabolic rate. The pelagic *Euterpina* was intermediate between these two meiobenthic copepods (Coull and Vernberg, 1970).

During the process of vertical migration, the question of energetic cost to a species is raised. Foulds and Roff (1976) examined this question for the freshwater mysid crustacean *Mysis relicta,* which undergoes vertical migration. Their general conclusion, that there is a negligible cost to this species, was based on the following experimental results. The mysid was induced to swim at speeds of 1 and 1.6 cm/sec, which was similar to observed vertical migration rates. Under these conditions, there was no significant increase in oxygen consumption over the routine metabolic rate of animals, which were approximately 5 ml dry weight. However, at higher swimming speeds (2.1–3.6 cm/sec) an increase in oxygen consumption, which was 1.2 times the routine rate, was reported.

The rate of aerobic metabolism increased linearly with the speed of locomotion of the land crab *Cardisoma guanhumi* and the recovery period exhibited a large oxygen debt (Herreid *et al.,* 1979).

Data indicate that both the method of extraction of meiofauna from mud or sand and the presence or absence of substrate can affect the outcome of respiratory determinations (F. J. Vernberg *et al.,* 1977). When the mud-dwelling copepod, *Nannopus palustris,* was extracted using the highly efficient sieving, centrifugation, and sucrose flotation method, measured respiration rates were significantly lower than when the copepods were laboriously hand-sorted from the sediment. Similarly, when metabolic determinations were made on the interstitial copepod, *Hastigerella leptoderma,* in the absence of sand grains, the rates were significantly higher than when sand grains were provided as substrate. The higher metabolic rate observed in those copepods lacking a substrate can undoubtedly be linked to increased activity, for in the presence of a substrate these animals curl around a particle and are relatively quiescent, whereas in the absence of a substrate they flail about vigorously.

Thermal acclimation influenced routine and hyperroutine respiration of the Dungeness crab (*Cancer magister*) (Prentice and Schneider, 1979). For example, routine respiration of crabs acclimated to 17.5°C showed temperature insensitivity between 7.5° and 17.5°C, but animals acclimated to 7.5°C showed thermal sensitivity. Hyperroutine rates were thermally insensitive from 7.5° to 17.5°C for both acclimation conditions.

McFarland and Pickens (1965) studied the standard metabolic rate of adult grass shrimp *Palaemonetes vulgaris* collected during the four seasons of the year and exposed to three different constant temperatures. They reported an increasing rate of oxygen consumption with increasing test temperature, but not significant seasonal effects on standard oxygen consumption rates. With the overhead lights on, oxygen consumption was 20–25% higher. This shrimp is a eurytherm and exhibits a high degree of independence of individual thermal history, except at extreme temperatures. Its standard oxygen consumption rate is, therefore, a predictable function of the test temperature over a wide range of thermal conditions. These authors also determined the active metabolic rate by making the shrimp swim against a water current (sustained work). This rate is significantly higher and is independent of thermal history only during acute exposure to temperatures between 15° and 30°C; active rates at lower or higher test temperatures are affected by the previous temperature regime.

B. Feeding

Feeding may also influence the rate of oxygen uptake. After feeding, *Carcinus maenas* consumed more oxygen than when starved for a prolonged period (Wallace, 1973). Initially the rate dropped by 60%, probably reflecting carbohydrate consumption, followed by a 40% decline that indicates utilization of lipids and proteins. Marsden *et al.* (1973) measured the O_2 consumption of crabs after various periods of starvation. Animals of a wide variety of sizes were placed at 15°C and one group was starved while the second was fed crushed shellfish. After 1 week there were no marked differences between the rates for starved and fed crabs, but after 2 weeks a clear difference was observed. At most temperatures the starved animals had a lower rate of O_2 consumption than organisms that were fed. Vernberg (1959b) demonstrated that starvation could influence the ability of fiddler crabs (*Uca pugnax*) to show metabolic compensation for changes in temperature acclimation levels. Starvation may result in a reduction of specific energy reserves, including carbohydrates and lipids, and may result in the change of metabolic substrates used. For example, the barnacle *Semibalanus balanoides* used carbohydrates first and then lipids (Barnes *et al.*, 1963a).

The rate of oxygen consumption of the benthic harpacticoid copepod, *Thompsonula hyaenae,* was significantly influenced by diet under certain conditions (Sellner, 1976). Nongravid females fed the diatom *Navicula pelliculosa* had a higher rate than animals fed different *Navicula* sp., apparently a response correlated with the smaller size and lower food energy content of *N. pelliculosa.* Ikeda (1977) reported that the respiration rates of five species of zooplankton (copepods, a euphausiid, and an amphipod) decreased progressively during a period of starvation, but remained at a constant level or increased during feeding.

Aldrich found that feeding and handling produced rises in the activity rates of oxygen consumption of *Cancer pagurus* and *Maia squinado* (Aldrich, 1975a). He also reported that the quiescent rates of these two species showed individual variability and that the rates of fed and starved animals exhibited continual cycles of rise and fall in metabolism independent of the feeding regime (Aldrich, 1975b).

C. Parasitism

The metabolism of an organism parasitized by another species may be influenced. For example, when the caridean shrimp *Palaemonetes pugio* is infected by the isopod *Probopyrus pandalicola,* its metabolic rate is dramatically reduced (Anderson, 1975b). This response may reflect changes in lipid metabolism as a host response to parasitism.

D. Energy Budgets

An increasing interest in bioenergetics of crustaceans has been shown in recent years. Energy budgets have been published for individual barnacles (Wu and Levings, 1978), grass shrimp, a parasitic isopod (Anderson, 1977), a popoulation of isopods (*Tylos punctatus:* Haynes, 1974; *Idotea baltica:* Strong and Daborn, 1979), a crab (*Carcinus maenas:* Eriksson and Edlund, 1977), a freshwater amphipod (*Gammarus pulex:* Nilsson, 1974), three species of euphausiids (Sameoto, 1976), and *Artemia* (Hentig, 1971).

Because respiration is an important component of the energy budget, all of the factors that influence oxygen uptake must be taken into account, an obviously complex process.

E. Respiratory Pigments

Respiratory pigments may play an important role in allowing crustaceans to meet the environmental problem of declining oxygen concentrations. Weber (1980) reviewed the literature on the functions of invertebrate hemoglobins with special attention to adaptations to ambient hypoxia. More detailed discussions dealing specifically with Crustacea are found in Volume 5 by Mangum, in Volume 9 by Ghidalia, and in Chapter 2 of this volume by Cameron and Mangum.

F. Season

The metabolic response of crustaceans may vary with the season of the year. Earlier examples are the studies of Edwards and Irving (1943a,b). The mole crab *Emerita talpoida,* which consumes oxygen at a much higher rate during the winter than in summer at temperatures below 20°C, is active during the winter and continues growing throughout the year. In contrast, the beach flea, *Talorchestia megalophthalma,* shows no metabolic acclimation with season, and it becomes dormant during the winter. Another example of metabolic diversity in response to temperature is illustrated by the enzymatic activity of the semiterrestrial ghost crab *Ocypode quadrata* and of the offshore spider crab *Libinia emarginata.* The activity of cytochrome c oxidase in the ghost crab, which is dormant in the winter, was relatively unaffected by acclimation temperature, while enzyme activity varied with thermal acclimation (Vernberg and Vernberg, 1968).

Seasonal differences in respiratory rates reported in barnacles (Barnes *et al.,* 1963a,b; Barnes and Barnes, 1964) may be correlated with the animal's biochemical composition during the various seasons. The low metabolic rate of *Semibalanus balanoides* in October may be the result of the increase in relatively inert semen, which may account for 50% of the total body weight. The scarcity of food and the increase in tissue water content during the winter months may account for the continued low rate of oxygen uptake after the barnacle has discharged its semen during copulation in or around November. The metabolic substrates for winter animals are proteins and lipids, but their greater dependence on carbohydrates during the summer when food is more abundant may account for the increase in metabolism during the summer.

Seasonal differences were evident in fiddler crabs even after animals were acclimated to seasonal changes in temperature (Vernberg, 1959b). However, seasonal differences in oxygen consumption of fiddler crabs (*Uca pugnax*) were eliminated upon eyestalk removal (Silverthorn, 1973). Silverthorn postulated the existence of an eyestalk factor that directly regulates respiration.

Juvenile *Carcinus maenas* consumed oxygen at a level 1.5 to 2.2 times higher in spring than in summer and autumn when measured at the same temperature (Breteler, 1975).

Seasonal changes in respiratory rates have also been observed in meiofauna. When oxygen uptake rates of the copepod *Thompsonula hyaenae* were measured at 10° and 25°C at different seasons of the year, the rate at 10°C was more than twice as high during the spring as it was during the summer. Conversely at 25°C, the rate was lowest during the spring (Vernberg and Coull, 1981). Seasonal changes in respiration have also been reported for a saltmarsh isopod (*Sphaeroma rugicauda*) by Marsden (1979) and four species of pelagic copepods by Gaudy (1973).

G. Other Cyclic Changes

In addition to seasonal changes in metabolic rate, short-term rhythmic respiratory cycles have been reported. A persistent oxygen consumption rhythm which was correlated with tidal frequency has been reported for fiddler crabs (Brown *et al.*, 1954; Barnwell, 1968) and for the crab *Carcinus maenas* (Arudpragasan and Naylor, 1964). Typically the oxygen consumption rate of these animals is highest during that phase of the tidal cycle when greatest locomotor activity occurs. For example, fiddler crabs are relatively inactive in their burrows when covered by seawater at high tide. Under this circumstance, the oxygen content within the burrows may drop to almost zero. Conceivably, a rhythmic drop in metabolism during high tide would have adaptive significance for these animals.

Both tidal and diurnal rhythms of oxygen consumption were reported for the pink shrimp (*Penaeus duorarum*) by Subrahmanyan (1976) and the sandy beach isopod (*Tylos granulatus*) by Marsh and Branch (1979). The respiration of the isopod was low in the day but increased markedly during nocturnal low tide, suggesting a circatidal rhythm. These rhythms, which persisted under constant light, probably serve to reduce metabolic energy losses during high diurnal temperatures. In contrast, no diurnal fluctuation in respiration rate was reported for a brine shrimp (*Parartemia zietziana*) inhabiting salt lakes in Victoria, Australia (Marchant and Williams, 1977).

The respiration rates of small, tropical copepods (species not listed) and a large copepod, *Undinula vulgaris,* were higher at night than during the day (Gerber and Gerber, 1979). Similar results were reported for *Orconectes nais* (a crayfish) by Rice and Armitage (1974).

The question of the existence of metabolic cycles independent of locomotory behavior has been raised in rhythm studies, since many investigators reported a simultaneous increase in both. Aldrich and McMullan (1979) investigated this problem and found circadian, circatidal, and other rhythms in immobilized crabs (*Carcinus maeans*) which were similar to those of unrestrained crabs. These findings suggest that rhythms of oxygen consumption can be detected without the contribution of muscle tissue.

Both Sutcliffe *et al.* (1975), working with crayfish, and Aldrich (1979), with marine crabs, have described methods for recording oxygen uptake over periods of time.

A comprehensive treatment of rhythmicity in crustaceans is beyond the scope of this chapter. More detailed discussion can be found in Aschoff (1964), Menaker (1971), Palmer (1975), DeCoursey (1976), Vernberg (1975), and Chapter 3 in Volume 7 of this series.

IV. CONCLUSION

A close relationship between the respiration and the ecology of crustaceans is well documented. Respiratory responses are influenced by exogenous and endogenous factors, but not always in a predictable manner. There are certain research trends relating to the physiological ecology of crustaceans that can be noted based on the review of the respiratory literature. In order to be ecologically meaningful, more investigators are measuring the oxygen uptake of organisms under experimental conditions that more nearly approximate conditions encountered in the field. Multifactor interactions are being studied with animals exposed (acclimated) to constant and fluctuating conditions.

Since the respiration rate of the intact organism reflects the sum of numerous suborganismic oxidative processes, more investigators are beginning to describe the relative contributions and the importance of these various processes to the total energetics. Which systems require more oxygen under what combination of environmental factors? How is energy usage partitioned and controlled in different environments?

The emergence of environmental scientists who bridge the gap between ecologists with "field" orientation and those with primarily "laboratory" orientation is a healthy trend. Obviously, sound field studies and realistic laboratory studies are essential to the understanding of how crustaceans get by in this world. Much progress has been made since Waterman's two-volume treatise "Physiology of Crustacea" appeared in 1960 and 1961. Rapid advances in technology (e.g., instrumentation and computers) and the perpetual interest of scientists will ensure that more comprehensive studies—which better explain the complex interactions of environment and respiration—are developed.

Acknowledgments

The excellent assistance of Ms. Nan Baker and Ms. Anne Miller is gratefully noted.

REFERENCES

Aldrich, J. C. (1975a). On the oxygen consumption of the crabs *Cancer pagurus* (L.) and *Maia squinado* (Herbst). *Comp. Biochem. Physiol.* **50,** 223–228.

Aldrich, J. C. (1975b). Individual variability in oxygen consumption rates of fed and starved *Cancer pagurus* and *Maia squinado. Comp. Biochem. Physiol. A* **51A,** 175–183.

Aldrich, J. C. (1979). *In situ* recordings of oxygen consumption rhythms in *Carcinus maenas. Comp. Biochem. Physiol. A* 64A, 279–283.

Aldrich, J. C., and McMullan, P. M. (1979). Observations on non-locomotory manifestations of biological rhythms and excitement in the oxygen consumption rates of crabs. *Comp. Biochem. Physiol. A* **64A,** 707–709.

Ambore, N. E., and Venkatachari, S. A. T. (1978) . Respiratory metabolism in relation to body size, sex and gill area of the freshwater crab *Barytelphusa guerini* Milne Edwards. *Indian J. Exp. Biol.* **16,** 465–472.

Anderson, G. (1975a). Larval metabolism of the epicardian isopod parasite *Probopyrus pandalicola* and metabolic effects of *P. pandalicola* on its copepod intermediate host *Acartia tonsa. Comp. Biochem. Physiol.* **50,** 747–751.

Anderson, G. (1975b). Metabolic response of the caridean shrimp *Palaemontes pugio* to infection by the adult epibranchial isopod parasite *Probopyrus pandalicola. Comp. Biochem. Physiol. A* **52A,** 201–207.

Anderson, G. (1977). The effects of parasitism on energy flow through laboratory shrimp population. *Mar. Biol. (Berlin)* **42,** 239–251.

Andryuschenko, V. V. (1972). Effect of DDT on respiration and survival rate of Black Sea shrimp. *Hydrobiol. J.* **8,** 39–44.

Armitage, K. B., and Lei, C.-H. (1979). Temperature acclimatization in the filtering rates and oxygen consumption of *Daphnia ambigua. Comp. Biochem. Physiol.* **62,** 807–812.

Arudpragasan, K. D., and Naylor, E. (1964). Gill ventilation volumes, oxygen consumption and respiratory rhythms in *Carcinus maenas* L. *J. Exp. Biol.* **41,** 309–321.

Aschoff, J., ed.. (1964). "Circadian Clocks." North-Holland Publ., Amsterdam.

Babula, A., Bielawski, A., and Sobieski, J. (1978). Oxygen consumption and gill surface area in *Mesidothea entomon* (L.) (Isopoda, Crustacea). *Comp. Biochem. Physiol. A* **61A,** 595–597.

Barnes, H., and Barnes, M. (1964). Some relations between the habitat, behavior, and metabolism on exposure to air of the high-level intertidal cirripede *Chthamalus depressus* (Poli). *Helgol. Wiss. Meeresunters.* **10,** 19–27.

Barnes, H., Barnes, M., and Finlayson, D. M. (1963a). The seasonal changes in body weight, biochemical composition and oxygen uptake of two common boreo-Arctic cirripedes, *Balanus balanoides.* (L) and *Balanus balanus* (L.). *J. Mar. Biol. Assoc. U.K.* **43,** 185–211.

Barnes, H., Barnes, M., and Finlayson, D. M. (1963b). The metabolism during starvation of *Balanus balanoides. J. Mar. Biol. Assoc. U.K.* **43,** 213–233.

Barnwell, F. H. (1968). The role of rhythmic systems in the adaptation of fiddler crabs to the intertidal zone. *Am. Zool.* **8,** 569–583.

Belman, B. W., and Childress, J. J. (1976). Circulatory adaptations to the oxygen minimum layer in the bathypelagic mysid *Gnathophausia ingens. Biol. Bull. (Woods Hole, Mass.)* **150,** 15–37.

Bishop, J. W. (1968). Respiratory rates of migrating zooplankton in natural habitat. *Limnol. Oceanogr.* **13,** 58–62.

Bliss, D. E., and Skinner, D. M. (1963). "Tissue Respiration in Invertebrates." Am. Mus. Nat. Hist., New York.

Boulton, A. P., and Huggins, A. K. (1977). Biochemical changes occurring during morphogenesis of the brine shrimp *Artemia salina* and the effect of alterations in salinity. *Comp. Biochem. Physiol. A* **57A,** 17–22.
Brauer, R. W., ed. (1972). "Barobiology and the Experimental Biology of the Deep Sea," N.C. Sea Grant Program, Spec. Publ. Univ. of North Carolina Press,
Breteler, W. C. M. K. (1975). Oxygen consumption and respiratory levels of juvenile shore crabs, *Carcinus maenas,* in relation to weight and temperature. *Neth. J. Sea Res.* **9,** 243–254.
Brown, F. A., Bennett, M. F., and Webb, H. M. (1954). Persistent daily and tidal rhythms of O_2 consumption in fiddler crabs. *J. Cell Comp. Physiol.* **44,** 477–506.
Buikema, A. L., Jr. (1972). Oxygen consumption of the cladoceran *Daphnia pulex,* as a function of body size, light and light acclimation. *Comp. Biochem. Physiol. A* **42A,** 877–888.
Buikema, A. L., Jr. (1975). Some effects of light on the energetics of *Daphnia pulex* and implications for the significance of vertical migration. *Hydrobiologia* **47,** 43–58.
Bullock, T. H. (1955). Compensation for temperature in the metabolism and activity of poikilotherms. *Biol. Rev. Cambridge Philos. Soc.* **30,** 311–342.
Bulnheim, H.-P. (1972). Vergleichende Untersuchungen zur Atmungsphysiologie euryhaliner Gammariden unter besonderer Berücksichtigung der Salzgehaltsanpassung. *Helgol. Wiss. Meeresunters.* **23,** 485–534.
Burke, E. M. (1979). Aerobic and anaerobic metabolism during activity and hypoxia in two species of intertidal crabs. *Biol. Bull. (Woods Hole, Mass.)* **156,** 157–168.
Cairns, J., Jr., ed. (1980). "The Recovery Process in Damaged Ecosystems." Ann Arbor Sci. Publ., Ann Arbor, Michigan.
Cantelmo, A. C., Conklin, P. J., Fox, F. R., and Rao, K. R. (1978). Effects of sodium pentachlorophenate and 2,4-dinitrophenol on respiration in crustaceans. *In* "Pentachlorophenol: Chemistry, Pharmacology and Environmental Toxicology" (K. R. Rao, ed.), pp. 251–263. Plenum, New York.
Childress, J. J. (1968). Oxygen minimum layer, vertical distribution and respiration of the mysid Gnathophausia ingens. *Science* **160,** 1242–1243.
Childress, J. J. (1971). Respiratory adaptations to the oxygen minimum layer in the bathypelagic mysid *Gnathophausia ingens. Biol. Bull. (Woods Hole, Mass.)* **141,** 109–121.
Childress, J. J. (1975). The respiratory rates of midwater crustaceans as a function of depth of occurrence and in relation to the oxygen minimum layer off southern California. *Comp. Biochem. Physiol.* **50,** 787–799.
Childress, J. J. (1976). Physiological approaches to the biology of midwater crustacean. *In* "Predictions of Sonic Scattering Layers" (N. Anderson, ed.), pp. 37–80. Plenum, New York.
Childress, J. J. (1977). The effects of pressure, temperature and oxygen on the oxygen consumption rate of the midwater copepod *Gaussia princeps. Mar. Biol. (Berlin)* **39,** 19–24.
Chinnayya, B. (1971). Effect of Heavy Metals on the oxygen consumption by the shrimp, *Caridina rajadhari* Bouvier. *Indian J. Exp. Biol.* **9,** 277–278.
Clegg, J. S. (1976). Interrelationships between water and metabolism in *Artemia* cysts. III. Respiration. *Comp. Biochem. Physiol. A* **53A,** 89–93.
Comita, G. W. (1968). Oxygen consumption in *Diaptomus. Limnol. Oceanogr.* **13,** 51–57.
Conover, R. J. (1959). Regional and seasonal variation in the respiratory rate of marine copepods. *Limnol. Oceanogr.* **4,** 259–268.
Conover, R. J. (1962). Metabolism and growth in *Calanus hyperboreus* in relation to its life cycle. *Rapp. P.-V., Reun., Cons. Int. Explor. Mer* **153,** 190–197.

Coull, B. C., and Vernberg, W. B. (1970). Harpacticoid copepod respiration: *Enhydrosoma propinquum* (Brady) and *Longipedia helgolandica* (Klie). *Mar. Biol. (Berlin)* **5,** 341–344.

Dame, R. F., and Vernberg, F. J. (1978). The influence of constant and cyclic acclimation temperature on the metabolic rate of *Panopeus herbstii* and *Uca pugilator*. *Biol. Bull. (Woods Hole, Mass.)* **154,** 188–197.

DeCoursey, P. J., ed. (1976). "Biological Rhythms in the Marine Environment." Univ. of South Carolina Press, Columbia.

Dehnel, P. A. (1958). Effect of photoperiod on the oxygen consumption of two intertidal crabs. *Nature (London)* **184,** 1415–1417.

Dehnel, P. A. (1960). Effect of temperature and salinity on the oxygen consumption of two intertidal crabs. *Biol. Bull. (Woods Hole, Mass.)* **118,** 215–249.

deZwann, A., and Skjoldal, H. R. (1979). Anaerobic energy metabolism of the scavenging isopod *Cirolana borealis* (Lilljeborg). *J. Comp. Physiol.* **129,** 327–331.

Dickson, G. W., and Franz, R. (1980). Respiration rates, ATP turnover and adenylate energy charge in excised gills of surface and cave crayfish. *Comp. Biochem. Physiol. A* **65A,** 375–380.

Diwan, A. D. (1975). Morpho-history of the respiratory system of the freshwater crab, *Barytelphusa cunicularis* (Westwood, 1836). *Indian J. Zoology* **16,** 11–20.

Duncan, A. (1966). The oxygen consumption of *Potamopyrgus jenkinsi* (Smith) (Prosobranchiata) in different temperatures and salinities. *Verh.—Int. Ver. Theor. Angew. Limnol.* **16,** 1739–1751.

Edwards, G. A., and Irving, L. (1943a). The influence of temperature and season upon the oxygen consumption of the sand crab. *Emerita talpoida* Say. *J. Cell. Comp. Physiol.* **21,** 169–182.

Edwards, G. A., and Irving, L. (1943b). The influence of season and temperature upon the oxygen consumption of the beach flea, *Talorchestia megalophthalma*. *J. Cell. Comp. Physiol.* **21,** 183–189.

Edwards, R. R. C. (1978). Effects of water-soluble oil fractions on metabolism growth and carbon budget of the shrimp *Crangon crangon*. *Mar. Biol. (Berlin)* **46,** 259–265.

Engel, D. W., and Eggert, L. D. (1974). The effect of salinity and sex on the respiration rates of excised gills of the blue crab, *Callinectes sapidus*. *Comp. Biochem. Physiol. A* **47A,** 1005–1011.

Engel, D. W., Ferguson, R. L., and Eggert, L. D. (1975). Respiration rates and ATP concentrations in the excised gills of the blue crab as a function of salinity. *Comp. Biochem. Physiol. A* **52A,** 669–673.

Epp, R. W., and Lewis, W. M., Jr. (1979). Metabolic responses to temperature change in a tropical freshwater copepod (*Mesocyclops brasilianus*) and their adaptive significance. *Oecologia* **42,** 123–138.

Eriksson, S., and Edlund, A.-M. (1977). On the ecological energetics of O-group *Carcinus maenas* (L.) from a shallow sandy bottom in Gullmar Fjord, Sweden. *J. Exp. Mar. Biol. Ecol.* **30,** 233–248.

Felder, D. L. (1979). Respiratory adaptations of the estuarine mud shrimp, *Callianassa jamaicense* (Schmitt, 1935) (Crustacea, Decapoda, Thalassinidea). *Biol. Bull. (Woods Hole, Mass.)* **157,** 125–137.

Findley, A. M., Belisle, B. W., and Stickle, W. B. (1978). Effects of salinity fluctuations on the respiration rate of the Southern oyster drill *Thais haemastoma* and the blue crab *Callinectes sapidus*. *Mar. Biol. (Berlin)* **49,** 59–67.

Fish, J. D., and Preece, G. S. (1970). The ecophysiological complex of *Bathyporeia pilosa* and *B. pelagica* (Crustacea: Amphipoda). I. Respiration rates. *Mar. Biol. (Berlin)* **5,** 22–28.

Flugel, H. (1972). Pressure: Animals. *In* "Marine Ecology" (O. Kinne, ed.), Vol. 1, pp. 1407–1450. Wiley, Chichester.

Fontaine, M. (1929a). De l'augmentation de la consommation d'O_2 des animaux marine sous l'influence des fortes pressions. Ses variations en fonction de l'intensité de la compression. *C.R. Hebd. Seances Acad. Sci.* **188,** 460–461.

Fontaine, M. (1929b). De l'augmention de la consummation d'oxygène des animaux marins sous l'influence des fortes pressions. Ses variations en fonction de la duréd de la compression. *C.R. Hebd. Seances Acad. Sci.* **188,** 662–663.

Foulds, J. B., and Roff, J. C. (1976). Oxygen consumption during simulated vertical migration in *Mysis relicta* (Crustacea, Mysidacea). *Can. J. Zool.* **54,** 377–385.

Frier, J. O. (1976). Oxygen consumption and osmoregulation in the isopods *Sphaeroma hookeri* Leach and *S. rugicauda* Leach. *Ophelia* **15,** 193–203.

Gamble, J. C. (1970). Effect of low dissolved oxygen concentrations on the ventilation rhythm of three tubicolous crustaceans, with special reference to the phenomenon of intermittent ventilation. *Mar. Biol. (Berlin)* **6,** 121–127.

Gaudy, R. (1973). Les variations saisonnières de la respiration chez quatre especès de copépodes pélagiques de Golfe de Marseille. *Neth. J. Sea Res.* **7,** 267–279.

George, R. Y. (1979a). What adaptive strategies promote immigration and speciation in deep sea environment. *Sarsia* **64,** 61–65.

George, R. Y. (1979b). Behavioral and metabolic adaptation of polar and deep sea crustaceans: A hypothesis concerning physiological basis for evolution of cold adapted crustaceans. *Bull. Biol. Soc. Wash* **3,** 283–296.

George, R. Y. (1981). Functional adaptations of deep-sea organisms. *In* "Functional Adaptations of Marine Organisms" (F. J. Vernberg and W. B. Vernberg, eds.), pp. 279–332. Academic Press, New York.

Gerber, R. P., and Gerber, M. B. (1979). Ingestion of natural particulate organic matter and subsequent assimilation, respiration and growth by tropical lagoon zooplankton. *Mar. Biol. (Berlin)* **52,** 33–43.

Gilfillan, E. S. (1975). The effects of water body characteristics other than temperature and salinity on respiration in *Euphausia pacifica*. *In* "Respiration of Marine Organisms" (J. J. Cech, Jr., D. W. Bridges, and D. B. Horton, eds.), pp. 24–40.

Goldberg, E. D. (1976). "Strategies for Marine Pollution Monitoring." Wiley-Interscience, New York.

Gophen, M. (1976). Temperature dependence of food intake, ammonia excretion and respiration in *Ceriodaphnia reticulata* (Jurine) (Lake Kinneret, Israel). *Freshwater Biol.* **6,** 451–455.

Gray, I. E. (1957). A comparative study of the gill area of crabs. *Biol. Bull. (Woods Hole, Mass.)* **112,** 34–42.

Green, J. D. (1975). Feeding and respiration in the New Zealand copepod *Calamoecia lucasi* Brady. *Oecologia* **21,** 345–358.

Gyllenberg, G., and Lundqvist, G. (1979). Oxygen consumption of *Eurytemora hirundoides* nauplii and adults as a function of salinity. *Ann. Zool. Fenn.* **15,** 328–330.

Hagerman, L. (1970). The oxygen consumption of *crangon vulgaris* (Fab.) (Crustacea, Natantia) in relation to salinity. *Ophelia* **7,** 283–292.

Halcrow, K. (1963). Acclimation to temperature in the marine copepod, *Calanus finmarchicus* (Gunneus). *Limnol. Oceanogr.* **8,** 1–8.

Halcrow, K., and Boyd C. M. (1967). The oxygen consumption and swimming activity of the amphipod *Gammarus oceanicus* at different temperatures. *Comp. Biochem. Physiol.* **23,** 233–242.

Hammen, C. S. (1972). Lactate oxidation in the upper-shore barnacle, *Chthamalus depressus* (Poli). *Comp. Biochem. Physiol. A* **43A,** 435–441.

Haq, S. M. (1965). Development of the copepod *Euterpina acutifrons* with special reference to dimorphism in the male. *Proc. Zool. Soc. London* **144,** 175–201.

Harding, G. C. H. (1977). Surface area of the euphausiid *Thysanoessa raschii* and its relation to body length, weight, and respiration. *J. Fish. Res. Board Can.* **34,** 225–231.

Haynes, W. B. (1974). Sand-beach energetics: Importance of the isopod *Tylos punctatus*. *Ecology* **55,** 838–847.

Hentig, R. (1971). Influence of salinity and temperature on the development, growth, reproduction, and energy budget of *Artemia salina*. *Mar. Biol.* **9,** 145–182.

Herreid, C. F., II, Lawrence, W. L., and Ghulam, M. S. (1979). Respiration and heart rate in exercising land crabs. *Respir. Physiol.* **36,** 109–120.

Ikeda, T. (1977). The effect of laboratory conditions on the extrapolation of experimental measurements to the ecology of marine zooplankton. IV. Changes in respiration and excretion rates of boreal zooplankton species maintained under fed and starved conditions. *Mar. Biol. (Berlin)* **41,** 241–252.

Khan, M. A. Q., and Bederka, J. P., Jr. (1974). "Survival in Toxic Environments." Academic Press, New York.

King, E. N. (1965). The oxygen consumption of intact crabs and excised gills as a function of decreased salinity. *Comp. Biochem. Physiol.* **15,** 93–102.

Kinne, O. (1964). The effects of temperature and salinity on marine and brackish water animals. II. Salinity and temperature-salinity combinations. *Oceanogr. Mar. Biol.* **2,** 281–339.

Kinne, O. (1970). Temperature: Invertebrates marine ecology. *In* "Marine Ecology" (O. Kinne, ed.), Vol. 1, Part 1, pp. 407–514. Wiley-Interscience, New York.

Kotaiah, K., and Rajabai, B. S. (1972). Influence of thermal acclimation on oxygen consumption in the tropical freshwater crab. *Paratelphusa hydrodromus* (Herbst), with reference to size, sex and sudden changes of temperature. *Indian J. Exp. Biol.* **10,** 375–378.

Kutty, M. N., Murugapoopathy, G., and Krishnan, T. S. (1971). Influence of salinity and temperature on the oxygen consumption in young juveniles of the Indian prawn *Penaeus indicus*. *Mar. Biol. (Berlin)* **11,** 125–134.

Laird, C. E., and Haefner, P. A., Jr. (1976). Effects of intrinsic and environmental factors on oxygen consumption in the blue crab, *Callinectes sapidus* Rathbun. *J. Exp. Mar. Biol. Ecol.* **22,** 171–178.

Lance, J. (1965). Respiration and osmotic behaviour of the copepod *Acartia tonsa* in diluted sea water. *Comp. Biochem. Physiol.* **14,** 155–165.

Lasserre, P., and Renaud-Mornant, J. (1973). Resistance and respiratory physiology of intertidal meiofauna to oxygen-deficiency. *Neth. J. Sea Res.* **7,** 290–302.

Leffler, C. W. (1973). Metabolic rate in relation to body size and environmental oxygen concentration in two species of xanthid crabs. *Comp. Biochem. Physiol. A* **44A,** 1047–1052.

Lockwood, A. P. M. (1967). "Aspects of the Physiology of Crustacea." Freeman, San Francisco, California.

Longhurst, A. R. (1967). Vertical distribution of zooplankton in relation to the eastern Pacific oxygen minimum. *Deep-Sea Res.* **14,** 51–63.

MacDonald, A. G. (1975). "Physiological Aspects of Deep Sea Biology." Cambridge Univ. Press, London and New York.

MacDonald, A. G., Gilchrist, I., and Teal, J. M. (1972). Some observations on the tolerance of oceanic plankton to high hydrostatic pressure. *J. Mar. Biol. Assoc. U.K.* **52,** 213–223.

McFarland, W. N., and Pickens, P. E. (1965). The effects of season, temperature, and salinity on standard and active oxygen consumption of the grass shrimp, *Palaemonetes vulgaris* (Say). *Can. J. Zool.* **43,** 571–585.

McMahon, B. R., and Burggren, W. W. (1979). Respiration and adaptation to the terrestrial habitat in the land hermit crab *Coenobita clypeatus. J. Exp. Biol.* **79,** 265–281.

MacMillen, R. E., and Greenaway, P. (1978). Adjustments of energy and water metabolism to drought in an Australian arid-zone crab. *Physiol. Zool.* **51,** 230–240.

McWhinnie, M. A. (1964). Temperature responses and tissue respiration in Antarctic Crustacea with particular reference to the krill, *Euphausia superba. Antarct. Res. Ser. Vol.* **1,** 63–72.

Madanmohanrao, G., and Rao, K. P. (1962). Oxygen Consumption in a brackish water crustacean, *Sesarma plicatum* (Latreille) and a marine crustacean, *Lepas anserifera* L. *Crustaceana* **4,** 75–81.

Mangum, C., and Van Winkle, W. (1973). Responses of aquatic invertebrates to declining oxygen conditions. *Am. Zool.* **13,** 529–541.

Marchant, R., and Williams, W. D. (1977). Field estimates of oxygen consumption for the brine shrimp *Parartemia zietziana* Sayce (Crustacea: Anostraca) in two salt lakes in Victoria, Australia. *Freshwater Biol.* **7,** 535–544.

Marsden, I. D. (1979). Seasonal oxygen consumption of the salt-marsh isopod *Sphaeroma rugicauda. Mar. Biol.* **51,** 329–338.

Marsden, I. D., Newell, R. C., and Ahsanullah, M. (1973). The effect of starvation on the metabolism of the shore crab *Carcinus meanas. Comp. Biochem. Physiol. A* **45A,** 195–213.

Marsh, B. A., and Branch, G. M. (1979). Circadian and circatidal rhythms of oxygen consumption in the sandy-beach isopod *Tylos Granulatus* Krauss. *J. Exp. Mar. Biol. Ecol.* **37,** 77–89.

Marshall, N. B. (1954. "Aspects of Deep Sea Biology." Philosophical Library, New York.

Marshall, S. M. (1973). Respiration and feeding in copepods. *Adv. Mar. Biol.* **11,** 57–120.

Marshall, S. M., Nicholls, A. G., and Orr, A. P. (1935). On the biology of *Calanus finmarchicus*. VI. Oxygen consumption in relation to environmental conditions. *J. Mar. Biol. Assoc. U.K.* **20,** 1–28.

Mauchline, J., and Fisher, L. R. (1969). The biology of euphausiids. *Adv. Mar. Biol.* **7,** 1–454.

Menaker, M. (1971). "Biochronometry." Nat. Acad. Sci., Washington, D.C.

Mickel, T. J., and Childress, J. J. (1978). The effect of pH on oxygen consumption and activity in the bathypelagic mysid *Gnathophausia ingens. Biol. Bull. (Woods Hole, Mass.)* **154,** 138–147.

Moreira, G. S., and Vernberg, W. B. (1968). Comparative thermal metabolic patterns in *Euterprina acutifrons* dimorphic males. *Mar. Biol. (Berlin)* **1,** 282–284.

Moshiri, G. A., Goldman, C. R., Godshalk, G. L., and Mull, D. R. (1970). The effects of variations in oxygen tension on certain aspects of respiratory metabolism in *Pacifastacus leniusculus* (Dana) (Crustacea: Decapoda). *Physiol. Zool.* **43,** 23–29.

Naroska, V. (1968). Vergleichende Untersuchungen über den Einfluß des hydrostatischen Druckes auf Überlebensfähigkeit und Stoffwechselintensität mariner Evertebraten und Telosteer. *Kiel. Meresforsch.* **24,** 95–123.

Nash, J. (1979). The effect of daily fluctuating temperatures on the oxygen consumption of *Oniscus asellus* L. *Comp. Biochem. Physiol. A* **62,** 983–988.

Newell, R. C. (1976). "Adaptations to Environment: Essays on the Physiology of Marine Animals." Butterworth, London.

Newrkla, P. (1978). The influence of ionic concentration on population parameters, development time, activity, and respiration rate of *Arctodiaptomus spinosus* (Daday) (Calanoida, Copepoda). *Oecologia* **33,** 87–99.

Nilsson, L. M. (1974). Energy budget of a laboratory population of *Gammarus pulex* (Amphipoda). *Oikos* **25,** 35–42.

Ostapenya, A. P., Pechen, G. A., Babitskiy, V. A., and Pavlyutin, A. P. (1969). Metabolic rate of *Diaptomus graciloides* (Lill.) at low temperature. *Hydrobiol. J.* **5,** 88–91.

Palmer, J. D. (1975). "Biological Clocks in Marine Organisms." Wiley-Interscience, New York.

Pamatmat, M. M. (1978). Oxygen uptake and heat production in a metabolic conformer (*Littorina irrorata*) and a metabolic regulator (*Uca pugnax*). *Mar. Biol. (Berlin)* **48,** 317–325.

Pearcy, W. G., and Small, L. F. (1968). Effects of pressure on the respiration of vertically migrating crustaceans. *J. Fish. Res. Board Can.* **25,** 1311–1316.

Pearse, A. S. (1929). Observations on certain littoral and terrestrial animals at Tortugas, Florida, with special reference to migrations from marine to terrestrial habitats. *Pap. Tortugas Lab.* (Reprinted from *Carnegie Inst. Washington Pub.*) **391,** 205–223).

Precht, H. (1958). Concepts of the temperature adaptation of unchanging reaction systems of cold-blooded animals. *In* "Physiological Adaptation" (C. L. Prosser, ed.), pp. 50–77. Am. Physiol. Soc., Washington, D.C.

Prentice, E. F., and Schneider, D. E. (1979). Respiration and thermal tolerance of the Dungeness crab. *Cancer magister. Comp. Biochem. Physiol. A* **63,** 591–598.

Prosser, C. L. (1955). Physiological variation in animals. *Biol. Rev. Cambridge Philos. Soc.* **30,** 229–262.

Prosser, C. L. (1958). The nature of physiological adaptation. *In* "Physiological Adaptation" (C. L. Prosser, ed.), pp. 167–180. Am. Physiol. Soc., Washington, D.C.

Prosser, C. L. (1973). "Comparative Animal Physiology." Saunders, Philadelphia, Pennsylvania.

Pruitt, N. L., and Dimock, R. V., Jr. (1979). The effects of temperature and eyestalk extracts on oxygen consumption of the crayfish *Cambarus acuminatus* (Faxon). *Comp. Biochem. Physiol. A* **62A,** 631–634.

Quetin, L. B., and Childress, J. J. (1976). Respiratory adaptations of *Pleuroncodes planipes* Stimpson to its environment off Baja, California. *Mar. Biol. (Berlin)* **38,** 327–334.

Quetin, L. B., Mickel, T. J., and Childress, J. J. (1978). A method for simultaneously measuring the oxygen consumption and activity of pelagic crustaceans. *Comp. Biochem. Physiol. A* **59A,** 263–266.

Rakusa-Suszczewski, S., McWhinnie, M. A., and Cahoon, M. O. (1976). Respiration of the antarctic copepod, *Rhincalanus gigas. (Limnol. Oceanogr.* **21,** 763–765.

Ranta, E., and Hakala, I. (1978). Respiration of *Mysis relicta* (Crustacea Malacostraca). *Arch Hydrobiol.* **83,** 515–523.

Rao, K. R., Fox, F. R., Conklin, P. J., Cantelmo, A. C., and Brannon, A. C. (1979). Physiological and biochemical investigations of the toxicity of pentachlorophenol to crustaceans. *In* "Marine Pollution: Functional Responses" (W. B. Vernberg, F. P. Thurberg, A. Calabrese, and F. J. Vernberg, eds.), pp. 307–339. Academic Press, New York.

Reeve, M. R., Grice, G. D., Gibson, V. R., Walter, M. A., Darcy, K., and Ikeda, T. (1976). A controlled environmental pollution experiment (CEPEX) and its usefulness in the study of larger marine zooplankton under toxic stress. *In* "Effects of Pollutants on Aquatic Organisms" (A. P. Lockwood, ed.), pp. 145–162. Cambridge, Univ. Press, London and New York.

Rice, P. R., and Armitage, K. B. (1974). The effect of photoperiod on oxygen consumption of the crayfish *Orconectes nais* (Faxon). *Comp. Biochem. Physiol. A* **47A,** 261–270.

Robert, M., and Gray, I. (1972). Enzymatic mechanisms during temperature acclimation of the blue crab, *Callinectes sapidus*. I. Oxygen consumption and activity of glucose-6-phosphate dehydrogenase and 6-phosphogluconate dehydrogenase. *Comp. Biochem. Physiol. B* **42B,** 377–387.

Roff, J. C. (1973). Oxygen consumption of *Limnocalanus macrurus* Sars (Calanoida, Copepoda) in relation to environmental conditions. *Can. J. Zool.* **51,** 877–885.

Roux, C. (1975). Les variations de la courbe métabolisme/température de *Gammarus pulex* (L.) et *Gammarus fossarum* Koch (Amphipodes) sous L'influence de divers facteures écologiques. *Crustaceana* **29,** 33–48.

Sameoto, D. D. (1976). Respiration rates, energy budgets, and molting frequencies of three species of euphausiids found in the Gulf of St. Lawrence. *J. Fish. Res. Bd. Can.* **33,** 2568–2576.

Sastry, A. N. (1979). Metabolic adaptations of *Cancer irroratus* developmental stages to cyclic temperatures. *Mar. Biol. (Berlin)* **51,** 243–250.

Schembri, P. J. (1979). Oxygen consumption and the respiratory responses to declining oxygen tension in the crab *Ebalia tuberosa* (Pennant) (Crustacea: Decapoda: Leucosiidae). *J. Exp. Mar. Biol. Ecol.* **41,** 133–142.

Schlieper, C. (1929). Uber die Einwirkung niederer Salzkonzentrationen auf marine Organismen. *Z. Vgl. Physiol.* **9,** 478–514.

Schneider, D. E. (1967). An evaluation of temperature adaptations in latitudinally separated populations of the xanthid crab, *Rhithropanopeus harrisii* (Gould) by laboratory rearing experiments. Ph.D. Dissertation, Duke University, Durham, North Carolina.

Sellner, B. W. (1976). Survival and metabolism of the harpacticoid copepod *Thompsonula hyaenae* (Thompson) fed on different diatoms. *Hydrobiologia* **50,** 233–238.

Shcherbakov, A. P. (1935). The oxygen consumption of certain plankton crustaceans. *Tr. Limnol. Stn. Kossino* **19,** 25–41.

Shumway, S. E. (1978). Osmotic balance and respiration in the hermit crab, *Pagurus bernhardus,* exposed to fluctuating salinities. *Mar. Biol. Assoc. U.K.* **58,** 869–876.

Silverthorn, S. U. (1973). Respiration in eyestalkless *Uca* (Crustacea: Decapoda) acclimated to two temperatures. *Comp. Biochem. Physiol. A* **45A,** 417–420.

Silverthorn, S. U. (1975a). Hormonal involvement in thermal acclimation in the fiddler crab *Uca pugilator* (Bosc). I. Effect of eyestalk extracts on whole animal respiration. *Comp. Biochem. Physiol. A* **50A,** 281–283.

Silverthorn, S. U. (1975b). Hormonal involvement in thermal acclimation in the fiddler crab, *Uca pugilator* (Bosc). II. Effects of extracts on tissue respiration. *Comp. Biochem. Physiol. A* **50A,** 285–290.

Small, L. F., and Hebard, J. F. (1967). Respiration of a vertically migrating crustacean *Euphausia pacifica* Hansen. *Limnol. Oceanogr.* **12,** 272–280.

Small, L. F., Hebard, J. F., and McIntire, C. D. (1966). Respiration in euphausiids (*Euphausia pacifica*). *Nature (London)* **211,** 1210–1211.

Smith, G. A., Fitzpatrick, L. C., and Pearson, W. D. (1978). Metabolic relations to temperatures in the copepods *Diaptomus dorsalis* and *Mesocyclops edax* from North-Central Texas. *Comp. Biochem. Physiol. A* **59A,** 325–326.

Smith, K. K., Jr., and Teal, J. M. (1973). Temperature and pressure effects on respiration of thecostomatous pteropods. *Deep-Sea Res.* **20,** 853–858.

Spaargaren, D. J. (1977). On the metabolic adaptation of *Carcinus maenas* to reduced oxygen tensions in the environment. *Neth. J. Sea Res.* **11,** 325–333.

Strong, K. W., and Daborn, G. R. (1979). Growth and energy utilization of the intertidal isopod *Idotea baltica* (Pallas) (Crustaceal Isopoda). *J. Exp. Mar. Biol. Ecol.* **41,** 101–124.

Subrahmanyam, C. B. (1962). Oxygen consumption in relation to body weight and oxygen tension in the prawn *Penaeus indicus* (Milne Edwards). *Proc. Indian Acad. Sci.* **55,** 152–161.

Subrahmanyam, C. B. (1976). Tidal and diurnal rhythms of locomotory activity and oxygen consumption in the pink shrimp, *Penaeus duroramm*. *Contri. Mar. Sci.* **20,** 123–132.

Sutcliffe, D. W., Carrick, T. R., and Moore, W. H. (1975). An automatic respirometer for determining oxygen uptake in crayfish (*Austropotamobius pallipes* (Lereboullet) over periods of 3–4 days. *J. Exp. Biol.* **63,** 673–688.

Taylor, A. C. (1976). The respiratory responses of *Carcinus maenas* to declining oxygen tension. *J. Exp. Biol.* **65,** 309–322.

Taylor, A. C. (1977). The respiratory responses of *Carcinus maenas* (L.) to changes in environmental salinity. *J. Exp. Mar. Biol. Ecol.* **29,** 197–210.

Teal, J. M. (1971). Pressure effects on the respiration of vertically migrating decapod crustacea. *Am. Zool.* **11,** 571–576.

Teal, J. M., and Carey, F. G. (1967a). The metabolism of marsh crabs under conditions of reduced oxygen pressure. *Physiol. Zool.* **40,** 83–91.

Teal, J. M., and Carey, F. G. (1967b). Effects of pressure and temperature on the respiration of euphausiids. *Deep-Sea Res.* **14,** 725–733.

Theede, H. (1973). Comparative studies on the influence of oxygen deficiency and hydrogen sulphide on marine bottom invertebrates. *Neth. J. Sea Res.* **7,** 244–252.

Thompson, R. K., and Pritchard, A. W. (1969). Respiratory adaptation of two burrowing Crustaceans, *Callianassa californiensis* and *Upogebia pugettensis* (Decapoda, Thalassinidae). *Biol. Bull. (Woods Hole, Mass.)* **136,** 274–287.

Thurberg, F. P., Sawson, M. A., and Collier, R. S. (1973). Effects of copper and cadmium on osmoregulation and oxygen consumption in two species of estuarine crabs. *Mar. Biol. (Berlin)* **23,** 171–175.

Torres, J. J., Gluck, D. L., and Childress, J. J. (1977). Activity and physiological significance of the pleopods in the respiration of *Callianassa californiensis* (Dana) (Crustacea: Thalassinidea). *Biol. Bull. (Woods Hole, Mass.)* **152,** 134–146.

Uglow, R. F. (1973). Some effects of acute oxygen changes on heart and scaphognathite activity in some portunid crabs. *Neth. J. Sea Res.* **7,** 447–454.

Van Weel, P. B., Randall, J. E., and Takata, M. (1954). Observations on the oxygen consumption of certain marine Crustacea. *Pac. Sci.* **8,** 209–218.

Veerannan, K. M. (1972). Respiratory metabolism of crabs from marine and estuarine habitats. I. *Scylla serrata. Mar. Biol. (Berlin)* **16,** 284–290.

Vernberg, F. J. (1956). Study of the oxygen consumption of excised tissue of certain marine decapod Crustacea in relation to habitat. *Physiol. Zool.* **29,** 227–234.

Vernberg, F. J. (1959a). Studies on the physiological variation between tropical and temperate zone fiddler crabs of the genus *Uca*. II. Oxygen consumption of whole organisms. *Biol. Bull. (Woods Hole, Mass.)* **117,** 163–184.

Vernberg, F. J. (1959b). Studies on the physiological variation between tropical and temperate zone fiddler crabs of the genus *Uca*. IV. The influence of temperature acclimation on oxygen consumption of whole organisms. *Biol. Bull. (Woods Hole, Mass.)* **117,** 582–593.

Vernberg, F. J. (1962). Latitudinal effects on physiological properties of animal populations. *Annu. Rev. Physiol.* **24,** 517–546.

Vernberg, F. J. (1969). Acclimation of intertidal crabs. *Am. Zool.* **9,** 333–341.

Vernberg, F. J. (1972a). Genetic and phenotypic plasticity of oceanic invertebrates. *Proc. Eur. Mar. Biol. Symp., 5th, 1972,* pp. 245–257.

Vernberg, F. J. (1972b). Dissolved gases: Animals. *In* "Marine Ecology" (O. Kinne, ed.), Vol. 1, pp. 1491–1526. Wiley, Chichester.

Vernberg, F. J., ed. (1975). "Physiological Adaptation to the Environment." Intext Educational Publishers, New York.

Vernberg, F. J., and Costlow, J. D., Jr. (1966). Studies on the physiological variation between

tropical and temperate zone fiddler crabs of the genus *Uca*. IV. Oxygen consumption of larval and young crabs reared in the laboratory. *Physiol. Zool.* **39,** 36–52.
Vernberg, F. J., and Vernberg, W. B., eds. (1974). "Pollution and Physiology of Marine Organisms." Academic Press, New York.
Vernberg, F. J., and Vernberg, W. B., eds. (1981). "Functional Adaptations of Marine Organisms." Academic Press, New York.
Vernberg, F. J., Calabrese, A., Thurberg, F. P., and Vernberg, W. B., eds. (1977). "Physiological Responses of Marine Biota to Pollutants." Academic Press, New York.
Vernberg, F. J., Guram, M. S., and Savory, M. A. (1978). Metabolic response to thermal changes of the adult fiddler crab *Uca pugilator* and the effect of PCBs. *Mar. Biol.* **48,** 135–141.
Vernberg, W. B., and Coull, B. C. (1981). Meiofauna. *In* "Functional Adaptations of Marine Organisms" (F. J. Vernberg and W. B. Vernberg, eds.), pp. 147–178. Academic Press, New York.
Vernberg, W. B., and Vernberg, F. J. (1968). Physiological diversity in metabolism in marine and terrestrial crustacea. *Am. Zool.* **8,** 449–458.
Vernberg, W. B., and Vernberg, F. J. (1972). "Environmental Physiology of Marine Animals." Springer-Verlag, Berlin and New York.
Vernberg, W. B., DeCoursey, P. J., and O'Hara, J. (1974). Multiple environmental factor effects on physiology and behavior of the fiddler crab, *Uca pugilator*. *In* "Pollution and Physiology of Marine Organisms" (F. J. Vernberg and W. B. Vernberg, eds.), pp. 381–425. Academic Press, New York.
Vernberg, W. B., Coull, B. C., and Jorgensen, D. D. (1977). Reliability of laboratory metabolic measurements of meiofauna. *J. Fish. Res. Board Can.* **34,** 164–167.
Vernberg, W. B., Thurberg, F. P., Calabrese, A., and Vernberg, F. J., eds. (1979). "Marine Pollution: Functional Responses." Academic Press, New York.
Vlasblom, A. G., Graafsma, S. J., and Verhoeven, J. T. A. (1977). Survival, osmoregulatory ability, and respiration of *Idotea chelipes* (Crustacea, Isopoda) from Lake Veere in different salinities and temperatures. *Hydrobiologia* **52,** 33–38.
Vos, J., Bernaerts, F., Gabriels, I., and Decleir, W. (1979). Aerobic and anaerobic respiration of adult *Artemia salina* L., acclimated to different oxygen concentrations. *Comp. Biochem. Physiol.* **62,** 545–548.
Wallace, J. C. (1972). Activity and metabolic rate in the shore crab, *Carcinus maenas* (L.). *Comp. Biochem. Physiol. A* **41A,** 523–533.
Wallace, J. C. (1973). Feeding, starvation and metabolic rate in the shore crab *Carcinus maenas*. *Mar. Biol. (Berlin)* **20,** 277–281.
Walshe-Maetx, B. M. (1956). Controle respiratoire et métabolisme chez les Crustacés. *Vie Milieu* **7,** 523–543.
Waterman, T. H. (1960). "The Physiology of Crustacea," Vol. 1. Academic Press, New York.
Waterman, T. H. (1961). "The Physiology of Crustacea," Vol. 2. Academic Press, New York.
Weber, R. E. (1980). Functions of invertebrate hemoglobins with special reference to adaptations to environmental hypoxia. *Am. Zool.* **20,** 79–101.
Wolvekamp, H. P., and Waterman, T. H. (1960). Respiration. *In* "The Physiology of Crustacea" (T. H. Waterman, ed.), Vol. 1, pp. 35–100. Academic Press, New York.
Wu, R. S. S., and Levings, C. D. (1978). An energy budget for individual barnacles (*Balanus gladula*). *Mar. Biol. (Berlin)* **45,** 225–235.
Wycliffe, M. J., and Job, S. V. (1977). Standard, routine and active oxygen consumption of a freshwater shrimp. *Hydrobiologia* **54,** 33–39.

2

Environmental Adaptations of the Respiratory System: Ventilation, Circulation, and Oxygen Transport

JAMES N. CAMERON and CHARLOTTE P. MANGUM

I. INTRODUCTION

As animals move in space or time through a range of environmental variables, adjustments of physiological systems often occur that enable the

THE BIOLOGY OF CRUSTACEA, VOL. 8

ISBN 0-12-106408-5

animals to function in the most advantageous or efficient manner. In the very short term, we tend to call these (nongenetic) changes in physiological systems "responses" or "acute responses." In the very long term, we speak of (genetic) interspecific differences resulting from generally very long periods of evolutionary change. Both of these kinds of change are relatively easy to study in the laboratory: the former because the time scale is similar to that of the investigator's working periods, and the latter because one can simply make similar measurements on a variety of species collected from the field. In between these two time scales lies a gray zone, a medium time scale in which changes occur over a period of days, weeks, or months (season to season), which is also important to an assessment of the animal's ability to cope. These types of changes are much more difficult to study, not only because of the inherent requirement for much longer experiments, but also because of such complicating factors as growth, molt cycle, and sexual maturation. The changes are often gradual, and they may be masked by individual variability among animals or by additional effects of the factors mentioned above. Even our terminology for these changes is less precise, involving the terms *acclimation, acclimatization, adjustment,* and various others. There is consequently little literature dealing with the occurrence, extent, and significance of these medium time scale changes in the mechanics of the respiratory systems of crustaceans.

Additional obstacles appear to have impeded progress in the respiratory physiology of the Crustacea. One problem area has been technical and consists principally of the difficulty of working with crustacean blood. The coagulation of the blood is not easily inhibited by a trace amount of a substance that does not affect blood-gas transport or other properties, as is the case with vertebrate blood and heparin or EDTA. Various substances, like oxalate, do inhibit coagulation, but they must be used at fairly high concentrations, and they do affect blood properties. There is also no substance that can be used to keep catheters patent. Whether this is as much of a problem as it is perceived to be by physiologists is arguable, since many investigators find that repeated syringe sampling of blood produces essentially the same results as chronic sampling from catheters (J. N. Cameron, C. P. Mangum, and B. R. McMahon, unpublished observations). A second technical problem is that the vessels in Crustacea are small and fragile, and the cardiovascular consequences of cannulation of these vessels are not known. Still a third problem consists of the small size of most of the members of the Crustacea; in fact, the vast majority of the group is far too small to allow studies of cardiovascular systems. In some species of very small crustaceans, such as cladocerans and amphipods, it may be possible to observe heart rates directly; in others, enough blood can be obtained by pooling samples from many individuals so that certain work on blood characteristics may be done (e.g., Hoshi and Inada, 1973). For other measurements such as blood and water flow, the small animals are not presently usable. As a result, our information is limited almost entirely to a few species of decapods, primarily the larger crabs, some lobsters, crayfish, and a handful of others. Throughout this discussion, then, we will draw heavily on a few studies we do have, we will try to generalize to the crustaceans from other groups of animals, and we will attempt to indicate where information is needed.

Crustaceans occur in an extraordinary range of environments, including virtually all aquatic habitats, and a broad range of terrestrial ones (wood lice, crabs). The class includes the antarctic zooplankton living at nearly −2°C and intertidal forms at temperatures exceeding 40°C; inhabitants of the oxygen-rich aerial world and dwellers in anoxic sediments; plankters of the soft fresh waters of the Canadian shield and the *Artemia* of hypersaline lagoons; fauna of alkaline lakes and acidic swamps; inhabitants of springs with constant high CO_2 and members of grass flat communities with productivity-driven fluctuations in CO_2; all of these situations are dealt with successfully. The basic machinery of the respiratory system is similar in many of the groups, and though it has been detailed elsewhere (see McMahon and Wilkens, Chapter 6, Volume 5), it is worth reviewing briefly.

The crustaceans have an "open" or "mixed" circulatory system. Although these terms conjure up visions of blood sloshing randomly around an unstructured hemocoel, the reality in larger numbers of the group is that arterial blood is quite effectively channeled in a variable number of well-defined vessels to various regions of the body, propelled by a more-or-less well-developed heart (see Maynard, 1960, for descriptions of various crustacean hearts), which may develop pressure as high as some vertebrate systems (50 or 60 mm Hg in *Birgus latro:* Cameron and Mecklenburg, 1973; and in *Panulirus interruptus:* Belman, 1975). At the ends of the vessels, the blood passes into body cavities that are continuous with interstitial spaces. Smaller sinuses often drain into larger ones, which usually then enter the gills and return by "venous" channels to the pericardium. The gills have a rather variable structure in the larger crustaceans and are often absent in the smaller members of the class. Where present, they may be simple tufts or filaments, or they may be lamellate structures enclosed within well-organized chambers and provided with countercurrent water flow to aid in the effectiveness of gas exchange (Hughes *et al.*, 1969). None of the crustaceans has its O_2 carrying pigment contained in corpuscles, as in the vertebrates, but most have significant quantities of hemocyanin, or occasionally hemoglobin, circulating in solution in the blood (see Chapter 7, Volume 5, Mangum). The blood pigment serves to increase the oxygen-carrying capacity of the blood and also functions as an important buffer protein. The respiratory characteristics of a large number of hemocyanins and hemoglobins have been studied and reviewed elsewhere (Bonaventura and Bonaventura, 1979; Mangum, 1979). The principal substances employed in crustacean blood to buffer CO_2 and other acid–base changes are proteins and the bicarbonate–carbonic acid buffer system, as in most other aquatic animals. The importance of phosphate has not been demonstrated, as it is present in small quantities similar to seawater concentrations. The following discussion will focus on the modifications of this basic machinery that have been made to deal with the broad range of habitats occupied and secondary modifications employed on the medium time scale (defined above) to "fine tune" the system to the environment.

II. TEMPERATURE ADAPTATIONS

The majority of crustaceans tolerate a wide range of environmental temperatures; consequently, the rate of oxidative metabolism in an individual may vary more than tenfold in nature (Prosser, 1973). These changes in metabolic rate may be matched by corresponding changes in the rates of oxygen delivery and carbon dioxide excretion by the gas exchange system. Adding to the problem of increased transport demand at high temperature is the reduction of gas solubility: oxygen solubility at 30°C is only 61% of the 5°C value, and there are comparable reductions in the amount of oxygen that can be carried in blood as the free molecule. These added physical loads are partly offset by a reduction in viscosity of water at high temperatures, which reduces the energy cost of ventilation of the gills.

In response to a moderate increase in temperature and a concomitant increase in metabolic oxygen demand of the tissues, the following adjustments of the gas exchange system may occur: (1) At the gills, the total gas exchange may increase due to increased flow of both blood and water and, possibly, to changes in the spatial and temporal patterns of fluid flow inside and outside of the gills. Increased ventilation does occur, although not at all temperatures (McMahon *et al.*, 1978); the possibility of a compensatory change in flow distribution has not yet been examined in crustaceans. (2) At the tissues, the overall rate of oxygen delivery by blood may rise because of increased flow. This adjustment could be achieved entirely by the increased heart rate that is invariably observed in crustaceans, or it may involve an increased stroke output by the heart as well, which has been suggested by McMahon and Wilkens (1977). In the species examined so far, the effect of temperature on heart rate is at least as great as that on total metabolic oxygen demand (DeFur and Mangum, 1979). (3) Also at the tissues, extraction of the oxygen contained in a unit volume of blood may increase, a frequently noted finding (Mauro and Mangum, 1979). This increase is due to several factors with separate but cumulative effects: a small enhancement of the molecular gradient driving O_2 from blood to mitochondrion; an intrinsic structural change in the hemoblogin or hemocyanin molecule that generally lowers its oxygen affinity; and an extrinsic structural change with the same result, which is due to a drop in blood pH (see below).

None of these adaptations achieves a truly homeostatic relationship between total oxygen uptake and environmental temperature over very large intervals, the reasons for which are fairly clear. The respiratory properties of the blood appear to be conservative in the crustaceans, and the oxygen transport system often performs at maximum efficiency in the realtively narrow interval of 15°–25°C, regardless of species range and thermal regime (Mauro, 1978). At lower temperatures, oxygen affinity becomes so high that delivery to the tissues is impaired, and at higher temperatures oxygen affinity is so low that oxygenation at the gill is incomplete (see example in Mangum, 1979). This limitation of the oxygen transport system is mitigated at very low temperature by a compensatory response of ventilation (McMahon *et al.*, 1978; Mauro and Mangum, 1979), which is so strong in *Cancer magister* that the frequency of scaphognathite beat is the same after acclimation to 7° and 12°C. In the several species studied, postbranchial blood P_{O_2} always increases at low temperature, often by a factor in excess of two over a 10°C interval; the transport of oxygen in the free molecular form becomes the dominant mechanism of supplying oxidative metaboism (see summary of data for *Carcinus maenas* in Table 3, Mangum, 1979; Mauro, 1978; McMahon *et al.*, 1978; Mauro and Mangum, 1979). In contrast to the response of the scaphognathites, the heart slows rapidly and blood P_{O_2} at the tissues decreases at low temperature (Mauro, 1978; Mauro and Mangum, 1979). Although the decrease in *Callinectes sapidus* is so great that the tissues are virtually anoxic at 5°C, it does not offset the greatly elevated $HcyO_2$ affinity; the blood becomes a substantial store of O_2 which is unavailable to the mitochondria, and the total volume of oxygen delivered by the blood sharply drops.

In spite of its failure outside a relatively narrow temperature range, the performance of the oxygen transport system is enhanced by its ability to acclimate to a temperature change within that range. Three independent studies of decapod crustaceans have clearly demonstrated a compensatory decrease in blood oxygen affinity at low temperature, which mitigates the effects on molecular structure. In *Carcinus maenas,* the acclimation occurs within a few days, but only at the upper temperature extremes (>25°C: Truchot, 1975). In the crayfish *Pacifastacus leniusculus,* the change appears to span the range of tolerance, and it occurs within 1 month (Rutledge, 1981). In *Callinectes sapidus,* a seasonal change also occurs well within the natural temperature range, but it cannot be induced by 2–3 weeks exposure to a new temperature in the laboratory (Mauro and Mangum, 1979). The mechanisms responsible for the acclimation are virtually unknown. The response of the blood of *C. maenas* depends upon the presence of both dialyzable and nondialyzable components (Truchot, 1975), raising the possibility of *de novo* synthesis of either parts or complete aggregates of the hemocyanin molecule. Rutledge (1981), however, was unable to demonstrate a difference in the electrophoretic behavior of cold- and warm-adapted crayfish hemocyanin. To complicate the matter, the seasonal character of the phenomenon in *C. sapidus* raises the possibility of a neuroendocrine control, although the influence of a seasonal variable other than temperature has not yet been dismissed (see Section III).

The ventilatory and circulatory systems also have the ability to acclimate to a temperature change, although this ability varies both with species and with the particular temperature. While DeFur and Mangum (1979) found no acclimation of heart rate in *C. sapidus* and *Libinia emarginata* exposed to a wide range of temperatures, McMahon *et al.* (1978) found a significant adjustment in *C. magister* after 12-hr exposure to a higher temperature, but only at 12° and not 17°C. The response of the ventilatory system in this species is similar and the compensation much stronger: 12 hr after the temperature is raised from 7° to 12°C, the scaphognathite beat frequency shows essentially perfect acclimation, but the response to further warming is insignificant (McMahon *et al.*, 1978).

In addition to its effect on the oxygen transport demand and delivery system, temperature exerts a profound effect on the carbon dioxide and acid–base system in all poikilotherms. As pointed out in an earlier chapter of this treatise (Truchot, Chapter 8, Volume 5), the blood pH varies inversely with temperature in such a way as to maintain a nearly constant relative alkalinity and a nearly constant ionization of proteins. Thus, at 24°C the pH of neutrality (defined as $[OH^-] = [H^+]$) is 7, and at 5°C it is 7.37. Since animals tend to maintain a constant difference between blood pH and the pH of neutrality, a pH of about 7.6 would be expected at 24°C and 8 at 5°C. In order for the pH to be controlled in this fashion, there must be changes in the bicarbonate buffer system. Specifically, for pH to be reduced when temperature is increased, either the partial pressure of CO_2 in blood must rise or the HCO_3^- mst fall.

Working from a background of mammalian physiology, early workers assumed that blood pH was controlled via ventilatory control of blood P_{CO_2}, as it is in most air-breathers. For this control to be effected, the ratio of ventilation to CO_2 production (and oxygen consumption) must fall at higher temperatures in order to raise the CO_2 gradient across the gills and to maintain a higher internal P_{CO_2}. In recent years, however, it has become increasingly clear that in water-breathing animals such a reduction of ventilation at high temperatures would cause unacceptable sacrifices in oxygen uptake, since oxygen solubility is about 1/30 of the CO_2 solubility in water and since ventilation is set at a relatively high level at all times in order to maintain oxygen delivery (Cameron and Batterton, 1978b). In fact, the partial pressure of CO_2 appears to rise in *Carcinus maenas* at high temperature according to one study (Truchot, 1973b), but not according to another (Howell *et al.*, 1973). Additionally, there is no significant rise (Howell *et al.*, 1973; Cameron and Batterton, 1978b) or significant decrease in ventilation relative to CO_2 production in either *C. sapidus* (Batterton and Cameron, 1978) or *C. magister* (McMahon *et al.*, 1978).

Since the appropriate changes of pH with temperature change are not brought about through manipulation of P_{CO_2}, clearly the HCO_3^- reserve is altered by some physiological process. It is easy to show that an insufficient change occurs in response to temperature, due to passive chemical changes in the blood equilibrium (Cameron and Batterton, 1978b); this is to be expected, since the carbonic acid system has a ΔH of 2200 kcal/mol, and the observed pH–temperature relationship acts like a system with a ΔH of about 7000 kcal/mol.

One obvious way of controlling the blood bicarbonate reserve would be the various ion exchange mechanisms, principally Cl^- for HCO_3^- and Na^+ for H^+ or NH_4^+ (Krogh, 1938), theoretically located at any external body surface. More than 90% of the total external surface is contained in the gills, which therefore become prime suspects, although antennal glands and gut are also possible control sites. In an investigation of hypercapnic responses, the antennal glands failed to show any appreciable acid–base regulating ability (Cameron and Batterton, 1978a), but otherwise no studies have been published on the organ system roles in thermal adjustment of the acid–base state.

The time course of the change in blood pH at a new temperature might provide a clue as to the control mechanism, but the available information is not conclusive. Mangum (1973) found that the response of blood pH to an increase in temperature was generally immediate, with no further change except in the air-breathing *Uca pugnax*. The adjustment to a large temperature decrease, however, often required 4–6 hr or more. Truchot (1978) found a partial response in *Carcinus maenas* that was immediate and a slower phase lasting many hours. He postulated that the immediate response was due to passive chemical changes seen in constant content systems *in vitro* (Cameron and Batterton, 1978b) and that the longer term response involved physiological alteration of the bicarbonate buffer pool. Since the blood volume in crustaceans is large, the total mass of ionic material that must be transported during such physiological regulation is relatively large, and a significant energy cost will probably be associated with it. An interesting question is how much regulation will take place in animals in cycling thermal regimes, and how much pH variation will simply be tolerated in the short term to avoid the energy costs of constant readjustment. These are problems for which we have little or no data at present, and ones which can now be expected to yield answers, since we have a firm theoretical and experimental base for the appropriate new experiments.

A final category of respiratory responses is behavioral: i.e., animals may respond to unfavorable gas exchange conditions by simply moving to an area where conditions are more favorable. This may frequently occur in grass flat environments, which become very hot at midday or hypoxic at night, whereupon animals may shift off to deeper adjacent channels (J. N. Cameron, unpublished observations). A more unusual response of *Callinectes sapidus* to very high temperatures is to emerge on land and breathe air for short periods when the water temperatures exceed air temperatures and oxygen availability becomes low (Abbot, 1967; Batterton and Cameron, 1978).

III. ADAPTATIONS TO AMBIENT OXYGEN CONCENTRATIONS

The concentration of dissolved oxygen in aquatic systems is always a critical factor, since oxygen is only about 1/30 as abundant in water as it is in air. Furthermore, the oxygen may fall to very low concentrations in many habitats due to biological or chemical action. One notable example is that large class of shallow lakes in the northern latitudes that "winterkill," resulting in a long period of complete anoxia. No doubt many of the crustacean zooplankters that enter various resting stages in these situations do so partly to escape the unfavorable oxygen conditions during winter.

One particularly spectacular example of the long-term response of a crustacean to lowered oxygen is the production of respiratory pigment in animals that otherwise have little or none. More than 25 years ago this phenomenon was discovered accidentally by H. Munro Fox, who noticed that a neglected culture of *Daphnia* left in a dark closet became red. His subsequent studies of the hypoxic induction of hemoglobin in cladocerans are reviewed thoroughly by Goodwin (1960). Although the biosynthetic control mechanism remains to be elucidated, a series of recent studies by Hoshi and his colleagues have clearly demonstrated the respiratory importance of the transient hemoglobin. In general, the hemoglobin in a variety of cladoceran species functions at maximum efficiency at water P_{O_2} of about 25–50 mm Hg, and it has little or no role in air-saturated water, which is not surprising in view of the small body size of cladocerans and of the high oxygen affinity

(P_{50} = ca. 2–4 mm Hg) of the carrier. In that P_{O2} range, however, the hemoglobin transports well over half of the oxygen consumed (e.g., Hoshi and Inada, 1973; see also Mangum, Chapter 7, Volume 5 of this treatise).

In vertebrates, there is now a considerable body of information on physiological modulation of oxygen affinity of the blood, which has been most widely studied in response to ambient oxygen concentrations. The cofactors, which are various organic phosphates (see review by Isaacks, 1979), are synthesized and their levels controlled by the metabolic machinery within the red cell. Although some of the same cofactors are present in the red cells of other animal phyla, the available evidence suggests that they do not have an appreciable influence on HbO_2 affinity (see review by Mangum, 1976). One would not necessarily expect to find analogous modulators in crustacean bloods, either from a phylogenetic point of view or from the extracellular location of their oxygen carriers. Yet the evidence in favor of the existence of cofactors grows. The temperature acclimation of $HcyO_2$ affinity, which in one species clearly entails a dialyzable as well as a nondialyzable component (Truchot, 1975), is discussed in Chapter 1 of this volume. The influence of inorganic ions on the respiratory properties of crustacean bloods is discussed briefly in Chapter 4 (see also Chapter 7 in Volume 5). Although the response to hypoxia has not been directly investigated, it is quite possible that the stimulus responsible for seasonal effect in *C. sapidus* (Mauro and Mangum, 1979) will prove to be low oxygen rather than low temperature; both variables characterize the environmental conditions to which the crabs were acclimatized.

The respiratory adjustments to hypoxia include adaptations of the circulatory and ventilatory systems, as well as the oxygen transport system. In both terrestrial and aquatic crabs, the acute ventilatory response is usually a pronounced increase in flow to compensate for the lower amount of oxygen present in a unit volume of water (early studies reviewed by Wolvekamp and Waterman, 1960; Larimer and Gold, 1961; Batterton and Cameron, 1978; Burnett, 1979; Herreid *et al.*, 1979). In contrast, the acute response of heart rate rarely involves a significant tachycardia, and bradycardia invariably occurs at only moderate low oxygen levels (reviewed by DeFur and Mangum, 1979). Whether stroke volume increases in hypoxia is not entirely clear, but the available information indicates once again that the cardiovascular system is capable of far less compensatory change than the ventilatory apparatus.

In *Daphnia magna* the acute response of ventilation to hypoxia is relatively small (viz., only a 10% increase in activity), but the difference between hemoglobin-poor and hemoglobin-rich individuals is notable. While the movements of the thoracic appendages begin a sharp decline to zero at

50–60 mm Hg in individuals with little or no hemoglobin in the blood, fully 75–100% of the activity continues in hemoglobin-rich animals in the range 15–60 mm Hg (Hoshi and Inada, 1973).

The responses of the crayfish, *Orconectes virilis,* and the lobster, *Homarus gammarus,* to relatively long-term hypoxia have now been studied by McMahon *et al.* (1974) and Butler *et al.* (1978). In both studies the results were in some respects similar in that they observed an initially large change in certain parameters in response to hypoxia, such as ventilation volume, and then a gradual amelioration of some of these changes over the next few days of hypoxia. In particular, ventilation volume declined over several days, but remained above control values, with some gradual rise in the percent extraction of oxygen from water in the gills. Both the beating of the scaphognathites and the heart become more frequent and more regular; periodic apnoea that is characteristic of well-rested animals tends to disappear during hypoxia (Fig. 1). Overall oxygen consumption is maintained in both species at ambient oxygen concentrations down to about one-third normal. The hyperventilation observed during hypoxia leads to a slight respiratory alkalosis (reduction in blood P_{CO_2} and increase in pH) and increases the oxygen affinity of the blood via the Bohr shift (McMahon *et al.*, 1974; Burnett, 1979).

In the gill-breathing vertebrates, particularly the teleosts, it is becoming increasingly clear that the functional gill surface area can be varied considerably, principally by perfusing a varying number of secondary lamellae (Davis, 1972; Smith and Johnson, 1977). A reduction of functional exchange area at rest is equally advantageous, since a large surface area increases the ionic and water flux problems encountered by all aquatic osmoregulators. During hypoxia in crustacean gill-breathers, it would obviously be advantageous to increase the functional exchange area, but so far nothing at all is known about whether this response indeed takes place. At this writing, the evidence suggests that crustacean gills lack the endogenous contractility found in the annelids and mollusks, and that they also lack the sphincterlike control points found in the lamellar circulation of fish (Vogel *et al.*, 1976). Various techniques are now available for investigation of such problems, including radioactively tagged microspheres, corrosion casting, and SEM methods, so it is not unreasonable to expect some interesting results in the near future.

As with temperature responses, a suite of behavioral responses to altered environmental oxygen is probably to be expected for crustaceans. Many cases of hypoxia are localized in such a way that relocation vertically or laterally eliminates the problem for the animal. The degree to which these responses are provoked is probably a function of the severity of the oxygen

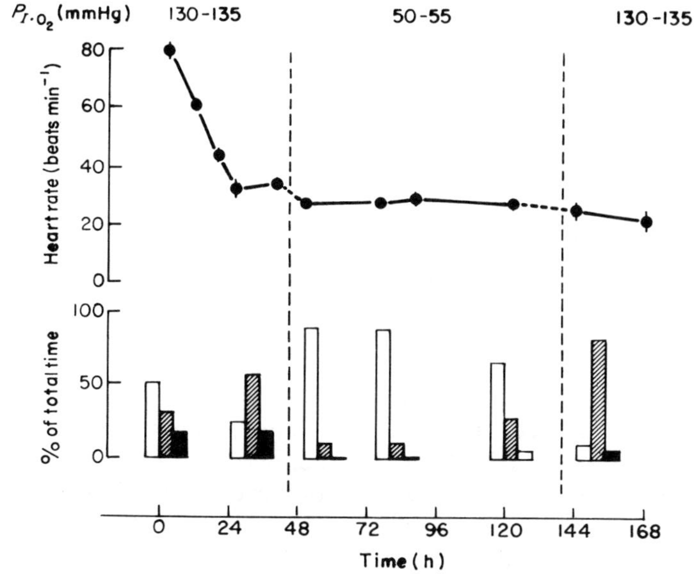

Fig. 1. The response of the lobster *Homarus gammarus* to long-term hypoxia. The upper panel shows heart rate for a 2-day control period, then a 4-day hypoxic period, followed by a further 2-day control period. The lower panel shows ventilatory activity during the same time, where the open bars represent ventilation of both gill chambers, the hatched bars ventilation on one side only, and the solid bars represent periods of apnoea. (Reprinted by permission from Butler *et al.*, 1978.)

depletion locally and of the energy costs associated with relocation. In a few specialized cases, such as the sessile barnacles, the result may instead be closure and the use of a form of anaerobic metabolism which is modified in a way that is not yet known (Jamieson *et al.*, 1977).

IV. ENVIRONMENTAL CARBON DIOXIDE AND pH

While the partial pressure of CO_2 in most waters remains very low at all times, in highly productive shallow environments, stagnant swamp waters, polluted waters, and a few other special situations (e.g., hot springs), the partial pressure of CO_2 can be high and highly variable (Kramer *et al.*, 1978). This will have two effects: the pH of the water will drop because of the equilibration of dissolved CO_2 with carbonic acid, which dissociates into HCO_3^- and H^+ ions; and the partial pressure of CO_2 (P_{CO_2}) will increase in the blood of crustaceans, with a corresponding drop in blood pH.

Unlike exercise, which can produce a rise in blood P_{CO2}, environmental hypercapnia cannot be alleviated quickly by ventilation, since a lower limit is imposed by the external partial pressure.

A. Effects of CO_2

To illustrate the effects of CO_2 on the blood of a crustacean, some typical acid–base variables are shown in Fig. 2 from a study of the blue crab (Cameron, 1978). This acid–base diagram, sometimes called a Davenport diagram, depicts combinations of HCO_3^- (sometimes total CO_2), pH, and P_{CO_2} according to the Henderson-Hasselbalch equation, $pH = pK' + \log(HCO_3^-/S_{CO_2})$, where $S_{CO_2} = P_{CO_2} \times$ solubility coefficient (α), and pK' is the operational pK for the carbonic acid system. The curved lines on the graph are so-called P_{CO_2} isopleths, which give combinations of pH and HCO_3^- for any given constant P_{CO_2}. A more complete discussion of acid–base concepts is presented in an earlier volume of this work (Truchot, Chapter 8, Volume 5). Shown at A on the diagram is a value for the resting blue crab, *Callinectes sapidus* Rathbun, at 26°C. Determination of the buffer properties of the blood removed from the animal would give a linear relationship, O–O′, called the *in vitro* buffer line; this line reflects the passive chemical buffering characteristics of the mixed bicarbonate-protein buffer system present. The greater the slope of this line, the greater the buffering power of the blood. If P_{CO_2} is increased outside the animal, a new steady state will be reached in which the positive P_{CO_2} gradient is reestablished from inside to outside; *i.e.*, the P_{CO_2} will rise inside until CO_2 excretion resumes down a diffusion gradient. While the difference between inside and outside may change slightly, the new steady state value will be higher than the resting value for any external P_{CO_2} value greater than 2 or 3 Torr. These values can often be exceeded in nature, as for example in the Amazon basin where P_{CO_2} values of 28–30 Torr have been measured (G. F. Holeton, personal communication).

If a change in P_{CO_2} occurs, the blood is shifted from its normal values upward and to the left along the *in vitro* buffer line, as from A to B in Fig. 2, increasing the HCO_3^- somewhat and depressing the pH. This change can take from a few minutes to a couple of hours, depending on the blood volume and circulation of the animal, and is purely passive. If changes are followed for a longer period of time, a secondary increase in HCO_3^- and pH will be observed, along the line B to E. This change represents an active readjustment of the HCO_3^- buffer reservoir, and almost certainly involves ionic exchanges across the gills (see below and Truchot, Chapter 8, Volume 5). The secondary, active change takes longer and may not be complete until 24–48 hr. It also involves transfer of an appreciable quantity of mate-

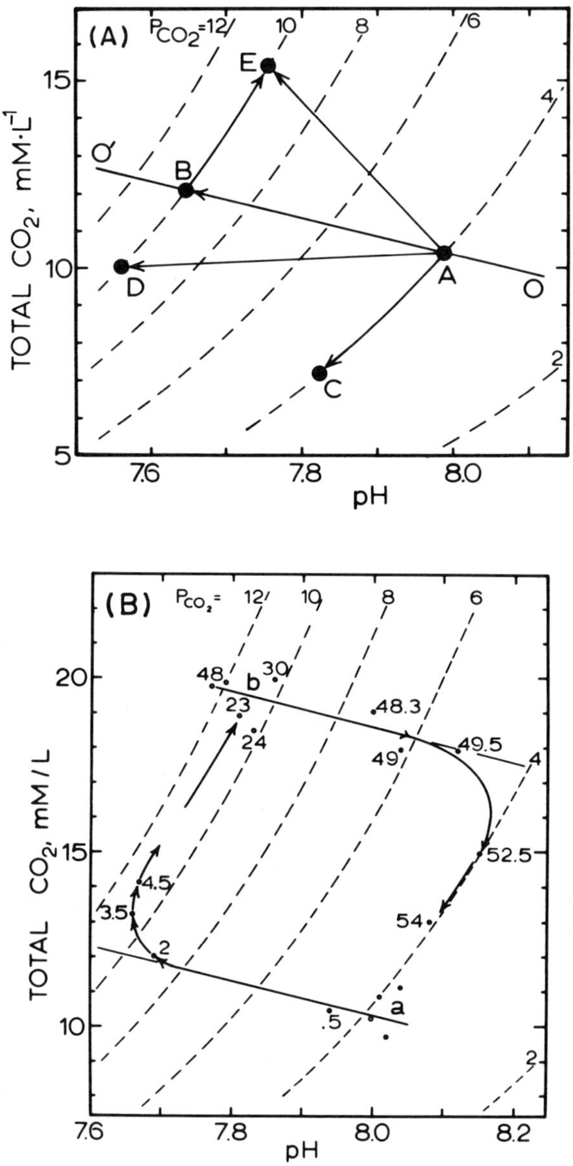

Fig. 2. Top: Acid–base diagram of blood of the blue crab, showing the normal *in vitro* buffer line (O–O′), the response to pure respiratory (CO_2) acidosis (A–B), a compensated respiratory acidosis (A–E), a pure metabolic acidosis (A–C), and a mixed acidosis (A–D). Bottom: Response of *Callinectes sapidus* to hypercapnia. Resting values at *a*, and numbers above points show times after start of hypercapnia. At 48 hr, the animal was returned to atmospheric CO_2 concentrations, and the new *in vitro* buffer line through the compensated points at *b* shows the course of the subsequent alkalosis, which by about 60 hr had disappeared. (Bottom redrawn from Cameron, 1978.)

rial and has an associated energy cost that has not been calculated thus far. Figure 2 also shows the time course of adjustment to a change in external P_{CO_2} in *Callinectes sapidus* (Cameron, 1978); note that the pH compensation is not complete and that, after restoring the animal to a normal ambient P_{CO_2}, the blood goes through a similar process in reverse with about the same time course. The antennal glands are apparently not involved in this adjustment (Cameron and Batterton, 1978a), which implicates gill processes.

Since these adjustments of the blood acid–base state are relatively slow and will require energy for transport of ions across membranes and epithelia, it would be interesting to see whether short-term, e.g., diurnal, fluctuations of external CO_2 would provoke the responses or whether the animal would simply tolerate short-term fluctuations of its internal P_{CO_2} and pH. We are not aware of any research to date that provides an answer.

Against a background of mammalian physiology, which had firmly established that ventilation was affected by both oxygen and CO_2 concentrations in the blood and in the medium breathed, early experimenters attempted to determine the response of ventilation in crustaceans to increased ambient CO_2 (Johnson, 1936; Wolvekamp and Waterman, 1960; Arudpragasam and Naylor, 1964). It is difficult in some cases to evaluate these studies, since CO_2 concentrations are often not convertible to partial pressure, being given in mg·liter^{-1} or ppm; in addition, concentrations have often been very high, say 5 and 10% (38 and 75 Torr), well beyond the physiological and ecological range. These high concentrations often provoke irritation reactions and may even partially anesthetize the animals. Nonetheless, since getting oxygen from water is so much greater a problem than getting rid of CO_2, ventilation is set at a high level to satisfy O_2 demands (Dejours, 1975; Cameron and Batterton, 1978b). Since ventilation is already much higher than necessary for the elimination of CO_2 because of oxygen scarcity, a further increase in response to increased CO_2, either in blood or in inspired water, is not to be expected on theoretical grounds, and is not in fact found when physiologically realistic concentrations of CO_2 are tested (Batterton and Cameron, 1978).

Much of the discussion of whether ventilation and CO_2 are closely linked has focused on the values of pre- or postbranchial P_{CO_2} in crabs and on the difference between them. For example, three recent papers examine the variation in P_{CO_2} as a function of temperature in order to determine whether ventilation has a significant role in controlling pH via control of P_{CO_2} (Howell *et al.*, 1973; Truchot, 1975; Cameron and Batterton, 1978b). What has been ignored in all of these is that since crab blood lacks any circulating carbonic anhydrase, and since the uncatalyzed equilibration between CO_2

and bicarbonate is a slow process (half-times of more than 30 sec at physiological temperatures: Edsall, 1969), the P_{CO_2} and pH *in vivo* will undoubtedly be much different from the value determined from a sample removed and allowed to come to equilibrium in an electrode measuring system.

For the blue crab, *Callinectes sapidus,* recent morphometric studies indicate that blood will spend about 6 sec in the respiratory lamellae (Aldridge and Cameron, 1979). Using this information and a computer simulation of circulation, a qualitative description of the *in vivo* values to be expected for P_{CO_2} and HCO_3^- has been diagramed in Fig. 3. These are qualitative because it is not known how long blood spends in the circulation between gills and tissue beds or how long it spends in contact with CO_2-producing tissues. What is clear is that the time will be short of what is required for equilibrium to be achieved, and that CO_2 diffusion out through the gill is much more rapid than the interconversion of HCO_3^- to CO_2. Consequently, the P_{CO_2} values measured from samples withdrawn in the usual fashion will reflect little more than the HCO_3^- concentration they are equilibrated with. This whole area suggests a number of productive lines of research: determining the actual *in vivo* P_{CO_2}'s and the times for circulatory transit between tills and tissues and return.

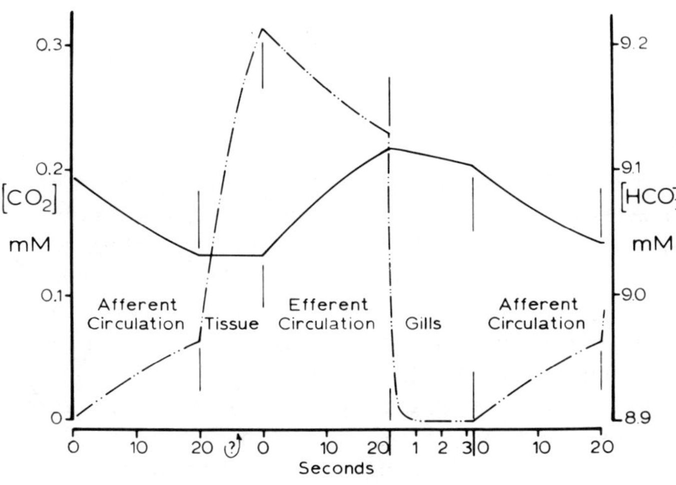

Fig. 3. Results of a simulation model of P_{CO_2} and bicarbonate during a transit of the circulation in the blue crab, *Callinectes sapidus*. Initial values from Aldridge and Cameron (1979), including a calculated 6-sec residence time in gills. Residence times for other portions of the circulation are estimates based on blood volume and calculated cardiac output. The graph is intended to show the disequilibrium at all points in the circulation.

B. Effects of pH

Changes in environmental pH, aside from changes in P_{CO2} that may or may not attend them, will have at least some known effects on gas exchange in crustaceans and possibly many more unknown ones. This is an area which has received little attention, however. Since 1938 it has been suspected (and was later shown) that freshwater crustaceans take up certain monovalent ions in electroneutral exchanges for internal ions. Specifically, Krogh (1938) proposed that Cl^- was taken up from the environment in exchange for internal HCO_3^-, and that Na^+ was taken up in exchange for either H^+ or NH_4^+. Changes in external pH will almost certainly affect the kinetics of these uptake mechanisms, since the electrochemical gradients across which they operate will be affected and since the energy requirements for transport are determined by those electrochemical gradients. Thus, one would predict that as pH drops externally, Na^+ uptake in exchange for H^+ or NH_4^+ would become more difficult, and that as pH increases, HCO_3^- excretion in exchange for Cl^- would become more difficult. These changes affect the respiratory system through their influence upon blood pH and secondarily through influence upon hemocyanin affinity.

V. RESPIRATORY ADAPTATIONS TO SALINITY

Variations in environmental salinity affect the respiratory system in four ways: by affecting gas solubility in the water pumped over the gills; by affecting solubility of gases in plasma; by directly affecting affinity of blood pigments; and by causing an altered pH in the blood, which also affects pigment affinity. The (short-term) responses to salinity in various contexts have been discussed in previous chapters of this work (see Volume 5), but a few longer term aspects have been pointed up in recent work.

Studies of both *Carcinus maenas* (Truchot, 1973a) and *Callinectes sapidus* (Weiland and Mangum, 1975; Cameron, 1978) have shown that the blood pH in waters of low salinity is considerably elevated over the blood pH found at seawater salinity and the same temperature. In both species, the increased pH leads to an increase in hemocyanin affinity, whereas the lower salt levels in the blood cause a reduction in affinity; the two factors acting together produce only a small change in P_{50} over a wide range of salinity, as shown in Fig. 4 (Truchot, 1973a). The exact time course and organ system involvement in this compensatory response have not been studied to date.

In response to reduced oxygen availability at higher salinities, crustaceans presumably increase the rate of ventilation and perfusion of the gills in proportion to the oxygen reduction. It is not presently known, however,

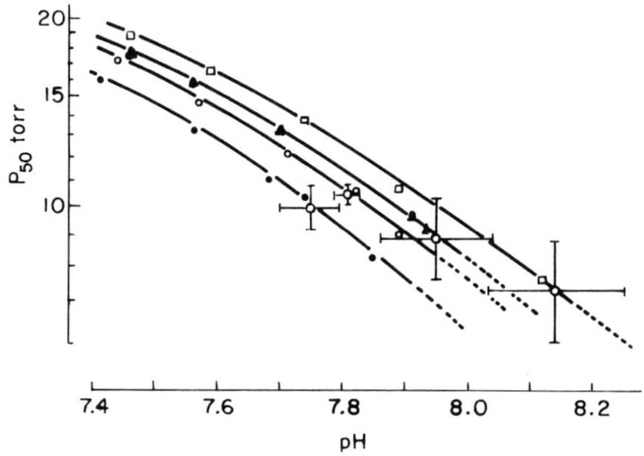

Fig. 4. Variations in P_{50} of *Carcinus maenas* adapted to 36 ppt (●), 30.2 ppt (○), 20.2 ppt (▲), 11.7 ppt (□), and a temperature of 15°C. The *in vivo* values are shown with error bars, illustrating the combined effect of pH and salinity in preventing large increases in P_{50}. (Reprinted by permission from Truchot, 1973a.)

whether there is a long-term adaptational response different in kind or degree from the acute response.

VI. EXERCISE

There are only a few studies of respiratory function of crustaceans during exercise, and most of them are quite recent. None so far has dealt with longer term adjustments to exercise, but has rather been concerned with the changes occurring during, and a few hours after, an exercise period of a few minutes to ca. 1 hour. It is known, however, that some crustaceans undertake considerable migrations; shrimp migrate from estuarine systems along the Atlantic and Gulf coasts of America out through tidal passes into the open ocean for a total voyage of many kilometers. Blue crab migrations in the Chesapeake Bay system have been well known for centuries (Warner, 1976) and may involve movements of 300 km or more. The hints we have from present studies of exercise are that laboratory experiments produce a "burst" type activity that is largely anaerobic and can only be sustained for short periods (Herreid *et al.*, 1977; McDonald *et al.*, 1979; Smatresk *et al.*, 1979). The animals so far employed for such studies, e.g., *Cardisoma guanhumi* and *Cancer magister*, are not animals that sustain long periods of vigorous swimming, as does *Callinectes*. The overland migrations of *Cardisoma* (Bliss, 1968) are probably undertaken at a sedate pace that allows aerobic metabolism to keep pace with energy expenditure. It would be very interesting to see if there are changes in the respiratory system that correspond to periods of sustained activity. An example might be the production of additional respiratory pigment at migration seasons.

VII. SUMMARY

Assessing the functioning of the respiratory system as an integrated whole is a fascinatingly difficult enterprise. So many vital processes are intertwined that it is very difficult to assess the effects of change on any one process without also discussing many others. The system consists of apparatus for delivering a high volume of water to an exchange surface, the exchange surface itself, another fluid system for pumping a transporting fluid (blood) from exchanger to tissues, and the exchange beds in the tissues. Gases are transported in opposite directions: i.e., oxygen in and CO_2 out. Control of the system seems to be geared to oxygen delivery, but any change necessitated by varying oxygen demand causes adjustments not only in CO_2 transport in the opposite direction, but also in other vital parameters such as blood pH, osmotic pressure, and gill transport of water. For an animal to be evolutionarily successful, the respiratory system must have evolved to provide a compromise between optimal efficiency at some sort of average demand and a high capacity for dynamic change in response to altered requirements for tissue metabolism or altered environmental conditions.

In studying the limits of adaptation of the system as it exists in any one animal, a time frame is imposed by the investigator, often in an implicit fashion, as discussed in the Section I. In this chapter, we have attempted to deal with a time scale intermediate between the immediate or acute responses often studied by physiologists and the evolutionary changes brought to light by comparative studies. The variables we have been most concerned with, then, have been those with intrinsically slow time courses *in vivo*. Thus, it is less likely that the long- and short-term responses of a factor like ventilation frequency will differ appreciably, since this is a process that can respond immediately through rather simple reflex arcs. On the intermediate scale, we have shown how readjustment of bicarbonate reserves takes a longer time, involving the transfer of a considerable quantity of material by active transport. Change in amount or type of respiratory pigment is another example of a medium timescale change. We did not discuss, due to lack of data, the very long term; for example, does the gill surface area of a population of crabs reared for several or many generations at low oxygen concentration increase? This might give us some idea of the genetic plasticity of populations, and it would not be surprising to find a large reservoir of genetic variability that could be expressed on this type of time scale.

REFERENCES

Abbot, W. (1967). Unusual climbing behavior by *Callinectes sapidus* Rathbun (Decapoda, Brachyura). *Crustaceana* **13,** 128.

Aldridge, J. B., and Cameron, J. N. (1979). CO_2 exchange in the blue crab, *Callinectes sapidus* (Rathbun). *J. Exp. Zool.* **207,** 321–328.

Arudpragasam, K. D., and Naylor, E. (1964). Gill ventilation volumes, oxygen consumption, and respiratory rhythms in *Carcinus maenas* (L.). *J. Exp. Biol.* **41,** 309–322.

Batterton, C. V., and Cameron, J. N. (1978). Characteristics of resting ventilation and response to hypoxia, hypercapnia and emersion in the blue crab, *Callinectes sapidus* (Rathbun). *J. Exp. Zool.* **203,** 403–418.

Belman, B. W. (1975). Some aspects of the circulatory physiology of the spiny lobster *Panulirus interruptus. Mar. Biol. (Berlin)* **29,** 295–305.

Bliss, D. E. (1968). Transition from water to land in decapod crustaceans. *Am. Zool.* **8,** 355–392.

Bonaventura, J., and Bonaventura, C. (1980). Hemocyanins: Relationships in their structure, function and assembly. *Am. Zool.* **20,** 7–17.

Burnet, L. E. (1979). The effects of environmental oxygen levels on the respiratory function of hemocyanin in the spider crab *Libinia emarginata,* and the ghost crab, *Ocypode quadrata. J. Exp. Biol.* **210,** 289–299.

Butler, P. J., Taylor, E. W., and McMahon, B. R. (1978). Respiratory and circulatory changes in the lobster (*Homarus vulgaris*) during long term hypoxia. *J. Exp. Biol.* **73,** 131–146.

Cameron, J. N. (1978). Effects of hypercapnia on blood acid-base status, NaCl fluxes, and trans-gill potential in freshwater blue crabs, *Callinectes sapidus. J. Comp. Physiol. B* **123,** 137–141.

Cameron, J. N., and Batterton, C. V. (1978a). Antennal gland function in the freshwater blue crab, *Callinectes sapidus:* Water, electrolyte, acid–base and ammonia excretion. *J. Comp. Physiol. B* **123,** 143–148.

Cameron, J. N., and Batterton, C. V. (1978b). Temperature and blood acid–base status in the blue crab, *Callinectes sapidus. Respir. Physiol.* **35,** 101–110.

Cameron, J. N., and Mecklenburg, T. A. (1973). Aerial gas exchange in the coconut crab, *Birgus latro,* with some notes on *Gecarcoidea lalandii. Respir. Physiol.* **19,** 245–261.

Davis, J. C. (1972). An infrared photographic technique useful for studying vascularization of fish gills. *J. Fish. Res. Board Can.* **29,** 109–111.

DeFur, P. L., and Mangum, C. P. (1979). The effects of environmental variables on the heart rates of invertebrates. *Comp. Biochem. Physiol. A* **62A,** 283–294.

Dejours, P. (1975). "Principles of Comparative Respiratory Physiology." Am. Elsevier, New York.

Edsall, J. T. (1969). Carbon dioxide, carbonic acid, and bicarbonate ion: Physical properties and kinetics of interconversion. *In* CO_2: Chemical, biochemical and physical aspects. *NASA [Spec. Publ.] SP* **NASA SP-188,** 15–27.

Goodwin, F. W. (1960). Biochemistry of pigments. *In* "The Physiology of Crustacea" (T. A. Waterman, ed.), Vol. 1, pp. 101–140. Academic Press, New York.

Herreid, C. F., II, Lee, L. W., Shah, G. M., O'Mahoney, P. M., and Murto, J. (1977). Exercise in land crabs. *Am. Zool.* **17,** 918.

Herreid, C. F., II, O'Mahoney, P. M., and Shah, G. M. (1979). Cardiac and respiratory responses to hypoxia in the land crab, *Cardisoma guanhumi* (Latrielle). *Comp. Biochem. Physiol.* **63A,** 145–152.

Hoshi, T., and Inada, Y.-Y. (1973). Studies on physiology and ecology of plankton. XXVII. O_2-

consumption, thoracic limb movement and O_2-dissociation from haemoglobin of *Daphnia magna in vivo*. *Sci. Rep. Niigata Univ., Ser. D* **10,** 79–86.

Howell, B. J., Rahn, H., Goodfellow, D., and Herreid, C. F. (1973). Acid–base regulation and temperature in selected invertebrates as a function of temperature. *Am. Zool.* **13,** 556–563.

Hughes, G. M., Knights, B., and Scammel, C. A. (1969). The distribution of P_{O2} and hydrostatic pressure changes within the branchial chambers in relation to gill ventilation of the shore crab *Carcinus maenas*. *J. Exp. Biol.* **51,** 203–220.

Isaacks, R. E. and Harkness, D. R. (1980). Erythrocyte organic phosphates and hemoglobin function in birds, reptiles, and fishes. *Am. Zool.* **20,** 115–129.

Jamieson, D. D., Taylor, R. S., and Taylor, K. M. (1977). Energy metabolism of the lateral depressor muscle of the barnacle *Balanus nigrescens*. *Comp. Biochem. Physiol. B* **58B,** 315–320.

Johnson, M. L. (1936). The control of respiratory movements in Crustacea by oxygen and carbon dioxide. *J. Exp. Biol.* **13,** 467–475.

Kramer, D. L., Lindsey, C. C., Moodie, G. E. E., and Stevens, E. D. (1978). The fishes and the aquatic environment of the central Amazon basin, with particular reference to respiratory patterns. *Can. J. Zool.* **56,** 717–729.

Krogh, A. (1938). The active absorption of ions in some freshwater animals. *Z. Vergl. Physiol.* **25,** 335–350.

Larimer, J. L., and Gold, A. H. (1961). Responses of the crayfish *Procambarus simulans* to respiratory stress. *Physiol. Zool.* **34,** 167–176.

McDonald, D. G., McMahon, B. R., and Wood, C. M. (1979). An analysis of acid–base disturbances in the haemolymph following strenuous activity in the Dungeness crab, *Cancer magister*. *J. Exp. Biol.* **79,** 47–58.

McMahon, B. R., and Wilkens, J. L. (1977). Periodic respiratory and circulatory performance in the red rock crab, *Cancer productus*. *J. Exp. Zool.* **202,** 363–374.

McMahon, B. R., Burggren, W. W., and Wilkens, J. L. (1974). Respiratory responses to long-term hypoxic stress in the crayfish *Orconectes virilis*. *J. Exp. Biol.* **60,** 195–206.

McMahon, B. R., Sinclair, F., Hassall, C. D., DeFur, P. L., and Wilkes, P. R. H. (1978). Ventilation and control of acid–base status during temperature acclimation in the crab, *Cancer magister*. *J. Comp. Physiol.* **128,** 109–116.

Mangum, C. P. (1973). Evaluation of the functional properties of invertebrate hemoglobins. *Neth. J. Sea Res.* **7,** 303–315.

Mangum, C. P. (1976). Primitive respiratory adaptations. *In* "Adaptation to Environment" (R. C. Newell, ed.), pp. 191–278. Butterworth, London.

Mangum, C. P. (1980). Respiratory function of the hemocyanins. *Am. Zool.* **20,** 19–38.

Mauro, N. A. (1978). The effects of temperature on O_2 transport in decapod crustaceans. *Am. Zool.* **18,** 639. Mauro, N. A., and Mangum, C. P. (1979). The role of the blood in the temperature dependence of oxidative metabolism in decapod crustaceans. (In prep.).

Maynard, D. M. (1960). Circulation and heart function. *In* "The Physiology of Crustacea" (T. H. Waterman, ed.), Vol. 1, pp. 161–226. Academic Press, New York.

Prosser, C. L. (1973). "Comparative Animal Physiology." Saunders, Philadelphia, Pennsylvania.

Rutledge, P. S. (1981). Effects of temperature acclimation on crayfish hemocyanin oxygen binding. *Am. J. Physiol.* **240,** R93–R98.

Smatresk, N. J., Preslar, A. J., and Cameron, J. N. (1979). Post-exercise acid–base disturbance in *Gecarcinus lateralis,* a terrestrial crab. *J. Exp. Zool.* **210,** 205–210.

Smith, D. G., and Johnson, D. W. (1977). Oxygen exchange in a simulated trout gill secondary lamella. *Am. J. Physiol.* **233,** R145–R161.

Truchot, J.-P. (1973a). Fixation et transport de l'oxygène par le sang de *Carcinus maenas:* Variations en rapport avec diverses conditions de température et de salinité. *Neth. J. Sea Res.* **7,** 482–495.

Truchot, J.-P. (1973b). Temperature and acid–base regulation in the shore crab *Carinus maenas. Respir. Physiol.* **17,** 11–29.

Truchot, J.-P. (1975). Factors controlling the *in vitro* and *in vivo* oxygen affinity of the hemocyanin in the crab *Carcinus maenas. Respir. Physiol.* **24,** 173–189.

Truchot, J.-P. (1978). Mechanisms of extracellular acid–base regulation as temperature changes in decapod crustaceans. *Respir. Physiol.* **33,** 161–176.

Vogel, W., Vogel, V., and Pfautsch, M. (1976). Arterio-venous anastomoses in rainbow trout gill filaments. *Cell Tissue Res.* **167,** 373–385.

Warner, W. (1976). "Beautiful Swimmers; Watermen, Crabs and the Chesapeake Bay." An Atlantic Monthly Press Book, Boston, Massachusetts.

Weiland, A. L., and Mangum, C. P. (1975). The influence of environmental salinity on hemocyanin function in the blue crab *Callinectes sapidus. J. Exp. Zool.* **193,** 265–274.

Wolvekamp, H. P., and Waterman, T. H. (1960). Respiration. *In* "The Physiology of Crustacea" (T. H. Waterman, ed.), Vol. 1, pp. 35–100. Academic Press, New York.

3

Adaptive Aspects of Feeding Mechanisms

JOHN GRAHAME

I. INTRODUCTION

An animal feeds to obtain energy and materials for growth, maintenance, and reproduction. It is a necessary component of fitness that adequate food should be obtained. Potential food items occur in considerable variety and range of suitability; proportions of different items will differ from time to time, and some times will be better for feeding than others. The time spent feeding diminishes that available for other essential activities. Natural selec-

THE BIOLOGY OF CRUSTACEA, VOL. 8

ISBN 0-12-106408-5

tion may not only have produced mechanisms for gathering and processing certain sorts of food, but also behavioral responses governing their use. The outcome will be a set of mechanisms and patterns of use that produce the optimal result that maximizes fitness, and in this sense feeding mechanisms may be said to show adaptive aspects. Animals select feeding times and sites, choose food items, and may under some circumstances not feed at all, even if suitable items are available. These observations have led to the concept of feeding strategies (Schoener, 1971). The number of items of a particular kind taken may be expressed as a function of their abundance; this is the functional response (Holling, 1959).

There is much information on the types of food eaten by crustaceans and the ways in which it is captured and brought to the mouth. Fewer workers have been concerned with the more subtle aspects of feeding strategies and functional responses. Here, virtually all the information is concerned with planktonic filter feeders.

This chapter is concerned with the ways in which crustaceans feed; these methods may be divided into the categories of deposit feeding and scavenging, grazing, predation, parasitism, and filtering. Very little will be said about parasitism, as it is being covered in Chapter 5, Volume 6 and in Chapter 4, Volume 7. It must be stressed that the divisions between these methods are not rigid, and many crustaceans employ more than one method of feeding. A principal aim is to explore the way in which crustacean feeding methods are tailored to suit the animal's environment. There is a section on assimilation efficiencies; this is the only part of the work that is concerned with what happens inside the animal. The interest of ecologists in energy budgets and the debate over "superfluous feeding" among herbivorous zooplankters have brought the question of assimilation, which might be thought properly to be a physiological topic, into the sphere of ecology.

A large part of this review is concerned with filter feeding. This is a reflection of the bias in the literature and the fact that the study of filter feeding has a relatively long tradition of quantitative work. It is here that students of crustaceans have progressed furthest in understanding the strategies involved in feeding.

II. FOOD AND FEEDING MECHANISMS (EXCEPT FILTERING)

A. Deposit Feeding and Scavenging

Deposit feeding is defined here to mean feeding on the organic matter in deposited sediments. This includes processes that involve sorting outside the mouth as well as those in which sediments are ingested with little or no sorting. Thus, distinction may be made between selective and nonselective deposit feeders.

The cephalocarid *Hutchinsoniella macracantha* is a small marine animal living in unconsolidated flocculent benthic material (Sanders, 1963). The thoracic limbs are all similar; a metachronal rhythm of limb movement passes forward, and the interlimb spaces are alternately enlarged and compressed. Water and detrital material are drawn up into the median space between limb pairs; the water is then pumped out posterolaterally between successive limbs leaving suspended material caught on setae on the limb bases. This is brushed back into the median space by setae and drawn up into the ventral food groove on the next suction phase. In the food groove, the material is transferred forward by spines on the limb endites. Limb activity resuspends settled material in the vicinity; this is then drawn toward the animal by the currents associated with the limb beat. *Hutchinsoniella* is a nonselective deposit feeder, ingesting diatoms, detritus, and inorganic particles (Sanders, 1963). Its feeding is simple and probably represents the primitive method of early crustaceans (Manton, 1977).

Deposit feeding has been retained with varying degrees of elaboration in many Crustacea and is often found as an alternate or secondary method of feeding. Many Cladocera are benthic deposit feeders (Fryer, 1968). Food particles are scraped up or resuspended by the trunk limbs, drawn into the carapace chamber, and filtered off on setae, from whence they are mechanically passed forward to the mouth. Often, sticky secretions produced from the trunk limb bases or labral glands are used to bind the particles together. In *Graptoleberis testudinaria,* particles are scraped up and passed to the mouthparts without any filtration being involved.

Among the Amphipoda, *Corophium volutator* is considered to be a selective deposit feeder (Hart, 1930; Goodhart, 1939; Meadows and Reid, 1966). Material is brought to the burrow and resuspended in the pleopod current, particles being caught on the setae of the gnathopods. Many other amphipods are also selective deposit feeders (Enequist, 1949; Mills, 1967). *Bathyporeia pilosa* and *Bathyporeia sarsi* burrow in sand with the ventral surface uppermost; sand grains pass from in front of the animal into a small cavity above the ventral surface where they are held against the mouthparts by the first pereopods. Organic matter adhering to the grains is scraped off and ingested (Nicolaisen and Kanneworff, 1969). Caprellids illustrate well the ability of many crustaceans to exploit several food sources by different methods. They may behave as deposit feeders, filter feeders, scavengers, or predators (Saunders, 1966; Caine, 1974, 1977). The freshwater *Hyalella azteca* feeds principally on bottom sediments (Hargrave, 1970), and since it assimilates diatoms and bacteria with high efficiency, it may be supposed that these are the main nutritional source.

Cumaceans also clean food off sand grains (Dixon, 1944). Larger crustaceans also employ methods that may be termed deposit feeding. *Galathea* uses the third maxillipeds to sweep the substrate, collecting diatoms and other small algae as well as detritus, which are passed on to the oral appendages (Nicol, 1932). Hermit crabs use the small chela to scrape up material (Orton, 1927; Roberts, 1968). The material is passed to the third maxillipeds, which pass it on to the other mouthparts where it is sorted; edible particles are ingested and inedible material either dropped or rejected on currents. Orton (1927) noted that in *Pagurus bernhardus*, feeding behavior could be stimulated by adding extracts of the scallop, *Pecten*, to the water. The prawn, *Penaeus merguiensis*, probes the substrate with the chelate pereopods, passing particles to the mouthparts (Hindley and Alexander, 1978). It will eat large particles, but small items from the substrate, such as bacterial colonies, microflora, and microfauna, are considered an important part of the diet of this and related penaeids (Moriarty, 1977). The atyid prawns, *Caridina, Jonga, Atya, Micratya,* and *Potimirim* (Fryer, 1977), possess long setae ("bristles") distally on the chelae that are used to brush up deposited material and also in passive filter feeding. The bristles are spread by blood pressure, which swings each individual bristle out against the springlike tension of the cuticle at the base. Collected material is stripped off the bristles by "grid setae" on the first maxilliped and passed on toward the mandibles (Fryer, 1977).

Crabs of the genus *Uca* (Ocypodidae) feed on organic matter in the sediments of the estuaries and beaches where they live. Special setae with excavate tips, the spoon-tipped hairs, are found to a varying degree on the mouthparts and chelae, their abundance being related to the type of sediment preferred by the species (Crane, 1941; Miller, 1961; Icely and Jones, 1978). Their development is greatest in those species living on sandy substrates. Similar setae are found in the ocypodid, *Scopimera* (Ono, 1965; Fielder, 1970), and in the mictyrid, *Mictyris longicarpus* (Cameron, 1966), but they are absent in *Ocypode*, itself (Crane, 1941).

The feeding of *Uca* has been described by Altevogt (1957) and Miller (1961). Surface material is scooped up by the chela (in males, the minor chela) and passed between the third maxillipeds to the buccal cavity. Here the material is sorted; heavy inorganic particles fall to the bottom of the buccal cavity and out between the third maxillipeds. The buccal cavity may be flooded with water from the branchial cavity. This floats the lighter particles out of the sediment, and the suspension is held between the mouthparts. On draining, fine particles are left on the plumose setae of the mouthparts. The need for sorting is greatest on coarse sediments; the sand-dwelling *Uca pugilator* uses the spoon-tipped hairs on the second maxillipeds in scraping sand grains across bristles on the first maxilliped endites, thus removing adhering food material (Miller, 1961). A process of flotation in the buccal cavity has also been suggested for *Scopimera inflata* (Fielder, 1970) and for *Mictyris* (Cameron, 1966). The need for water used in sorting sediment is a factor affecting the distribution of ocypodes. In *Uca* there is strong sexual dimorphism: the male has one chela hypertrophied for social behavior, which is useless for feeding. As the nutritional physiology of males and females is similar, the males extend the duration of feeding, so that although they make fewer feeding movements per unit time than do the females, the number of feeding movements over a long period is the same (Valiela *et al.*, 1974).

It seems reasonable also to consider scavenging as it also involves the taking of dead organic matter. The individual food portions will, however, be much larger than in the case of deposit feeders. Many crustaceans scavenge to provide at least part of their diet, the most obvious examples being among the decapods, e.g., *Ocypode* (Hughes, 1966), isopods (Naylor, 1955), ostracodes (Lochhead, 1968), and copepods (Alldredge, 1972). The ostracode, *Conchoecia,* uses sticky secretions from glands in the carapace margins to entangle food particles (Lochhead, 1968). *Hippa pacifica* is recorded by Matthews (1955) as scavenging stranded *Physalia,* ingesting the zooids, including nematocysts. Among Cladocera, *Pseudochydorus globosus* feeds on small dead crustaceans (Fryer, 1968). The food is seized, dragged, and pushed toward the median chamber, then to the food groove, and on to the mouthparts.

B. Grazing

In terrestrial terms, grazing refers to the removal of living vegetation from a surface. The term has been widely applied to crustacean feeding types (see papers in Crisp, 1964). Here it is restricted to those animals which take their food from a solid surface which is large relative to themselves. The feeding of *Bathyporeia* referred to above (Section II,A) could legitimately be regarded as grazing; however, it seems more like the method employed by *Uca* and has accordingly been treated as deposit feeding.

It is difficult to classify the feeding method of harpacticoid copepods. They are often said to feed on diatoms (e.g., see Jewett and Feder, 1977). Marcotte (1977) considers that *Tisbe furcata* is a raptorial feeder. Food is grasped and manipulated by the antennules and second maxillae, and to a lesser extent by the maxillipeds. A particle is orientated longitudinally along the long axis of the body when the first maxillae push it forward to the mandibles, which crush it preparatory to ingestion. Other authors have shown that harpacticoids feed on general particulate matter (Harris, 1973), perhaps digesting mainly the associated bacteria (Brown and Sibert, 1977; Rieper, 1978), and also on blue-green algal colonies (Roman, 1978). Hicks and Grahame (1979) have suggested that *Diarthrodes nobilis* grazes on microorganisms that colonize a mucus bag secreted by the copepod.

Because of the size of the animal relative to the food particles, it is possible to agree with Marcotte (1977) or to regard benthic harpacticoids as grazers or deposit feeders.

Grazing has been described in isopods and amphipods. Jones (1972) gave an account of the mouthparts of *Jaera,* concluding that for the four species investigated, all of them fed by scouring and abrading algal or bacterial films from rocks and the alga, *Fucus*. Carefoot (1973) found that *Ligia pallasii* fed on encrusting and colonial benthic diatoms, on young shoots of such algae as *Enteromorpha, Cladophora,* and *Bangia,* and some animal matter. He considered that the choice of food was based not only on the calorific value of the food, but also on composition, availability, manipulability, and efficiency of digestion.

Freshwater species of *Gammarus* prefer fungal colonies to fresh angiosperm leaf material (Bärlocher and Kendrick, 1973a), although fungi are apparently not an obligate component of the diet (Bärlocher and Kendrick, 1973b). Moore (1975) found that *Asellus aquaticus* and *Gammarus pulex* ingested algae after both mechanical and chemical selection, excluding particularly filamentous blue-green algae. However, *Talorchestia longicornis* has been found to ingest and assimilate blue-green algae in its salt marsh habitat (Brenner *et al.*, 1976).

Terrestrial crustaceans feed in ways which may be variously regarded as scavenging, grazing, or predation, as in the case of terrestrial isopods.

An interesting case is reported by Herreid (1963) for the land crab, *Cardisoma guanhumi*. This animal lives in a burrow from which it emerges to eat fallen leaves, fruits, and grass. It will respond to the sight of a falling leaf at up to 3 m, and the sound of falling leaves or fruit landing on the ground will bring crabs from their burrows to search for food.

C. Predation and Parasitism

1. PREDATION

Many crustaceans will take live animal food without apparently having any particular structures or behavior patterns definitely associated with predation, e.g., the ostracode, *Conchoecia spinirostris* (Angel, 1970), and the crab, *Pachygrapsus crassipes* (Hiatt, 1948). Many calanoid copepods are considered to be carnivorous to a greater or lesser degree; and here, while there may be raptorial appendages, prey capture seems to be a matter of chance, the prey being apprehended by physical contact or at very close proximity (Wickstead, 1962; Ambler and Frost, 1974; Alvarez and Mat-

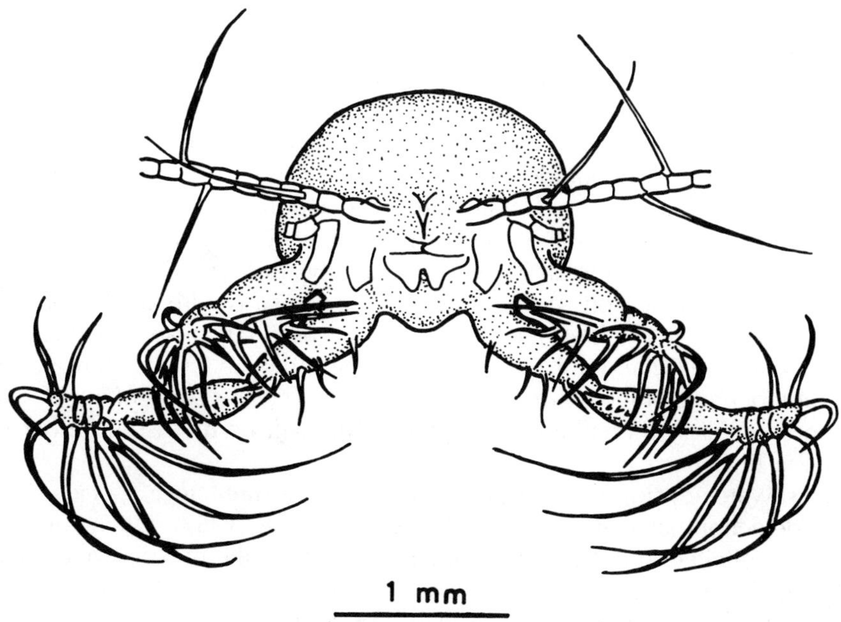

Fig. 1. Female *Pareuchaeta* with second maxilla and maxilliped in raptorial posture. (After Wickstead, 1962.)

thews, 1975). Some deep-water calanoids in particular have carnivorous tendencies (Wickstead, 1962), and the second maxilla and maxilliped are clearly raptorial (Fig. 1). The deployment of these appendages in a posture where they are ready to grab small prey once it is apprehended is an adaptation for predation. The calanoid, *Candacia,* is a carnivore preying on chaetognaths (Wickstead, 1959).

Experimental studies have shown that some copepods prey actively on fish or *Artemia* larvae in the laboratory. Lillelund and Lasker (1971) studied two species of *Labidocera* and one of *Pontellopsis,* with larvae of the anchovy, *Engraulis,* and *Artemia* as prey. The copepods were attracted by an active larva if it was in their immediate vicinity, ignoring eggs or motionless larvae. The prey was grasped and sometimes bitten without being eaten. Presence of *Artemia* reduced the predation of *Labidocera trispinosa* on the fish larvae. Density of fish larva prey had no apparent effect on the kill rate except at very low levels, when there was a drastic decline. Båmstedt and Holt (1978) found that Stage V copepodids of *Pareuchaeta norvegica* showed a clear preference for small prey over large when offered *Pseudocalanus* and *Calanus*. Landry (1978) showed that *L. trispinosa* creates a swimming current using the antennae and maxillipeds, with a funnel of water being drawn in from in front of the animal. The copepods detected prey (copepod larvae) remotely, lunged at it, and captured it with the second maxillae. Capture rates of different sized prey reflected the ability of the copepod to detect prey (this increased with prey size) versus the ability of the prey to escape, also increasing with prey size and abruptly becoming much greater in copepodids compared with nauplii. Thus, with *Acartia* as food, capture rates increased with later stage nauplii, decreasing again with copepodids. Of special interest is the shape of the functional response curve of *Labidocera* predation: there was apparently no lower threshold below which predation ceased, and capture increased linearly with number of prey up to 100 nauplii per liter, the maximum level used. The results of Lillelund and Lasker (1971) and of Landry (1978) suggest that the functional response of *Labidocera* is of type 1 or 2 (Holling, 1959); i.e., the curve of attack rate on prey density is rectilinear or convex. This will be discussed further in Section III,D.

Freshwater cyclopoid copepods include many predatory species, and prey detection has been said to be by contact (Fryer, 1957). Upon contact, attack is elicited and the prey grasped and held by the maxillae and maxillipeds. Fryer (1957) suggested that an abundance of food had obviated the need for efficient mechanisms and habits of predation, which consequently had not evolved. The anostracan, *Branchinecta gigas,* preys on other small Anostraca and copepods (White *et al.,* 1969). The animal has a hunting posture in which the abdomen is held over the trunk; while swimming ventral side up, prey encountering the abdomen dart away, sometimes colliding with and being trapped by the trunk appendages. The animal is said to rely solely on chance encounters for its food.

Strickler and Bal (1973) have shown that in *Cyclops scutifer,* there are mechanoreceptors on the first antennae. These could sense immobile particles at close range (1–5 mm) by detecting "fluid mechanical signals" generated by the animal and presumably reflected from the inert object, or by detecting fluid mechanical disturbances generated by active prey. This capacity, if widespread in small Crustacea, may be the basis for remote nonvisual prey detection. Kerfoot (1978) has shown that *Cyclops vernalis* has a well-defined attack pattern when preying on the cladoceran, *Bosmina*. Recognition of a subject as suitable (or otherwise) for attack can occur over a distance of at least two or three body lengths (≈3 mm). The attack field stretches out in a cone-shaped zone in front of the *Cyclops;* most attacks occur in two regions, one in front of each antennule. The *Cyclops* reorients if necessary and leaps on the prey, attempting to grasp it with the mouthparts. The attack sequences may be repeated several times. Of short duration (≈75 msec), they must be studied by high-speed cinematography. The findings of Strickler and Bal (1973) and Kerfoot (1978) suggest that in fact carnivorous copepods may locate prey remotely, albeit over fairly short distances.

While carnivorous isopods and amphipods may rely on chance encounters for food (e.g., D. A. Jones, 1968; von Westernhagen, 1976), some are visual predators. Sheader and Evans (1975) have shown that *Parathemisto gaudichaudi* hunts moving prey visually.

Decapods respond to chemical cues which initiate hunting and feeding behavior (McLeese, 1970; Mackie, 1973), and Fuzessery and Childress (1976) consider chemosense to be important among crustaceans in general, especially those living in deep water.

Brachyuran crabs are well known as predators on mollusks. Zipser and Vermeij (1978) report that crushing involves holding the shell in the master chela and applying force for a prolonged period. Elner (1978) found that *Carcinus maenas* preying on *Mytilus* used several methods to open the shell, the most usual being to crush the shell with the master chela. The crab sought weak spots in the shell by trial and error, applying series of crushing pulses in succession to one spot. If these were unsuccessful, the shell was slightly reoriented, and another series of pulses followed. Elner and Hughes (1978) showed that for *C. maenas* of a given size, there existed an optimal sized mussel for which energy gain was maximized. If prey were unlimited, the crabs chose mussels close to the predicted optimum; as these became depleted, they took mussels above and below the optimum size. Suboptimal prey were readily taken as a result of rapid adjustment to the availability of different quality prey. Elner and Hughes (1978) considered this to be an adaptation to heterogeneous prey size by a nonvisual predator. They showed that recognition and assessment of the prey items were achieved by manipulation with the chelae.

2. PARASITISM

If an animal feeds on another without killing it, it may be considered to be parasitizing its "prey." Indeed, parasitism has apparently often evolved from predation. Goldberg (1971) reported that the snapping shrimp, *Synalpheus fritzmuelleri,* is omnivorous and feeds on the snail *Coralliophila* in an interesting way. The shrimp inserts the dactylus of the large chela under the siphonal canal, lifts the operculum, and tears off pieces of the foot, which it eats. After a time the snail is released, having suffered minimal damage. The copepod, *Sapphirina angusta,* is said to be parasitic as a juvenile and predaceous as an adult (Heron, 1973), although both attack the same "prey," the salp, *Thalia democratica.* Heron (1973) distinguished between parasitic and predaceous phases on the basis of the damage done to the salp: juvenile copepods did not cause death, while the adult females did and were thus regarded as predators. Pinnotherid crabs are variously regarded as commensals or parasites, often of lamellibranchs (Pearce, 1966; Kruczynski, 1975); they feed on the food-laden mucus strings of the host. Kruczynski (1975) has shown that ^{14}C from labeled phytoplankton is taken up by female *Pinnotheres maculatus* living in bivalves, but the precise pathway of this uptake is still unclear.

Coelenterates provide hosts in many commensal or parasitic relationships. The cladoceran, *Anchistropus emarginatus,* parasitizes *Hydra* and feeds on the tissues (Fryer, 1968). The barnacle *Hoekia monticulariae* lives in coral skeletons, using the mandibles and maxillae to rasp away coral tissue as it grows over the small aperture of the chamber (Ross and Newman, 1969). The Pacific xanthid crabs, *Trapezia* and *Tetralia,* are considered by Knudsen (1967) to be ectoparasites of corals, using the walking legs to excite mucus production; the mucus is then eaten by the crab. Patton (1967) believed that the xanthid, *Domecia acanthophora,* living on the coral *Acropora* in the Caribbean, fed on organic detritus filtered from the water with the second maxillipeds. In this case, coral mucus was thought to be only a secondary food source, if used at all.

III. FILTER FEEDING

A. Mechanisms

Filter feeding is widespread among Crustacea and has received much attention. Marshall and Orr (1960) distinguished five methods of filter feeding, depending on which limbs bore the filtratory setae. These were (1) trunk limb filters (Anostraca, Cladocera, and Cirripedia); (2) thoracic limb filters (some Amphipoda and Decapoda); (3) maxillary filters (many Copepoda); (4) maxillular and mandibular filters (some Ostracoda); and (5) antennal and antennular filters (some Amphipoda and Decapoda).

There is a degree of overlap between what some authors consider to be filter feeding and others to be deposit feeding. This is especially true when an animal resuspends sediments and filters food from the suspension, as do several of the crustaceans considered in Section II,A. Branchiopods, mysids, and euphausiids are all known to have some members which may either filter food from the water column or descend to the bottom and resuspend sediment (Cannon and Manton, 1927; Lowndes, 1933; Mauchline and Fisher, 1969). The categorization as deposit feeder or a filter feeder depends, to an extent, on whether the feeding method or the source of food is held to be paramount.

1. FILTER FEEDING IN NONPLANKTONIC CRUSTACEA

Among amphipods, *Haustorius arenarius* has a maxillary filter (Dennell, 1933). *Melphidippella* "stands" ventral side up on the bottom, catching falling detritus with its antennules, antennae, and third and fourth pereopods (Enequist, 1949). Passive filter feeding also occurs in atyid prawns (Fryer, 1977).

Pedunculate barnacles are "captorial" feeders, curling the cirri (thoracic legs) often independently around relatively large items of food (Barnes and Rees, 1959; Howard and Scott, 1959; E. C. Jones, 1968). The acorn barnacle, *Verruca stroemia,* is also capable of withdrawing a single cirrus, as well as the whole net (Stone and Barnes, 1973). In low water velocities this barnacle beats rhythmically; as water velocities increase, beating gives way to prolonged extension of the cirral net, which curls when food is captured. Both the "fast beat" and "extension" feeding activities of acorn barnacles are stimulated by water currents, faster flow being needed to stimulate activity in species from turbulent habitats (Southward and Crisp, 1965). *Balanus* extends the cirral net into the water, withdrawing it again, and scraping off trapped particles (Crisp and Southward, 1961). The large cirri (fourth, fifth, and sixth pairs) are combed by the third pair, which in turn are combed by the second, and then the second by the first. Thus, food is passed toward the mouth. The fine comb setae of the first, second, and possibly third pairs of cirri filter the respiratory flow of water passing into the mantle cavity, allowing for feeding on smaller particles than are taken by the large cirri. Barnes (1959) suggested that glandular secretions from the bases of the cirri enmeshed fine food particles.

Crisp (1967) found that the cirri and mouthparts of pedunculate barnacles were sensitive to a range of substances, including amino acids. This capacity was thought to reinforce initial capture responses and cause living prey to be eaten.

Filter feeding occurs quite widely among decapods. The anomuran, *Porcellana longicornis,* sweeps the third maxillipeds back and forth through the water, alternately spreading and flexing the setae (Nicol, 1932). Captured food is brushed off the setae by the second maxillipeds. *Petrolisthes cinctipes* feeds in a very similar way and is strongly stimulated to feed by amino acids, trehalose, and glucose (Hartman and Hartman, 1977). Feeding is also stimulated by newly hatched *Artemia* nauplii or by dialysate of the hatch solution. *Emerita analoga* filters with the antennae, extending them from the sand into the backwash of a wave and filtering the water flowing from behind the animal (Efford, 1966). The rate at which the antennae are withdrawn to be cleaned by the mouthparts is related to water velocity, being more rapid in faster flowing water. In the presence of food, the antennal cleaning rate increases markedly; this response is mediated by the arrival of food in the mouthparts. Both tactile and chemical stimuli are apparently involved.

The hermit crab, *Pagurus bernhardus,* can filter both *Artemia* nauplii and the flagellate, *Dunaliella,* and can evidently satisfy its metabolic requirements by filtering zooplankton (Gerlach *et al.,* 1976). Filtering occurs on the setae of the mouthparts. Scully (1978) has reported that *Pagurus longicarpus* will roll onto its back and collect foam and floating detritus from the surface of very shallow water (1–3 cm depth). *Diogenes brevirostris* is usually a deposit feeder, but can employ the antennae as a "cast net" (Boltt, 1961). The antennae may be held passively in a current or swept to and fro, twisting so that the oblique angle between the setae faces the direction of each sweep. The setae are cleaned between the third maxillipeds, and food particles are eaten. This behavior provides a secondary food source for the hermit crab. *Stratiotes setosus* uses antennary filter feeding as a primary feeding mechanism; the antennae are specially jointed at the base to facilitate turning at the end of each sweep (Greenwood, 1972).

Filter feeding is important in larval decapods (e.g., see Factor, 1978). Budd *et al.* (1978) have suggested that filter feeding occurs not only in the larvae, but also in the adults of the crayfish, *Orconectes immunis.* In adults, filtering may be used to supplement the diet or as a food source immediately after ecdysis when the animal is vulnerable and unable to forage widely.

2. FILTER FEEDING IN PLANKTONIC CRUSTACEA

Many authors, when discussing the feeding of copepods and branchiopods in particular, refer to the filtering rate or the volume swept clear per unit time. It is generally assumed that during the experiment, a certain particle category has decreased in number because it was filtered and ingested by the experimental animals. This means that ingestion rate and filtration rate are linked, the latter being calculated from a measure of the former. The use of filtration rate as a measure of feeding has two potentially confusing consequences. First, there may be different apparent filtration rates in the same experiment, due to differential capture of different particle categories. Second, over a range of particle concentrations where filtration rate remains constant (a common feature), ingestion is increasing linearly with particle concentration. It is therefore important to distinguish carefully between filtration rate and ingestion rate.

The anostracan, *Chirocephalus diaphanus,* has a filtering mechanism (Cannon, 1928a) which seems easily derived, in evolutionary terms, from the feeding mechanism of *Hutchinsoniella* (Section II,A). *Chirocephalus* swims on its back, with a metachronal rhythm of leg movements, the waves of which pass forward. Alternate expansion and compression of the interlimb spaces draws a current down into the food groove, where particles are trapped by setae on the basal endites of the limbs. While the water is pumped out laterally, the trapped particles are flushed back into the food groove and carried forward to the mouth in a water current. Labral gland secretions bind the food before swallowing. Lowndes (1933) considered that *C. diaphanus* fed principally on detritus and not on already suspended particles. It took larger food as well, pushing this into the food groove with the limb endites and forward by means of spines and setae on the gnathobases.

While Anostraca filter with all the trunk limbs, in Cladocera this function may be concentrated on a few. This is seen in *Daphnia* where the third and fourth trunk limbs are specialized as filters (Cannon, 1933).

In mysids (Cannon and Manton, 1927) and euphausiids (Mauchline and Fisher, 1969; Berkes, 1975), the exopodites of the thoracic limbs create a feeding current passing along the rotating exopodite and into the ventral groove between the limb bases. Food is retained on the basal endites of the limbs and on the maxillae.

Filter feeding in calanoid copepods was first studied by Esterly (1916) and Cannon (1928b) using *Calanus*. The feeding current is generated by the antennae, first maxillae, and maxillipeds; the filter is formed by setae on the second maxillae. Esterly (1916) suggested that locomotion and feeding in *Calanus* were independent. Cannon (1928b) considered that filter feeding was "automatic," the feeding current being a necessary resultant of the swimming current. Lowndes (1935) criticized Cannon's use of the term "automatic," finding that there was no evidence that calanoid feeding was either automatic or nonselective. However, Cannon (1928b) did point out that material was sometimes vigorously rejected from the second maxillae, a possibility suggested by Esterly (1916) to explain his observation that feeding did not always occur.

Filtration is usually regarded as a process of sieving. Other filtration mechanisms are possible (Rubenstein and Koehl, 1977), but there seems to be no direct evidence that any of these may be used by copepods. Boyd (1976) considers that the filtering mouthparts should be regarded as "leaky sieves." In this case, the lower limit of filtration would be determined by the distance between filtering setules and the upper by screening setae on the maxillae rejecting large particles (e.g., in *Calanus:* Gauld, 1964). Both of these limits are subject to some variation in the working filter, due to distension by impinging particles. Nival and Nival (1973) have determined the proportion of the total filtering area covered by meshes of different sizes in several species of copepods. The curves they plot (Fig. 2) can be interpreted as representing the efficiency with which different sized particles will be retained. From Fig. 2 it can be estimated that *Acartia clausi* will retain 50% of

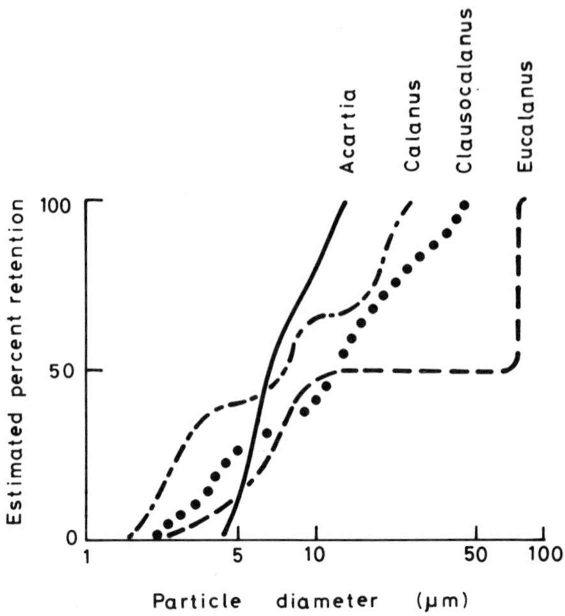

Fig. 2. Cumulative retention efficiency for different sized particles, estimated from intersetule distances on the second maxilla and maxilliped for *Acartia clausi, Calanus helgolandicus, Clausocalanus arcuicornis,* and *Eucalanus elongatus* (Dana). (After Nival and Nival, 1973, omitting two species for clarity.)

particles of 6 μm diameter, reaching 100% retention at about 14 μm, whereas *Clausocalanus arcuicornis* reaches these retention efficiencies at about 14 μm and 45 μm, respectively. These curves (Fig. 2) are based on intersetule distances for both the second maxilla and maxilliped. It is clear that if they do reflect the filtering ability of calanoids, different species will retain the same size particle with different efficiencies. Nival and Nival (1976) tested predictions based on their filtration efficiency curves for *A. clausi* and found that the curves gave a good approximation of the experimental results. There were some discrepancies which the authors attributed to the complexity of the processes of capture and ingestion.

Calanoids not only filter feed, but also grasp larger particles with their mouthparts. Petipa (1960) observed *Calanus helgolandicus* capturing the dinoflagellate, *Noctiluca,* and Conover (1966a) studied the capture of the diatom, *Coscinodiscus,* by *Calanus hyperboreus.* Capture was successful only upon physical contact of a cell with the mouthparts when the maxillipeds brought the cell to within reach of the second maxillae. The first and second maxillae and maxillipeds were used to manipulate the cell and press it to the mandibles, where it was shredded and the cell fragments eaten.

Artemia nauplii were apparently sensed from a distance of a few millimeters, and the tips of the pereopods and maxillipeds were lowered in anticipation of contact. Conover (1956) described filtration by "seining" in *Acartia,* where the second maxillae are active rather than passive filters. *Acartia* works its second maxillae by spreading the setae and then quickly drawing them together, capturing suspended particles. Gauld (1964) referred to this process as "scooping," and suggested that *Centropages, Temora, Pseudocalanus,* and *Calanus* also captured large particles by scooping with the second maxillae. Feeding calanoids may cause fragmentation of diatom chains (O'Connors *et al.*, 1976) and destruction of large particles (e.g., see Poulet and Marsot, 1978) by the activities of their mouthparts.

It has been apparent, since the work of Harvey (1937), that calanoids can detect the nature of the particles they encounter, and several recent workers confirm this (e.g., Conover, 1966a). Particles may be rejected after "tasting." Petipa (1960) suggested that *Calanus* might in some way be attracted by the bioluminescence of *Noctiluca*. However, Esaias and Curl (1972) and White (1979) have shown the opposite effect with *Acartia* and *Calanus* grazing on bioluminescent dinoflagellates. Grazing was reduced at higher levels of bioluminescence, apparently as a result of "startle" or confusion of the copepods by the flashing light of the cells.

Ong (1969) found sensilla on the labrum and mandibles of the estuarine copepod, *Gladioferens pectinatus,* which evidently allowed it to discriminate between detritus and phytoplankton, the latter being selected. Friedman and Strickler (1975) characterized the receptors on the mouthparts of *Diaptomus pallidus* as chemoreceptors; mechanoreceptors were lacking. They suggested that herbivorous copepods selected food on the basis of olfaction rather than size, but that size might be appreciated by determining the area of stimulation from a particle. Poulet and Marsot (1978) found that *Acartia clausi* and *Eurytemora herdmani* discriminated between synthetic particles of two sizes on a chemical basis, selecting those which included a freeze-dried water soluble fraction from a phytoplankton bloom. Substantial destruction of large nonenriched particles occurred, and the authors concluded that the copepods were discriminating between the particles on the basis of "scent," regardless of size.

B. Filtration by Planktonic Crustacea in Relation to Environmental Factors

1. FOOD AVAILABILITY AND TYPE

The supposition of automatic feeding and the belief of some workers that phytoplankton was too sparse to support such forms as *Calanus* (e.g., see

Fuller, 1937) were consistent with results indicating that filtration was independent of cell concentration (Fuller, 1937; Gauld, 1951) and, presumably, always maximal. Using *Daphnia magna,* Ryther (1954) found that there was evidence for dependence of filtration rate on cell concentration. In particular, filtration seemed to be inhibited by senescent cultures of *Chlorella.* Rigler (1961) showed that filtration by *D. magna* declined in phytoplankton concentrations above a certain level, which was called the "critical concentration," or "incipient limiting level" (McMahon and Rigler, 1965). Rigler proposed that this might be a general phenomenon. He noted that Marshall and Orr (1955) had found that *Calanus finmarchicus* increased the rate of fecal pellet production with food concentration up to a certain level, beyond which there was little further increase. Rigler (1961) proposed that in low concentrations of food, feeding rate was limited by the ability to filter, while in high concentrations it was limited by the ability to ingest or digest the filtered food. Thus, filtration was independent of concentration up to the critical concentration while ingestion increased; above the critical concentration, filtration declined while ingestion was constant. McMahon and Rigler (1963) showed that *Daphnia* regulated its feeding rate above the critical concentration by two mechanisms, i.e., by rejection of food from the food groove and by reduction of the pumping rate of the filtering legs. In marine copepods, similar responses have been found (e.g., see Frost, 1972). However, Mullin (1963) and Haq (1967) reported curves of ingestion rate on food concentration for *Calanus* and *Metridia,* respectively; in these experiments the ingestion rate peaked and then declined, with further increases in food concentration. Conover (1978b) thought that this form of response might result from an "overshoot" in maximum ingestion combined with incomplete adjustment of feeding rate to the experimental food level over the duration of the experiment ($\approx$24 hr). This raises the question of the length of time for which filter-feeding copepods may expect, in nature, to experience a particular food level. A number of authors consider that an upper critical concentration is rarely if ever encountered by copepods in the sea, either because absolute concentrations of food are too low (Adams and Steele, 1966; Reeve and Walter, 1977) or because of rapid adaptation of the copepods to changing conditions (Conover, 1978a,b). However, Parsons *et al.* (1967) and Gamble (1978) believe that an upper critical concentration may at times be important for copepods in the sea. There is evidence that in freshwater habitats, natural algal populations often exceed the critical concentration for *Daphnia* species over long periods of time (Burns and Rigler, 1967; Burns, 1968a; Hayward and Gallup, 1976).

The behavior of filter feeders in low food concentrations has also attracted attention, and there is evidence that copepods may reduce or cease feeding when food is sparse (Adams and Steele, 1966; Parsons *et al.*, 1967, 1969; Frost, 1975). This will be discussed further in Section III,D.

The nature of the food available (plant or animal and the kind involved) may modify the behavior of filter-feeding Crustacea. In particular, blue-green algae may be unsuitable as food. Burns (1968a) found that filamentous blue-green algae caused the pumping of thoracic appendages in *Daphnia* to be interrupted, and any filaments reaching the labral region were rejected. However, if other food was sparse, *Daphnia galeata* did ingest cells or short fragments of *Anabaena. Daphnia schoedleri* does not feed on the blue-greens, *Anabaena* and *Aphanizomenon* (Hayward and Gallup, 1976). Arnold (1971) fed *Daphnia pulex* on a variety of blue-green algae, as well as on *Ankistrodesmus, Chlorella,* and *Chlamydomonas,* for comparison. The cyanophytes were in general less valuable as food, supporting lower survivorship and reproduction than other foods.

Crowley (1973) found that seston limited the filtering of *Daphnia pulex* and so did unspecified dissolved inhibitors. O'Brien and deNoyelles (1974) found, with *Ceriodaphnia reticulata,* that particle concentration alone affected filtering rates. In particular, they failed to find any effects due to the water chemistry, and the blue-green alga, *Microcystis,* did not have any effect.

Of particular relevance to the design of experiments on filtering is the fact that filtering rates are sensitive to the age of phytoplankton cultures (e.g., Conover, 1956; Mullin, 1963), being depressed in old or senescent cultures.

2. DIURNAL CHANGES

Gauld (1953) concluded that *Calanus* might show an alternation of feeding and nonfeeding periods, but that this depended on vertical migration and was not an independent behavior pattern. If the animals did not migrate down in the day, they continued to feed. At the same time, he noted that *Calanus* could spend periods of several days in dense food suspensions without feeding. Petipa (1958) found that adults of *Acartia clausi* fed around the clock, but the gut was most full at 0100, with a secondary peak at 0700; nauplii fed only during the day. Mackas and Bohrer (1976) examined the guts of zooplankters by a fluorescence technique and concluded that many marine copepods show a diurnal periodicity of feeding. There might be either one or two peaks of activity, and they suggested that individual animals migrated down as soon as the gut was full. Nauplii of *Semibalanus balanoides* also showed a diurnal periodicity of feeding activity. Hu (1978) found that euphausiids fed most at night if they were vertical migrators, while those that did not migrate had a constant stomach content.

Chisholm *et al.* (1975) have shown that *Daphnia middendorffiana* has a bimodal rhythm of filtering activity. Peaks of activity were at 2400 and 1400 in the late summer (Barrow, Alaska), and the authors noted that at these times, the temperature in the habitat was near the optimum for filtering activity (11°C). Haney and Hall (1975) showed a bimodal periodicity of feeding in *D. pulex,* with both peaks at night. The first was after upward migration and the second before the morning descent; 86% of filter feeding was found to take place at night. In contrast, *Diaptomus pallidus* was found to migrate, but not to show any diurnal variation in filter-feeding activity. The diel activity rhythms of *Daphnia* can be shown to persist under laboratory conditions lacking environmental cues (Duval and Geen, 1976). Starkweather (1978) concluded that the periodicity of feeding in *Daphnia* was sensitive to a number of environmental factors, including nutritional history and food availability.

Temperature changes are known to affect filtering rates (as would be expected), and temperature optima are found. These may have ecological significance in the timing of maximal filtering effect (Chisholm *et al.,* 1975) and in the distribution of *Daphnia* species (Burns, 1969).

C. Sources of Food for Planktonic Filter Feeders

Some early workers doubted that there was sufficient particulate matter in seawater to sustain filter feeders, and that dissolved organic matter must make up the difference. The problem was discussed by Esterly (1916) with respect to calanoid copepods. From examination of gut contents, he showed that diatoms and sometimes crustaceans were taken as food, but that there were insufficient of these in the sea to meet the animals' needs. He postulated that they relied in part on nannoplankton, which might leave no recognizable remains in the gut. This view was repeated by Fuller and Clarke (1936), whose measurements of filtration rate in *Calanus finmarchicus* suggested that it could not obtain enough food from larger phytoplankton. Gauld (1951) considered that filtration might be adequate to meet the needs of calanoids, but that data were still insufficient for proof of this. Corner (1961) reviewed the literature on the subject. As a result of experiments using *Calanus helgolandicus,* he concluded that filtration, at an average of 21.5 ml/animal/day, supplied adequate particulate matter for the copepod's needs. Corner *et al.* (1967) reached the same conclusion for *C. finmarchicus*. It is now widely thought that small planktonic Crustacea do meet their nutritional needs by filtering or capturing particles from the water, and the supposed reliance on dissolved organic matter originally suggested by Putter (quoted in Esterly, 1916) is no longer believed to be the case. Nevertheless, dissolved organic substances can be taken up from the water, and should they be present in sufficient concentration might play a part in nutrition (Gyllenberg and Lundqvist, 1978).

It has long been apparent that marine calanoids will partake of a variety of foods (see Esterly, 1916), and attempts have been made to classify them as herbivores or carnivores. Anraku and Omori (1963) examined the mouthparts of several species and concluded that there was a relationship between form of mouthparts (especially the mandibles) and the main food items. Herbivores (e.g., *Calanus*) had short, grinding mandibular teeth; predators (e.g., *Tortanus*) had long, sharp mandibular teeth; omnivorous species (e.g., *Acartia*) had teeth intermediate between these. Itoh (1970) attempted to put this on a quantitative basis by deriving an "edge index," which increased in value from herbivores through omnivores to carnivores. Arashkevich (1968) distinguished the same groups as had Anraku and Omori (1963), extending these to include deep-sea copepods.

Sullivan *et al.* (1975) examined the mandibles of several herbivorous copepods by scanning electron microscopy (SEM). They confirmed the presence of siliceous crowns on the teeth, originally described by Beklemishev (quoted by Sullivan *et al.*, 1975), and commented that the form of the teeth suggested a cracking rather than a grinding function. Turner (1978) has questioned the classification of the Pontellidae as carnivores. He considered that diatom remains in the guts of *Pontella meadii, Anomalocera ornata,* and *Labidocera aestiva* were too numerous to be the result of ingestion of herbivorous copepods. He pointed out that, whereas the teeth appeared long and pointed under the light microscope, SEM examination of the mandibles revealed that some possessed blunted molariform tips, suggesting a crushing role. Moreover, setules were present on the setae of the second maxillae and capable of retaining particles down to 10–20 μm. Turner concluded that while some pontellids were indisputably capable of predation, they would be more accurately classified as omnivores.

Corner *et al.* (1974) have shown that *Calanus* must be considered omnivorous. The turnover of nitrogen and phosphorous over winter indicates that feeding occurs at a time when only microzooplankton and detritus are readily available. They showed that *Calanus* can feed on the nauplii of the barnacle *Elminius*.

Nonliving particulate organic matter (detritus) may also be a food source for filter-feeding Crustacea. Baylor and Sutcliffe (1963) found that *Artemia* grew when fed on particles produced by foaming seawater in the laboratory and suggested that particulate organic matter (POM) arising from natural foams could be an important food source for zooplankton. Paffenhöfer and Strickland (1970) found that *Calanus helgolandicus* would feed on dead phytoplankton cells and on their own fragmented feces. They would also ingest polystyrene spheres of suitable size. However, the copepods would not feed on "marine snow," i.e., amorphous natural detritus. Corner *et al.* (1974) found that both natural detritus and laboratory foam tower POM, although eaten, were inadequate diets for overwintering *C. helgolandicus*. The laboratory POM was made by foaming seawater enriched by soluble extracts of the diatom, *Biddulphia*. Poulet (1976) estimated the proportion of living particulate matter in Bedford Basin, Canada, by determining the amount of adenosine triphosphate and using a conversion factor to obtain values for living organic carbon. He found that *Pseudocalanus minutus* consistently ingested a higher proportion of nonliving than living carbon, reflecting the greater abundance of nonliving particulate carbon. Poulet concluded that detritus was a basic food for *P. minutus*, not merely a supplementary one.

Studies by Heinle *et al.* (1977) and Roman (1977) indicated that detritus derived from a variety of plant sources with associated bacteria and ciliate protozoans were, by themselves, nutritionally inadequate for copepods. These authors showed that while detrital fractions were ingested and digested, they could not by themselves support normal growth and activity. Chervin (1978) showed that the copepods *Acartia* and *Eurytemora* in the River Hudson Estuary selected against detritus, while in neritic waters *Temora, Pseudocalanus,* and *Centropages* selected for detritus. Chervin suggested that this might reflect the origin and palatability of the detritus fractions in the two locations, the estuarine detritus (with a large terrigenous component) being relatively unacceptable.

Coral mucus suspended in the water is ingested by the copepod, *Acartia negligens,* and this material may represent a significant food source for reef-dwelling planktonic copepods (Richman *et al.,* 1975).

Bacteria are often considered to be a potential food source for filter feeders, and *Daphnia* has been shown to be able to retain and ingest very small particles (Gellis and Clark, 1935; Tezuka, 1971; Lampert, 1974). Peterson *et al.* (1978) have shown that *Daphnia* can feed on the very small, naturally occurring bacteria from an Alaskan lake. The actual contribution of bacteria to the total requirements of the animals is likely to be small (Tezuka, 1971).

D. Strategies of Filter Feeders

1. PARTICLE SELECTION

A great deal has been written on particle selection by filter feeders. There are two sets of criteria by which potential food particles might be assessed: mechanical, involving size, shape, and possibly texture; and chemical, i.e., "taste" or "scent."

Harvey (1937) showed that *Calanus* filtered large cells more readily than small cells. There was some evidence of preconditioning in the results of Harvey's experiments, although he was cautious about drawing such a conclusion from the limited data. The greater retention of larger particles within the range taken by filter feeders is a feature pointed out by many authors (e.g., Mullin, 1963; Frost, 1972, 1977; Poulet, 1973; Paffenhöffer and Knowles, 1978). Larger animals and later developmental stages may take larger cells (Burns, 1968b; Nival and Nival, 1976; Poulet, 1977; Paffenhöfer and Knowles, 1978). Wilson (1973) conducted a series of experiments in which *Acartia tonsa* was allowed to feed on plastic beads 7–70 μm in diameter, in varying size-frequency distributions. He found that there was apparent selection, with the animals preferentially ingesting those beads that were just larger than the largest peak in the size-frequency distribution. Wilson concluded that *A. tonsa* was selecting on the basis of abundance and size, and he hypothesized that the copepod adjusted the attitude of its "grasping appendages" so as to catch items just larger than the last one ingested. The actual range of ingested bead sizes was 13.9–59 μm. In grazing experiments using natural seawater and particle populations with several species of copepod and their development stages, Poulet (1973, 1974, 1976, 1977, 1978), Poulet and Chanut (1975), Allan *et al.* (1977), Richman *et al.* (1977), and Gamble (1978) have all concluded that grazing is maximal in the peaks of the particle biomass distribution spectra. Moreover, these authors conclude that there is at least some degree of flexibility, and that copepods track shifting peaks of biomass and so adjust to changing food availability. How this tracking is achieved is obscure, and there is semantic confusion. Allan *et al.* (1977) and Richman *et al.* (1977) refer to tracking as "selective feeding," whereas Poulet and co-workers maintain that it is "nonselective opportunistic" feeding. Poulet and Marsot (1978) found that *Acartia clausi* and *Eurytemora herdmani* discriminated between synthetic food particles on the basis of "scent" quite independently of size (see also Friedman and Strickler, 1975). Piontkovskii and Petipa (1975) considered that size, shape, cell wall configuration, and cell mobility were all involved in selection of food by *Acartia*. Allan *et al.* (1977) have made the interesting suggestion that selectivity of grazing (with respect to particle size) appears with metamorphosis from the nauplius VI to copepodid I, the nauplii being less able to show a tracking response.

Boyd (1976) has used the retention efficiency curves proposed by Nival and Nival (1973) as a probability function showing the likelihood of retention of a particular size particle by a given species. This approach predicts the ingestion of particles in the manner shown in the experiments of Wilson (1973). Many of the electivity indices shown by Wilson (1973) and Poulet (1973) can be generated in this way. Boyd (1976) concludes that many of the features of copepod filtration can in fact be explained in simple mechanistic terms by treating the filtration process as one of sieving. Frost (1977) also considers that filtration is a simple mechanical process. From experiments with *Calanus pacificus,* Frost (1977) concluded that the copepod filtered suspensions of food particles indiscriminately, but that capture, handling, and ingestion were more efficient for larger particles. At the larger end of the particle spectrum, copepods may grasp their food rather than filter it (e.g., Conover, 1966a). It has been noted that the process of filtration and raptorial feeding on large cells or animal prey are different and may interfere with each another (Conover, 1966a; Lonsdale *et al.*, 1979).

There may be some scope for altering the mechanical nature of the filter, depending on particle size and abundance. Frost (1977) and Cowles (1979) have discussed this possibility, with different conclusions. The possibility that mucin (e.g., from the labral glands) might be used to modify the surface properties of the filtering setules does not seem to have been explored. It is indisputable that once particles have been captured, considerable choice may be exercised over whether they are ingested or not (Conover, 1966a; Burns, 1968a; Frost, 1977; Poulet and Marsot, 1978). From a consideration of the work of the authors cited above, it seems plausible to separate the process of "filter feeding" into at least two phases: particle capture, which is probably indiscriminate within the mechanical limits of the filter, and particle handling preparatory to ingestion, when choice is exercised. Lehman (1976) has pointed out that if the cost of rejection is high, nutritionally unsuitable material may be ingested rather than rejected.

2. THE FUNCTIONAL RESPONSE OF FILTER FEEDERS

Holling (1959) proposed that there were three basic types of functional responses, i.e., of the relationship of attack rate to density of food items (prey). These are graphically represented in Fig. 3. In type 1, search is random and search rate is constant at all prey densities, leading to a linear increase to the maximum amount of food items taken with increasing density. In type 2, the search rate decreases at higher prey densities, and the curve of prey taken on density becomes convex. In type 3, the curve is sigmoid because not only is there an upper threshold, but also a lower one, beyond both of which the search rate decreases. Holling (1965) proposed that a type 2 response was typical of invertebrate predators and type 3 of vertebrates, but Hassell *et al.* (1977) have argued that the type 3 response may also be common among invertebrates.

Ingestion rates of filter feeders may be thought of in terms of functional response curves. Initial work (e.g., Fuller, 1937), which suggested that filtration and cell concentration were independent and that copepods were perpetually short of food, indicated a type 1 or 2 response. The demonstration

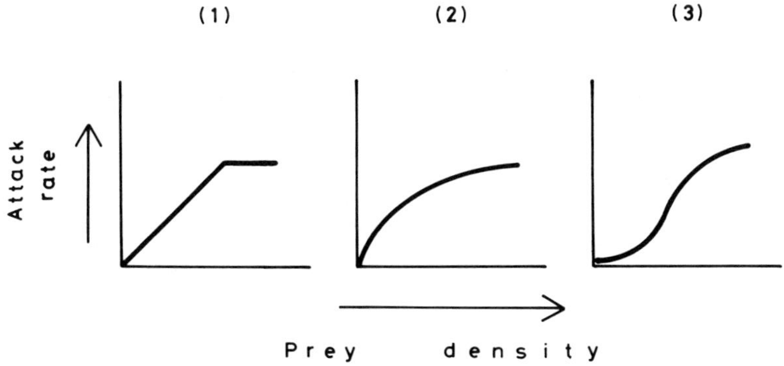

Fig. 3. The three types of funtional response of a predator to prey density. (After Holling, 1959.)

of a "critical concentration" (Rigler, 1961) revealed an upper threshold beyond which ingestion was constant. Adams and Steele (1966) first produced evidence that there might be a lower threshold response as well. They found that *Calanus finmarchicus* reduced its filtering effort at low food concentrations and noted that this was quite the contrary of expectations based on the "encounter theory" of grazing developed by Cushing (1959). Adams and Steele (1966) pointed out that the lower threshold was of potential ecological importance, since it would cause reduced phytoplankton mortality at low food densities in common with the general effect of sigmoid responses in predator–prey systems (Hassell *et al.,* 1977). A type 3 response for filtering calanoids is also indicated by the work of Parsons *et al.* (1967, 1969), Corner *et al.* (1972), Frost (1975), and Reeve and Walter (1977). Frost (1975) specifically designed experiments to test for the existence of a lower threshold in the feeding activity of *Calanus pacificus.* He found that, while there was measurable feeding at the lowest cell concentrations used, below about 200 cells/ml filtration was reduced (Fig. 4). These data, when combined with the critical concentration upper threshold, imply a type 3 response for *C. pacificus.* Other authors have pointed out that it is impossible to statistically reject the fit of either a curvilinear or rectilinear plot of the functional response, and they retain one of these in preference to a sigmoid response expression (Mullin *et al.,* 1975).

Frost (1972, 1975) has shown that as food particles get bigger, the density at which the lower threshold is found fails; thus with larger food particles, filtration is maintained to lower food densities. With a large *Coscinodiscus* of cell diameter 87 μm, the threshold was not reached until a density of only 4 cells/ml was attained (Frost, 1975). Boyd (1976) suggests that *Calanus* treats particles larger than 100 μm raptorially rather than by filtration. It is

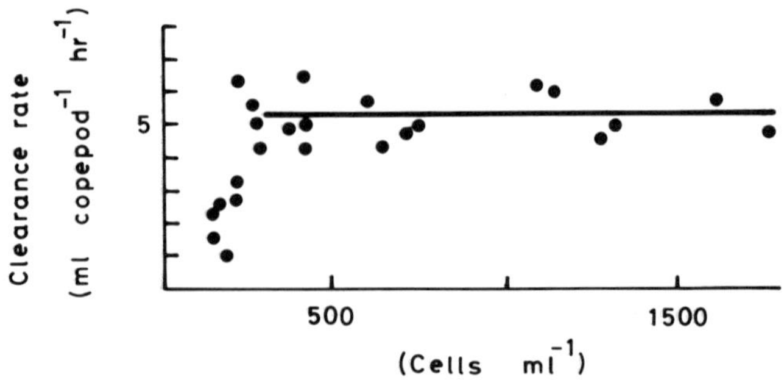

Fig. 4. Clearance rate of adult female *Calanus pacificus* feeding at various cell densities of *Thalassiosira fluviatilis*. Line fitted by least squares regression for points at cell densities >200 cells/ml. (After Frost, 1975.)

interesting to note that in the case of omnivorous or carnivorous copepods feeding on animal prey, i.e., by raptorial feeding, a lower threshold is generally not seen; instead the response is of type 2 (Ambler and Frost, 1974; Corner *et al.*, 1976; Lampitt, 1978; Landry, 1978). The only exception seems to be in the work of McQueen (1969), where type 3 responses are evident for cyclopoid predators. Observations of other carnivorous copepods do suggest changes in their behavior at very low prey densities (Lillelund and Lasker, 1971; Kerfoot, 1978) such as would indicate a type 3 functional response, even when it is difficult to demonstrate this.

IV. ASSIMILATION EFFICIENCIES

A. Assimilation

Assimilated food is that fraction of ingested food which is incorporated by the animal into new tissue or used in respiration. In the terminology of the International Biological Programme (see Petrusewicz and Macfadyen, 1970):

$$\text{Assimilation} = \text{Consumption} - (\text{Feces} + \text{Excreta})$$

where "Excreta" includes all such losses as nitrogenous waste, mucus, and shed cuticle (Crisp, 1971). This definition is not consistent with frequent usage, where assimilation is often taken to mean simply "Consumption minus Feces" (Crisp, 1971; Conover, 1978b). The quantities involved in the category of "Excreta" are often ignored and not measured. It is convenient to express assimilation as a ratio or efficiency by dividing the assimilation figure by the figure for consumption. The utility of this measure is that it allows an estimate of the usefulness of a food source to an animal and enables comparisons to be made between the digestive performances of different animals or the values of different foods.

Several methods of determining assimilation have been proposed. If food eaten and feces produced are known, then for that time the assimilation is simply $C - F$, and the efficiency of assimilation is $(C - F)/C$. If this is expressed gravimetrically, it may be necessary to take account of the ash (inorganic) content of the food and feces. If the food is rich in ash, the gravimetric assimilation estimate may be biased, and comparisons with other values become difficult. Alternatively, C and F may be expressed in terms of joules or calories, overcoming this problem and providing a more satisfactory estimate, since the feces may have a different caloric content from the food, even if the ash content is the same in both.

Direct quantitative methods have the disadvantage that consumption and defecation must be accurately measured. Conover (1966b) proposed a simple, indirect method which avoided this problem. By determining the organic and ash fractions in food and feces, and by making the assumption that the ash fraction is unchanged by the digestive processes, the assimilation efficiency (U') may be calculated by

$$U' = \frac{F' - E'}{(1 - E')(F')}$$

where F' is the ash-free dry weight: total dry weight in the food, and E' is the same ratio for the feces. Conover (1966b) found that this method gave results which compared well with those from direct gravimetric determinations when both were applied to copepods feeding on algal cells. Corner *et al.* (1967) found the method acceptable when compared with assimilation measured by nitrogen uptake, and it has been used successfully for amphipods by Hargrave (1970). However, if the ash fraction of the gut contents is altered by the animal, this invalidates the assumptions of the method, and the results become misleading. Such alteration is known to occur in some crustaceans (Prus, 1971) as well as in some mollusks (Grahame, 1973).

Assimilation may also be measured by labeling the food with a radioactive tracer and measuring its uptake. An early application of this technique to crustaceans was by Marshall and Orr (1955), who used the incorporation of ^{32}P to measure assimilation by *Calanus finmarchicus*. They commented on the high efficiency of assimilation (>50%, sometimes >90%), which was contrary to generally accepted opinion. Choice of tracers may be important: Zimmerman (1973) considered that low values for assimilation by *Lucifer chacei* were a result of using ^{35}S, which was thought to be poorly incorporated. The use of tracers is complicated by the fact that many materials (including ^{32}P) may be rapidly turned over by the experimental animals. Loss of dissolved organic matter originally taken up through the gut is a potentially confounding factor for any determinations of assimilation, however made (Johannes and Satomi, 1967).

Assimilation may also be estimated if a budget of nitrogen or carbon turnover has been compiled using biochemical methods. This approach has been taken by Corner *et al.* (1967, 1976), Johannes and Satomi (1967), and Dagg (1976).

In many instances, assimilation efficiencies quoted have been arrived at by difference or summation. For example, if the energy used in growth and respiration are known, assimilation may be taken as the sum of these. While it is sometimes necessary to use the expedient of such estimates, they are unsatisfactory as independent values and have in general been excluded from this discussion. Selected data on assimilation efficiencies in Crustacea are given in Table I.

The type of food clearly affects assimilation; when a crustacean is tested with a variety of food species, there is usually a wide spread of assimilation values (e.g., Schindler, 1971). This presumably reflects the differing suitability of different foods. Blue-green algae are often poorly assimilated, although some [such as *Anabaena* fed to *Daphnia* (Arnold, 1971)] may be assimilated with very high efficiency indeed. Some Crustacea show higher assimilation efficiencies of animal versus plant food. *Calanus helgolandicus,* when fed on the diatom, *Biddulphia,* had an inexplicably low assimilation efficiency (Corner *et al.,* 1972). The same species, when fed on *Elminius* nauplii, assimilated them very well, apparently because of the high quality of this food (Corner *et al.,* 1976). In many other cases, assimilation efficiencies are apparently similar for animal and plant foods, or the variation is so great that differences are probably masked. Some of the variability is doubtless due to the differences in methods between investigators, or even within a particular study. However, there is evidence of a trend toward higher assimilation of animal versus plant foods.

Other factors that have been found to affect assimilation efficiency are sex, reproductive condition (Prus, 1971; Hayward and Gallup, 1976), and development stage (Mootz and Epifanio, 1974). In the case of *Daphnia,* temperature, crowding, light intensity, and animal size all affect assimilation (Hayward and Gallup, 1976).

"Assimilation" still lacks a rigorously applied definition in practice, and satisfactory generalizations are difficult on the basis of the available data. It seems clear that preferred foods are assimilated with high efficiency. However, in some instances (e.g., animals feeding on decaying leaves or sediments) the nutritional value of a substrate lies in one component which may be only a small part of the whole bulk, and the overall assimilation will be low, even when the nutritional fraction is efficiently absorbed.

TABLE I

Selected Values of Assimilation Efficiency among Crustacea

Taxa	Food	Method of determination	Assimilation efficiency (%)[a]	Reference
Branchiopoda				
Daphnia pulex	Cyanophyceae (7 species)	^{14}C	15.8–100, average for 7 food species = 71.1	Arnold, 1971
Daphnia schoedleri	*Ankistrodesmus; Chlamydomonas; Frustulia*	^{14}C	78–94, higher values at 10°C than at 22°C and among gravid females	Hayward and Gallup, 1976
Daphnia longispina	Selected algae (11 species)	^{14}C	10.5–100, average for 11 food species = 48.9	Schindler, 1971
Maxillopoda				
Copepoda				
Chiridius armatus	*Artemia* nauplii	Conover ash ratio	89.8	Alvarez and Matthews, 1975
	Mixed zooplankton	Conover ash ratio	95.3	Alvarez and Matthews, 1975
Pareuchaeta norvegica	Small crustacean prey	Conover ash ratio	91–94.1	Båmstedt and Holt, 1978
Calanus hyperboreus	*Rhizosolenia setigera*	Conover ash ratio	18.8–73.3, mean 56.8	Conover, 1966c
	Dunaliella sp.	Conover ash ratio	75.1–89.9, mean 86.7	Conover, 1966c
Calanus finmarchicus	*Skeletonema costatum*	Nitrogen uptake and Conover ash ratio	53.8–67.5, mean 61.7	Corner *et al.*, 1967
Calanus helgolandicus	*Biddulphia sinensis*	Nitrogen and phosphorus uptake	34.1 (N), 40.4 (P)	Corner *et al.*, 1972

(continued)

TABLE I *(Continued)*

Taxa	Food	Method of determination	Assimilation efficiency (%)[a]	Reference
Calanus helgolandicus	*Elminius* nauplii	Nitrogen uptake	92.6	Corner *et al.*, 1976
Metridia longa and *Metridia lucens*	*Artemia* nauplii, various algae	Conover ash ratio	59 (nauplii); 35–94 (algae)	Haq, 1967
Tigriopus brevicornis	Natural particulate material	Nitrogen uptake and Conover ash ratio	75.4 ± 11.8 (mean and standard deviation)	Harris, 1973
Acartia negligens	Coral mucus	^{14}C	18.3 (of complete material); 50 (of organic fraction)	Richman *et al.*, 1975
Diaptomus gracilis	Selected algae (11 species)	^{14}C	13.7–100, average for 11 food species = 41.5	Schindler, 1971
Cyclops strenuus strenuus	Selected algae (11 species)	^{14}C	3.7–38, average for 11 food species = 14.9	Schindler, 1971
Cyclops strenuus strenuus	*Daphnia*	Estimated	50	Schindler, 1971
Cirripedia				
Balanus improvisus	*Cladophora*	Direct calorific and gravimetric	68 ± 7.9 (calories); 66 ± 7.7 (gravimetric)	Kuznetsova, 1973
	Balanus nauplii	Direct calorific and gravimetric	95 ± 3.8 (calories); 94 ± 3.8 (gravimetric)	Kuznetsova, 1973
	Asterionella	Direct gravimetric	86 ± 3.6 (gravimetric)	Kuznetsova, 1973
Semibalanus balanoides	*Artemia* nauplii	Conover ash ratio	68.2	Ritz and Crisp, 1970
	Artemia nauplii	Direct calorific and gravimetric	86.5 (calories); 80.3 (gravimetric)	Ritz and Crisp, 1970
Balanus glandula	*Skeletonema costatum*	Direct gravimetric, calories estimated	99.3	Wu and Levings, 1978

Eumalacostraca				
Peracarida				
Amphipoda				
Gammarus pseudolimnaeus	Leaves	Direct calorific	17–19	Bärlocher and Kendrick, 1975
	Fungi	Direct calorific	67.9–83.2	Bärlocher and Kendrick, 1975
Talorchestia longicornis	Cyanophyte mat	^{14}C	66.7 ± 2.5	Brenner *et al.*, 1976
Calliopius laeviusculus	Phytoplankton or zooplankton	Carbon and nitrogen uptake	90	Dagg, 1976
Hyalella azteca	Elm leaves	Conover ash ratio	8.5	Hargrave, 1970
	Sediment	Conover ash ratio	14.9 (surface); 6.5 (subsurface)	Hargrave, 1970
	Chara epiphytes	Conover ash ratio	72.8	Hargrave, 1970
	Algae (5 species)	^{14}C	5.5–75, average for 5 food species = 39.1	Hargrave, 1970
	Bacteria (3 species)	^{14}C	60–82.5, average for 3 food species = 73.5	Hargrave, 1970
Isopoda				
Ligia pallasii	*Ulva*	Direct calorific	78 ± 13	Carefoot, 1973
	Nereocystis	Direct calorific	74 ± 9	Carefoot, 1973
Tylos punctatus	*Macrocystis*	Direct gravimetric	64	Hayes, 1974
Armadillidium vulgare	Thistle leaves	Direct gravimetric and ^{85}Sr	50–75, "true value" 55	Hubbell *et al.*, 1965
Asellus aquaticus	Decaying alder leaves	Direct gravimetric and Conover ash ratio	30	Prus, 1971

(*continued*)

TABLE I (*Continued*)

Taxa	Food	Method of determination	Assimilation efficiency (%)[a]	Reference
Eucarida				
Decapoda				
Libinia emarginata	Fish, mussel, and alga	Direct gravimetric	95 (alga); 96.5–99.3 (fish or mussel)	Aldrich, 1974
Palaemonetes pugio	*Nitzschia closterium*	Carbon uptake	78–79; 47–49 when corrected for loss of dissolved organic matter from the animals	Johannes and Satomi, 1967
Homarus americanus (larvae and juveniles)	*Artemia* adults	Direct calorific	82 ("absorption efficiency"); 81 ("assimilation efficiency")	Logan and Epifanio, 1978
Menippe mercenaria (larvae)	*Artemia* nauplii	Direct calorific	41.7 in stage 1 zoea, rising to 85.4 in megalopa	Mootz and Epifanio, 1974
Lucifer chacei	*Dunaliella tertiolecta*	^{35}S	7.7–10.4	Zimmerman, 1973
	Artemia nauplii	^{35}S	21.4–22.5	Zimmerman, 1973

[a] Data for percentage assimilation are quoted with ranges where appropriate and with standard errors, except where stated otherwise.

B. Superfluous Feeding

Observations that some copepods were responding to abundant phytoplankton by greatly increasing their ingestion led Beklemishev (1962) and Cushing and Vucetic (1963) to propose that assimilation was lower at such times, i.e., that the animals were eating at such a rate that digestion became inefficient. Cushing and Vucetic (1963) also considered that some food was wasted outside the animal by not being ingested. They pointed out that reproduction in *Calanus finmarchicus* was limited to the period when this "superfluous feeding" was expected to be occurring.

There is little direct evidence from crustaceans to support the inverse relation between ingestion and assimilation, which is central to the hypothesis of superfluous feeding. Schindler (1968) found that the assimilation efficiency of *Daphnia magna* decreased with increasing food concentration. Schindler (1971) showed that for *Diaptomus gracilis,* ingestion and assimilation were inversely related over a range of 11 different food algae, and he suggested that *Diaptomus* might regulate its ingestion according to the assimilability of the food. This must be distinguished from superfluous feeding, where (implicitly) assimilation is supposed to decline with increasing ingestion of a single food species.

Conover (1966c) failed to find a significant inverse relationship between ingestion and assimilation for *Calanus hyperboreus* feeding on *Thalassiosira fluviatilis,* even at food levels well above those at which, as proposed by Beklemishev (1962), superfluous feeding should start. Corner *et al.* (1967) found no inverse relationship between assimilation and availability of nitrogen for *Calanus finmarchicus.*

More recently, Gaudy (1974) has found evidence that the assimilation efficiency of copepods declines initially when low rations increase, stabilizing at relatively low levels (<50%) at higher rations. Unfortunately, the assimilation figures quoted by Gaudy (1974) are the result of summation and were not independently determined. It was considered that the pattern of assimilation efficiencies found did not constitute superfluous feeding in the sense of Beklemishev (1962). It is noteworthy that in another filter-feeding animal, the mussel *Mytilus edulis,* there is an exponential decline in assimilation efficiency with increase in food concentration (Widdows, 1978).

The present consensus is that the considerable variations in daily rations (see Gaudy, 1974, for review) are an opportunistic response by copepods (and other small, herbivorous Crustacea) to unpredictable fluctuations in food availability. When large quantities of food are available, rapid adaptation occurs (Fryer, 1968; Conover, 1978a; Mayzaud and Poulet, 1978) so that not only are large daily rations ingested (≈100% of body weight/day), but most of this is assimilated and used for growth, storage, or (especially) reproduction. This assumes that the food available is suitable. Under some circumstances, unsuitable algae may be ingested and poorly assimilated, even to the extent of algal cells passing unharmed through the gut (Moore, 1975; Porter, 1976).

V. CONCLUSIONS AND PROSPECTS

Crustaceans display a variety of feeding mechanisms, and many of these are very elegant. As Marshall and Orr (1960) pointed out, few are really specialized feeders, and most will ingest a variety of foods. Although there is considerable morphological complexity associated with the sorting of food from deposits in *Uca,* the animals will readily scavenge (Crane, 1941; Altevogt, 1957). Hermit crabs will feed on dead animal matter and on fine material from sediments, and they are capable of filtration. They employ different methods, depending on the type of food available, although there is evidence of preference of one method over another in different species (Boltt, 1961; Greenwood, 1972). Among calanoid copepods, many will take small animal or plant food, ranging in size over at least an order of magnitude, although there are some which are evidently specialized carnivores (e.g., *Candacia:* Wickstead, 1959). Many crustaceans combine filtration of already suspended particles with resuspension of deposits, which are then filtered.

Progress continues to be made in describing the structure and use of feeding appendages. For example, most atyid prawns possess remarkable and unique arrangements of setae on their chelipeds, using these for scraping up deposits and in passive filtration (Fryer, 1977).

Interesting recent developments have arisen concerning feeding strategies, particularly among filter feeders. While filtration in copepods has been said to be automatic, it is now understood to be under some degree of control. Control is also exercised over the rate of feeding, and it has been suggested that there is a lower threshold below which filter-feeding effort is reduced. Such a reduction of effort would be required by an energy gain maximization model of filter feeding, as has been proposed by Lam and Frost (1976) and Lehman (1976). This threshold decreases as food particles get larger, until the threshold virtually disappears in raptorially feeding copepods. This has interesting implications concerning the relative costs and benefits of filtering versus raptorial feeding; it implies that the costs of the former may be higher. This could arise simply if filtration demands a sustained effort, which if unrewarded would be expensive, while raptorial feeding is carried out over a relatively short interval of time after the "prey" has been sensed. However, we as yet know nothing of the relative costs of the two activities.

Filter feeders other than copepods also respond to cues related to food availability, e.g., barnacles (Southward and Crisp, 1965), mole crabs (Efford, 1966), and the anomuran, *Petrolisthes* (Hartman and Hartman, 1977).

An energy gain maximization model has also been successfully used by Elner and Hughes (1978) in understanding the pattern of predation by *Carcinus* on mussels. They found that *Carcinus* shows the features expected of a predator that forages optimally, using only tactile cues in assessing the prey.

There are challenging fields for the future. Further exploration of morphology and feeding mechanisms promises to reveal more elegant detail of the relation between structure and function and of both to the environment. The details of copepod filter feeding remain obscure, and this leads to a second field: that of foraging strategies. It is apparent that crustacean foraging strategies are worthy of study, but they have yet to play a full role in the development of this body of knowledge.

REFERENCES

Adams, J. A., and Steele, J. H. (1966). Shipboard experiments on the feeding of *Calanus finmarchicus* (Gunnerus). *In* "Some Contemporary Studies in Marine Science" (H. Barnes, ed.), pp. 19–35. Allen & Unwin, London.

Aldrich, J. C. (1974). Allometric studies on energy relationships in the spider crab *Libinia emarginata* (Leach). *Biol. Bull. (Woods Hole, Mass.)* **147,** 257–273.

Allan, J. D., Richman, S., Heinle, D. R., and Huff, R. (1977). Grazing in juvenile stages of some estuarine calanoid copepods. *Mar. Biol. (Berlin)* **43,** 317–331.

Alldredge, A. L. (1972). Abandoned larvacean houses: A unique source of food in the pelagic environment. *Science* **177,** 885–887.

Altevogt, R. (1957). Untersuchungen zur biologie, ökologie und physiologie indischer winkerkrabben. *Z. Morphol. Oekol. Tiere* **46,** 1–110.

Alvarez, V., and Matthews, J. B. L. (1975). Experimental studies on the deep-water pelagic community of Korsfjorden, western Norway. Feeding and assimilation by *Chiridius armatus* (Crustacea, Copepoda). *Sarsia* **58,** 67–78.

Ambler, J. W., and Frost, B. W. (1974). The feeding behavior of a predatory planktonic copepod, *Tortanus discaudatus*. *Limnol. Oceanogr.* **19,** 446–451.

Angel, M. V. (1970). Observations on the behaviour of *Chonchoecia spinirostris*. *J. Mar. Biol. Assoc. U.K.* **50,** 731–736.

Anraku, M., and Omori, M. (1963). Preliminary survey of the relationship between the feeding habit and the structure of the mouth-parts of marine copepods. *Limnol. Oceanogr.* **8,** 116–126.

Arashkevich, Ye. G. (1968). The food and feeding of copepods in the northwestern Pacific. *Oceanology (Moscow)* **9,** 695–709.

Arnold, D. E. (1971). Ingestion, assimilation, survival and reproduction by *Daphnia pulex* fed seven species of blue-green algae. *Limnol. Oceanogr.* **16,** 906–920.

Båmstedt, U., and Holt, M. R. (1978). Experimental studies on the deep-water pelagic community of Korsfjorden, western Norway. Prey-size preference and feeding of *Euchaeta norvegica* (Copepoda). *Sarsia* **63,** 225–236.

Bärlocher, F., and Kendrick, B. (1973a). Fungi and food preferences of *Gammarus pseudolimnaeus*. *Arch. Hydrobiol.* **72,** 501–516.

Bärlocher, F., and Kendrick, B. (1973b). Fungi in the diet of *Gammarus pseudolimnaeus* (Amphipoda). *Oikos* **24,** 295–300.

Bärlocher, F., and Kendrick, B. (1975). Assimilation efficiency of *Gammarus pseudolimnaeus* (Amphipoda) feeding on fungal mycelium or autumn-shed leaves. *Oikos* **26,** 55–59.

Barnes, H. (1959). Stomach contents and microfeeding of some common cirripedes. *Can. J. Zool.* **37,** 231–236.

Barnes, H., and Rees, E. S. (1959). Feeding in the pedunculate cirripede *Pollicipes polymerus* J. B. Sowerby. *Proc. Zool. Soc. London* **132,** 569–585.

Baylor, E. R., and Sutcliffe, W. H. (1963). Dissolved organic matter in sea water as a source of particulate food. *Limnol. Oceanogr.* **8,** 369–371.

Beklemishev, C. W. (1962). Superfluous feeding of marine herbivorous zooplankton. *Rapp. P.-V. Reun., Cons. Int. Explor. Mer* **153,** 108–113.

Berkes, F. (1975). Some aspects of feeding mechanisms of euphausiid crustaceans. *Crustaceana* **29,** 266–270.

Boltt, R. E. (1961). Antennary feeding of the hermit crab *Diogenes brevirostris* Stimpson. *Nature (London)* **192,** 1099–1100.

Boyd, C. M. (1976). Selection of particle sizes by filter-feeding copepods: A plea for reason. *Limnol. Oceanogr.* **21,** 175–180.

Brenner, D., Valiela, I., van Raalte, C. D., and Carpenter, E. J. (1976). Grazing by *Talorchestia longicornis* on an algal mat in a New England salt marsh. *J. Exp. Mar. Biol. Ecol.* **22,** 161–169.

Brown, T. J., and Sibert, J. R. (1977). Food of some benthic harpacticoid copepods. *J. Fish. Res. Board Can.* **34,** 1028–1031.

Budd, T. W., Lewis, J. C., and Tracey, M. L. (1978). The filter-feeding apparatus in crayfish. *Can. J. Zool.* **56,** 695–707.

Burns, C. W. (1968a). Direct observations of mechanisms regulating feeding behavior of *Daphnia,* in lakewater. *Int. Rev. Gesamten Hydrobiol. Hydrogr.* **53,** 83–100.

Burns, C. W. (1968b). The relationship between body size of filter-feeding Cladocera and the maximum size of particle ingested. *Limnol. Oceanogr.* **13,** 675–678.

Burns, C. W. (1969). Relation between filtering rate, temperature and body size in four species of *Daphnia. Limnol. Oceanogr.* **14,** 693–700.

Burns, C. W., and Rigler, F. H. (1967). Comparison of filtering rates of *Daphnia rosea* in lake water and in suspensions of yeast. *Limnol. Oceanogr.* **12,** 492–502.

Caine, E. A. (1974). Comparative functional morphology of feeding in three species of caprellids (Crustacea: Amphipoda) from the northwestern Florida Gulf coast. *J. Exp. Mar. Biol. Ecol.* **15,** 81–96.

Caine, E. A. (1977). Feeding mechanisms and possible resource partitioning of the Caprellidae (Crustacea: Amphipoda) from Puget Sount, USA. *Mar. Biol. (Berlin)* **42,** 331–336.

Cameron, A. M. (1966). Some aspects of the behaviour of the soldier crab, *Mictyris longicarpus. Pac. Sci.* **20,** 224–234.

Cannon, H. G. (1928a). On the feeding mechanism of the fairy shrimp, *Chirocephalus diaphanus* Prévost. *Trans. R. Soc. Edinburgh* **55,** 807–822.

Cannon, H. G. (1928b). On the feeding mechanism of the copepods, *Calanus finmarchicus* and *Diaptomus gracilis. Br. J. Exp. Biol.* **6,** 131–144.

Cannon, H. G. (1933). On the feeding mechanism of the Branchiopoda. *Philos. Trans. R. Soc. London, Ser. B* **222,** 267–352.

Cannon, H. G., and Manton, S. M. (1927). On the feeding mechanism of the mysid crustacean *Hemimysis lamornae. Trans. R. Soc. Edinburgh* **55,** 219–253.

Carefoot, T. H. (1973). Feeding, food preferences, and uptake of food energy by the supralittoral isopod *Ligia pallasii*. *Mar. Biol. (Berlin)* **18,** 228–236.

Chervin, M. B. (1978). Assimilation of particulate organic carbon by estuarine and coastal copepods. *Mar. Biol. (Berlin)* **49,** 265–275.

Chisholm, S. W., Stross, R. G., and Nobbs, P. A. (1975). Environmental and intrinsic control of filtering and feeding rates in Arctic *Daphnia*. *J. Fish. Res. Board Can.* **32,** 219–226.

Conover, R. J. (1956). Oceanography of Long Island Sound, 1952–1954. VI. Biology of *Acartia clausi* and *A. tonsa*. *Bull. Bingham Oceanogr. Collect.* **15,** 156–223.

Conover, R. J. (1966a). Feeding on large particles by *Calanus hyperboreus* (Kröyer). *In* "Some Contemporary Studies in Marine Science" (H. Barnes, ed.), pp. 187–194. Allen & Unwin, London.

Conover, R. J. (1966b). Assimilation of organic matter by zooplankton. *Limnol. Oceanogr.* **11,** 338–354.

Conover, R. J. (1966c). Factors affecting the assimilation of organic matter by zooplankton and the question of superfluous feeding. *Limnol. Oceanogr.* **11,** 346–354.

Conover, R. J. (1978a). Feeding interactions in the pelagic zone. *Rapp. P.-V. Reun. Cons. Int. Explor. Mer.* **173,** 66–76.

Conover, R. J. (1978b). Transformation of organic matter. *In* "Marine Ecology" (O. Kinne, ed.), Vol. 4, pp. 221–499. Wiley, Chichester.

Corner, E. D. S. (1961). On the nutrition and metabolism of zooplankton. I. Preliminary observations on the feeding of the marine copepod, *Calanus helgolandicus* (Claus). *J. Mar. Biol. Assoc. U.,.* **41,** 5–16.

Corner, E. D. S., Cowey, C. B., and Marshall, S. M. (1967). On the nutrition and metabolism of zooplankton. V. Feeding efficiency of *Calanus finmarchicus*. *J. Mar. Biol. Assoc. U.K.* **47,** 259–270.

Corner, E. D. S., Head, R. N., and Kilvington, C. C. (1972). On the nutrition and metabolism of zooplankton. VIII. The grazing of *Biddulphia* cells by *Calanus helgolandicus*. *J. Mar. Biol. Assoc. U.K.* **52,** 847–861.

Corner, E. D. S., Head, R. N., Kilvington, C. C., and Marshall, S. M. (1974). On the nutrition and metabolism of zooplankton. IX. Studies relating to the nutrition of overwintering *Calanus*. *J. Mar. Biol. Assoc. U.K.* **54,** 319–331.

Corner, E. D. S., Head, R. N., Kilvington, C. C., and Pennycuick, L. (1976). On the nutrition and metabolism of zooplankton. X. Quantitative aspects of *Calanus helgolandicus* feeding as a carnivore. *J. Mar. Biol. Assoc. U.K.* **56,** 345–358.

Cowles, T. J. (1979). The feeding response of copepods from the Peru upwelling system: Food size selection. *J. Mar. Res.* **37,** 601–622.

Crane, J. (1941). Eastern Pacific expeditions of the New York Zoological Society. XXVI. Crabs of the genus *Uca* from the west coast of Central America. *Zoologica (N.Y.)* **26,** 145–208.

Crisp, D. J., ed. (1964). "Grazing in Terrestrial and Marine Environments." Blackwell, Oxford.

Crisp, D. J. (1967). Chemoreception in cirripedes. *Biol. Bull. (Woods Hole, Mass.)* **133,** 128–140.

Crisp, D. J. (1971). Energy flow measurements. *In* "Methods for the Study of Marine Benthos" (N. A. Holme and A. D. McIntyre, eds.), IBP Handb. No. 16, pp. 197–279. Blackwell, Oxford.

Crisp, D. J., and Southward, A. J. (1961). Different types of cirral activity in barnacles. *Philos. Trans. R. Soc. London, Ser. B.* **243,** 271–307.

Crowley, P. H. (1973). Filtering rate inhibition of *Daphnia pulex* in Wintergreen Lake water. *Limnol. Oceanogr.* **18,** 394–402.

Cushing, D. H. (1959). On the nature of production in the sea. *Fish. Invest., London, Ser. 2* **22** (6), 1–40.

Cushing, D. H., and Vucetic, T. (1963). Studies on a *Calanus* patch. III. The quantity of food eaten by *Calanus finmarchicus*. *J. Mar. Biol. Assoc. U.K.* **43,** 349–371.

Dagg, M. J. (1976). Complete carbon and nitrogen budgets for the carnivorous amphipod, *Calliopius laeviusculus* (Kröyer). *Int. Rev. Gesamten Hydrobiol. Hydrogr.* **61,** 297–357.

Dennell, R. (1933). The habits and feeding mechanism of the amphipod *Haustorius arenarius* Slabber. *J. Linn. Soc. London, Zool.* **38,** 363–388.

Dixon, A. F. (1944). Notes on certain aspects of the biology of *Cumopsis goodsiri* (Van Beneden) and some other cumaceans in relation to their environment. *J. Mar. Biol. Assoc. U.K.* **26,** 61–71.

Duval, W. S., and Geen, G. H. (1976). Diel feeding and respiration rhythms in zooplankton. *Limnol. Oceanogr.* **21,** 823–829.

Efford, I. A. (1966). Feeding in the sand crab, *Emerita analoga* (Stimpson) (Decapoda, Anomura). *Crustaceana* **10,** 167–182.

Elner, R. W. (1978). The mechanisms of predation by the shore crab, *Carcinus maenas* (L.) on the edible mussel, *Mytilus edulis* L. *Oecologia* **36,** 333–344.

Elner, R. W., and Hughes, R. N. (1978). Energy maximization in the diet of the shore crab, *Carcinus maenas*. *J. Anim. Ecol.* **47,** 103–116.

Enequist, P. (1949). Studies on the soft bottom amphipods of the Skagerak. *Zool. Bidr. Uppsala* **28,** 297–492.

Esaias, W. E., and Curl, H. C. Jr. (1972). Effect of dinoflagellate bioluminescence on copepod ingestion rates. *Limnol. Oceanogr.* **17,** 901–906.

Esterly, C. O. (1916). The feeding habits and food of pelagic copepods and the question of nutrition by organic substances in solution in the water. *Univ. Calif., Berkeley, Publ. Zool.* **16,** 171–184.

Factor, J. R. (1978). Morphology of the mouthparts of larval lobsters, *Homarus americanus* (Decapoda: Nephropidae), with special emphasis on their setae. *Biol. Bull (Woods Hole, Mass.)* **154,** 383–408.

Fielder, D. R. (1970). The feeding behaviour of the sand crab *Scopimera inflata* (Decapoda, Ocypodidae). *J. Zool.* **160,** 35–49.

Friedman, M. M., and Strickler, J. R. (1975). Chemoreceptors and feeding in calanoid copepods (Arthropoda: Crustacea). *Proc. Natl. Acad. Sci. U.S.A.* **72,** 4185–4188.

Frost, B. W. (1972). Effects of size and concentration of food particles on the feeding behavior of the marine planktonic copepod *Calanus pacificus*. *Limnol. Oceanogr.* **17,** 805–815.

Frost, B. W. (1975). A threshold feeding behavior in *Calanus pacificus*. *Limnol. Oceanogr.* **20,** 263–266.

Frost, B. W. (1977). Feeding behavior of *Calanus pacificus* in mixtures of food particles. *Limnol. Oceanogr.* **22,** 472–491.

Fryer, G. (1957). The feeding mechanisms of some fresh water cyclopoid copepods. *Proc. Zool. Soc. London* **129,** 1–25.

Fryer, G. (1968). Evolution and adaptive radiation in the Chydoridae (Crustacea: Cladocera): A study in comparative functional morphology and ecology. *Philos. Trans. R. Soc. London, Ser. B* **254,** 221–385.

Fryer, G. (1977). Studies on the functional morphology and ecology of the atyid prawns of Dominica. *Philos. Trans. R. Soc. London, Ser. B.* **277,** 57–128.

Fuller, J. L. (1937). Feeding rate of *Calanus finmarchicus* in relation to environmental conditions. *Biol. Bull. (Woods Hole, Mass.)* **72,** 233–246.

Fuller, J. L., and Clarke, G. L. (1936). Further experiments on the feeding of *Calanus finmarchicus*. *Biol. Bull. (Woods Hole, Mass.)* **70,** 308–320.

Fuzessery, Z. M., and Childress, J. J. (1976). Comparative chemosensitivity to amino acids and their role in the feeding activity of bathypelagic and littoral crustaceans. *Biol. Bull. (Woods Hole, Mass.)* **149,** 522–538.

Gamble, J. C. (1978). Copepod grazing during a declining spring phytoplankton bloom in the northern North Sea. *Mar. Biol. (Berlin)* **49,** 303–315.

Gaudy, R. (1974). Feeding four species of pelagic copepods under experimental conditions. *Mar. Biol. (Berlin)* **25,** 125–141.

Gauld, D. T. (1951). The grazing rate of planktonic copepods. *J. Mar. Biol. Assoc. U.K.* **29,** 695–706.

Gauld, D. T. (1953). Diurnal variations in the grazing rate of planktonic copepods. *J. Mar. Biol. Assoc. U.K.* **31,** 461–474.

Gauld, D. T. (1964). Feeding in planktonic copepods. *In* "Grazing in Terrestrial and Marine Environments" (D. J. Crisp, ed.), pp. 239–245. Blackwell, Oxford.

Gellis, S. S., and Clarke, G. L. (1935). Organic matter in dissolved and in colloidal form as food for *Daphnia magna. Physiol. Zool.* **8,** 127–137.

Gerlach, S. A., Ekstrøm, D. K., and Eckardt, P. B. (1976). Filter feeding in the hermit crab, *Pagurus bernhardus. Oecologia* **24,** 257–264.

Goldberg, W. M. (1971). A note on the feeding behavior of the snapping shrimp *Synalpheus fritzmuelleri* Coutiére (Decapoda, Alpheidae). *Crustaceana* **21,** 318–320.

Goodhart, C. B. (1939). Notes on the bionomics of the tube-dwelling amphipod, *Leptocheirus pilosus* Zaddach. *J. Mar. Biol. Assoc. U.K.* **23,** 311–325.

Grahame, J. (1973). Assimilation efficiency of *Littorina littorea* (L.) (Gastropoda: Prosobranchiata). *J. Anim. Ecol.* **42,** 383–389.

Greenwood, J. G. (1972). The mouthparts and feeding behaviour of two species of hermit crabs. *J. Nat. Hist.* **6,** 325–337.

Gyllenberg, G., and Lundqvist, G. (1978). Utilization of dissolved glucose by two copepod species. *Ann. Zool. Fenn.* **15,** 323–327.

Haney, J. F., and Hall, D. J. (1975). Diel vertical migration and filter-feeding activities of *Daphnia. Arch. Hydrobiol.* **75,** 413–441.

Haq, S. M. (1967). Nutritional physiology of *Metridia lucens* and *M. longa* from the Gulf of Maine. *Limnol. Oceanogr.* **12,** 40–51.

Hargrave, B. T. (1970). The utilization of benthic microflora by *Hyalella azteca* (Amphipoda). *J. Anim. Ecol.* **39,** 427–437.

Harris, R. P. (1973). Feeding, growth, reproduction and nitrogen utilization by the harpacticoid copepod, *Tigriopus brevicornis. J. Mar. Biol. Assoc. U.K.* **53,** 785–800.

Hart, T. J. (1930). Preliminary notes on the bionomics of the amphipod *Corophium volutator* Pallas. *J. Mar. Biol. Assoc. U.K.* **16,** 761–789.

Hartman, H. B., and Hartman, M. S. (1977). The stimulation of filter feeding in the porcelain crab *Petrolisthes cinctipes* Randall by amino acids and sugars. *Comp. Biochem. Physiol. A* **56,** 19–22.

Harvey, H. W. (1937). Note on selective feeding by *Calanus. J. Mar. Biol. Assoc. U.K.* **22,** 97–100.

Hassell, M. P., Lawton, J. H., and Beddington, J. R. (1977). Sigmoid functional responses by invertebrate predators and parasitoids. *J. Anim. Ecol.* **46,** 249–262.

Hayes, W. B. (1974). Sand beach energetics: Importance of the isopod *Tylos punctatus. Ecology* **55,** 838–847.

Hayward, R. S., and Gallup, D. N. (1976). Feeding, filtering and assimilation in *Daphnia schoedleri* Sars as affected by environmental conditions. *Arch. Hydrobiol.* **77,** 139–163.

Heinle, D. R., Harris, R. P., Ustach, J. F., and Flemer, D. A. (1977). Detritus as food for estuarine copepods. *Mar. Biol. (Berlin)* **40,** 341–353.

Heron, A. C. (1973). A specialised predator-prey relationship between the copepod *Sapphirina angusta* and the pelagic tunicate *Thalia democratica*. *J. Mar. Biol. Assoc. U.K.* **53,** 429–435.

Herreid, C. F., II (1963). Observations on the feeding behavior of *Cardisoma guanhumi* (Latreille) in southern Florida. *Crustaceana* **5,** 176–180.

Hiatt, R. W. (1948). The biology of the lined shore crab, *Pachygrapsus crassipes* Randall. *Pac. Sci.* **3,** 135–213.

Hicks, G. R. F., and Grahame, J. (1979). Mucus production and its role in the feeding behaviour of *Diarthrodes nobilis* (Copepoda: Harpacticoida). *J. Mar. Biol. Assoc. U.K.* **59,** 321–330.

Hindley, J. P. R., and Alexander, C. G. (1978). Structure and function of the chelate pereiopods of the banana prawn *Penaeus merguiensis*. *Mar. Biol. (Berlin)* **48,** 153–160.

Holling, C. S. (1959). The components of predation as revealed by a study of small-mammal predation of the European pine sawfly. *Can. Entomol.* **91,** 293–320.

Holling, C. S. (1965). The functional response of predators to prey density and its role in mimicry and population regulation. *Mem. Entomol. Soc. Can.* **45,** 1–60.

Howard, G. K., and Scott, H. C. (1959). Predaceous feeding in two common gooseneck barnacles. *Science* **129,** 717–718.

Hu, V. J. H. (1978). Relationships between vertical migration and diet in four species of euphausiids. *Limnol. Oceanogr.* **23,** 296–306.

Hubbell, S. P., Sikora, A., and Paris, O. H. (1965). Radiotracer, gravimetric and calorimetric studies of ingestion and assimilation rates of an isopod. *Health Phys.* **11,** 1485–1501.

Hughes, D. A. (1966). Behavioural and ecological investigations of *Ocypode ceratophthalmus* (Crustacea: Ocypodidae). *J. Zool.* **150,** 129–143.

Icely, J. D., and Jones, D. A. (1978). Factors affecting the distribution of the genus *Uca* (Crustacea: Ocypodidae) on an East African shore. *Estuarine Coastal Mar. Sci.* **6,** 315–325.

Itoh, K. (1970). A consideration on feeding habits of planktonic copepods in relation to the structure of their oral parts. *Bull. Plankton Soc. Jpn.* **17,** 1–10.

Jewett, S. C., and Feder, H. M. (1977). Biology of the harpacticoid copepod, *Harpacticus uniremis* Kröyer on Dayville Flats, Port Valdez, Alaska. *Ophelia* **16,** 111–129.

Johannes, R. E., and Satomi, M. (1967). Measuring organic matter retained by aquatic invertebrates. *J. Fish. Res. Board Can.* **24,** 2467–2471.

Jones, D. A. (1968). The functional morphology of the digestive system in the carnivorous intertidal isopod *Eurydice*. *J. Zool.* **156,** 363–376.

Jones, E. C. (1968). *Lepas anserifera* Linné (Cirripedia Lepadomorpha) feeding on fish and *Physalia*. *Crustaceana* **14,** 312–313.

Jones, M. B. (1972). The mouthparts of the members of the *Jaera albifrons* group of species (Crustacea: Isopoda). *Mar. Biol. (Berlin)* **14,** 264–270.

Kerfoot, W. C. (1978). Combat between predatory copepods and their prey: *Cyclops, Epischura* and *Bosmina*. *Limnol. Oceanogr.* **23,** 1081–1102.

Knudsen, J. W. (1967). *Trapezia* and *Tetralia* (Decapoda, Brachyura, Xanthidae) as obligate ectoparasites of pocilloporid and acroporid corals. *Pac. Sci.* **21,** 51–57.

Kruczynski, W. L. (1975). A radiotracer study of food uptake by *Pinnotheres maculatus* in molluscan hosts. *Biol. Bull. (Woods Hole, Mass.)* **148,** 60–67.

Kuznetsova, I. A. (1973). Utilization of several kinds of feed by cirriped crustaceans. *Hydrobiol. J. (USSR)* **9,** 33–40.

Lam, R. K., and Frost, B. W. (1976). Model of copepod filtering response to changes in size and concentration of food. *Limnol. Oceanogr.* **21,** 490–500.

Lampert, W. (1974). A method for determining food selection by zooplankton. *Limnol. Oceanogr.* **19,** 995–998.
Lampitt, R. S. (1978). Carnivorous feeding by a small marine copepod. *Limnol. Oceanogr.* **23,** 1228–1231.
Landry, M. R. (1978). Predatory feeding behavior of a marine copepod, *Labidocera trispinosa. Limnol. Oceanogr.* **23,** 1103–1113.
Lehman, J. T. (1976). The filter-feeder as an optimal forager, and the predicted shapes of feeding curves. *Limnol. Oceanogr.* **21,** 501–516.
Lillelund, K., and Lasker, R. (1971). Laboratory studies of predation by marine copepods on fish larvae. *Fish. Bull.* **69,** 655–667.
Lochhead, J. H. (1968). The feeding and swimming of *Conchoecia* (Crustacea, Ostracoda). *Biol. Bull.* (*Woods Hole, Mass.*) **134,** 456–464.
Logan, D. T., and Epifanio, C. E. (1978). A laboratory energy balance for the larvae and juveniles of the American lobster *Homarus americanus. Mar. Biol.* (*Berlin*) *47,* 381–389.
Lonsdale, D. J., Heinle, D. R., and Siegfried, C. (1979). Carnivorous feeding behavior of the adult calanoid copepod *Acartia tonsa* Dana. *J. Exp. Mar. Biol. Ecol.* **36,** 235–248.
Lowndes, A. G. (1933). The feeding mechanism of *Chirocephalus diaphanus* Prévost, the fairy shrimp. *Proc. Zool. Soc. London* pp. 1093–1118.
Lowndes, A. G. (1935). The swimming and feeding of certain calanoid copepods. *Proc. Zool. Soc. London* pp. 687–715.
McLeese, D. W. (1970). Detection of dissolved substances by the American lobster (*Homarus americanus*) and olfactory attraction between lobsters. *J. Fish. Res. Board Can.* **27,** 1371–1378.
Mackas, D., and Bohrer, R. (1976). Fluorescence analysis of zooplankton gut contents and an investigation of diel feeding patterns. *J. Exp. Mar. Biol. Ecol.* **25,** 77–85.
Mackie, A. M. (1973). The chemical basis of food detection in the lobster *Homarus gammarus. Mar. Biol.* (*Berlin*) **21,** 103–108.
McMahon, J. W., and Rigler, F. H. (1963). Mechanisms regulating the feeding rate of *Daphnia magna* Straus. *Can. J. Zool.* **41,** 321–332.
McMahon, J. W., and Rigler, F. H. (1965). Feeding rate of *Daphnia magna* Straus in different foods labeled with radioactive phosphorus. *Limnol. Oceanogr.* **10,** 105–113.
McQueen, D. J. (1969). Reduction of zooplankton standing stocks by predaceous *Cyclops bicuspidatus thomasi* in Marion Lake, British Columbia. *J. Fish. Res. Board Can.* **26,** 1605–1618.
Manton, S. M. (1977). "The Arthropoda. Habits, Functional Morphology, and Evolution." Oxford Univ. Press, London and New York.
Marcotte, B. M. (1977). An introduction to the architecture and kinematics of harpacticoid (Copepoda) feeding: *Tisbe furcata* (Baird, 1837). *Mikrofauna Meeresboden* **61,** 183–196.
Marshall, S. M., and Orr, A. P. (1955). On the biology of *Calanus finmarchicus.* VIII. Food uptake, assimilation and excretion in adult and stage V *Calanus. J. Mar. Biol. Assoc. U.K.* **34,** 495–529.
Marshall, S. M., and Orr, A. P. (1960). Feeding and nutrition. *In* "The Physiology of Crustacea" (T. H. Waterman, ed.), Vol. I, pp. 227–258. Academic Press, New York.
Matthews, D. C. (1955). Feeding of the sand crab *Hippa pacifica* (Dana). *Pac. Sci.* **9,** 382–386.
Mauchline, J., and Fisher, J. R. (1969). The biology of euphausiids, *Adv. Mar. Biol.* **7,** 1–454.
Mayzaud, P., and Poulet, S. A. (1978). The importance of the time factor in the response of zooplankton to varying concentrations of naturally occurring particulate matter. *Limnol. Oceanogr.* **23,** 1144–1154.
Meadows, P. S., and Reid, A. (1966). The behaviour of *Corophium volutator* (Crustacea: Amphipoda). *J. Zool.* **150,** 387–399.

Miller, D. C. (1961). The feeding mechanism of fiddler crabs, with ecological considerations of feeding adaptations. *Zoologica (N.Y.)* **46,** 89–100.
Mills, E. L. (1967). The biology of an ampeliscid amphipod crustacean sibling species pair. *J. Fish. Res. Board Can.* **24,** 305–355.
Moore, J. W. (1975). The role of algae in the diet of *Asellus aquaticus* L. and *Gammarus pulex* L. *J. Anim. Ecol.* **44,** 719–730.
Mootz, C. A., and Epifanio, C. E. (1974). An energy budget for *Menippe mercenaria* larvae fed *Artemia* nauplii. *Biol. Bull. (Woods Hole, Mass.)* **146,** 44–55.
Moriarty, D. J. W. (1977). Quantification of carbon, nitrogen and bacterial biomass in the food of some penaeid prawns. *Aust. J. Mar. Freshwater Res.* **28,** 113–118.
Mullin, M. M. (1963). Some factors affecting the feeding of marine copepods of the genus *Calanus. Limnol. Oceanogr.* **8,** 239–250.
Mullin, M. M., Stewart, E. F., and Fuglister, F. J. (1975). Ingestion by planktonic grazers as a function of concentration of food. *Limnol. Oceanogr.* **20,** 259–262.
Naylor, E. (1955). The diet and feeding mechanism of *Idotea. J. Mar. Biol. Assoc. U.K.* **34,** 347–355.
Nicol, E. A. T. (1932). The feeding habits of the Galatheidea. *J. Mar. Biol. Assoc. U.K.* **18,** 87–106.
Nicolaisen, W., and Kanneworff, E. (1969). On the burrowing and feeding habits of the amphipods *Bathyporeia pilosa* Lindström and *Bathyporeia sarsi* Watkin. *Ophelia* **6,** 231–250.
Nival, P., and Nival, S. (1973). Efficacité de filtration des copépodes planctoniques. *Ann. Inst. Oceanogr. (Paris)* **49,** 135–144.
Nival, P., and Nival, S. (1976). Particle retention efficiencies of an herbivorous copepod, *Acartia clausi* (adult and copepodite stages): effects on grazing. *Limnol. Oceanogr.* **21,** 24–28.
O'Brien, W. J., and deNoyelles, F., Jr. (1974). Filtering rate of *Ceriodaphnia reticulata* in pond waters of varying phytoplankton concentrations. *Am. Midl. Nat.* **91,** 509–512.
O'Connors, H. B., Small, L. F., and Donaghay, P. L. (1976). Particle-size modification by two size classes of the estuarine copepod *Acartia clausi. Limnol. Oceanogr.* **21,** 300–308.
Ong, J. E. (1969). The fine structure of the mandibular sensory receptors in the brackish water calanoid copepod *Gladioferens pectinatus* (Brady). *Z. Zellforsch. Mikrosk. Anat.* **97,** 178–195.
Ono, Y. (1965). On the ecological distribution of ocypoid crabs in the estuary. *Mem. Fac. Sci., Kyushu Univ., Ser. E* **4,** 1–60.
Orton, J. H. (1927). On the mode of feeding of the hermit-crab, *Eupagurus bernhardus,* and some other decapoda. *J. Mar. Biol. Assoc. U.K.* **14,** 909–921.
Paffenhöfer, G.-A., and Knowles, S. C. (1978). Feeding of marine planktonic copepods on mixed phytoplankton. *Mar. Biol. (Berlin)* **48,** 143–152.
Paffenhöfer, G. A., and Strickland, J. D. H. (1970). A note on the feeding of *Calanus helgolandicus* on detritus. *Mar. Biol. (Berlin)* **5,** 97–99.
Parsons, T. R., LeBrasseur, R. J., and Fulton, J. D. (1967). Some observations on the dependence of zooplankton grazing on the cell size and composition of phytoplankton blooms. *J. Oceanogr. Soc. Jpn.* **23,** 10–17.
Parsons, T. R., LeBrasseur, R. J., Fulton, J. D., and Kennedy, O. D. (1969). Production studies in the Strait of Georgia. Part II. Secondary production under the Fraser River plume, February to May, 1967. *J. Exp. Mar. Biol. Ecol.* **3,** 39–50.
Patton, W. K. (1967). Studies on *Domecia acanthophora,* a commensal crab from Puerto Rico, with particular reference to modifications of the coral host and feeding habits. *Biol. Bull. (Woods Hole, Mass.)* **132,** 56–67.

Pearce, J. B. (1966). The biology of the mussel crab, *Fabia subquadrata,* from the waters of the San Juan Archipelago, Washington. *Pac. Sci.* **20,** 3–35.

Peterson, B. J., Hobbie, J. E., and Haney, J. F. (1978). *Daphnia* grazing on natural bacteria. *Limnol. Oceanogr.* **23,** 1039–1043.

Petipa, T. S. (1958). The diurnal feeding rhythm of the copepod crustacean *Acartia clausi* Giesbr. *Dokl. Biol. Sci. (Engl. Transl.)* **120,** 435–437.

Petipa, T. S. (1960). Role of *Noctiluca miliaris* Sur. in the feeding of *Calanus helgolandicus* Claus. *Dokl. Biol. Sci. (Engl. Transl.)* **132,** 961–963.

Petrusewicz, K., and Macfadyen, A. (1970). "Productivity of Terrestrial Animals. Principles and Methods," IBP Handb. No. 13. Blackwell, Oxford.

Piontkovskii, S. A., and Petipa, T. S. (1975). Elektivnost' v pitanii *Acartia clausi* Giesbr. *Biol. Morya (Kiev)* **33,** 3–11.

Porter, K. G. (1976). Enhancement of algal growth and productivity by grazing zooplankton. *Science* **192,** 1332–1334.

Poulet, S. A. (1973). Grazing of *Pseudocalanus minutus* on naturally occurring particulate matter. *Limnol. Oceanogr.* **18,** 564–573.

Poulet, S. A. (1974). Seasonal grazing of *Pseudocalanus minutus* on particles. *Mar. Biol. (Berlin)* **25,** 109–123.

Poulet, S. A. (1976). Feeding of *Pseudocalanus minutus* on living and non-living particles. *Mar. Biol. (Berlin)* **34,** 117–125.

Poulet, S. A. (1977). Grazing of marine copepod developmental stages on naturally occurring particles. *J. Fish. Res. Board Can.* **34,** 2381–2387.

Poulet, S. A. (1978). Comparison between five coexisting species of marine copepods feeding on naturally occurring particulate matter. *Limnol. Oceanogr.* **23,** 1126–1143.

Poulet, S. A., and Chanut, J. P. (1975). Nonselective feeding of *Pseudocalanus minutus. J. Fish. Res. Board Can.* **32,** 706–713.

Poulet, S. A., and Marsot, P. (1978). Chemosensory grazing by marine calanoid copepods (Arthropoda: Crustacea). *Science* **200,** 1403–1405.

Prus, T. (1971). The assimilation efficiency of *Asellus aquaticus* L. (Crustacea, Isopoda). *Freshwater Biol.* **1,** 287–305.

Reeve, M. R., and Walter, M. A. (1977). Observations on the existence of lower threshold and upper critical food concentrations for the copepod *Acartia tonsa* Dana. *J. Exp. Mar. Biol. Ecol.* **29,** 211–221.

Richman, S., Loya, Y., and Slobodkin, L. B. (1975). The rate of mucus production by corals and its assimilation by the coral reef copepod *Acartia negligens. Limnol. Oceanogr.* **20,** 918–923.

Richman, S., Heinle, D. R., and Huff, R. (1977). Grazing by adult estuarine calanoid copepods of the Chesapeake Bay. *Mar. Biol. (Berlin)* **42,** 69–84.

Rieper, M. (1978). Bacteria as food for marine harpacticoid copepods. *Mar. Biol. (Berlin)* **45,** 337–345.

Rigler, F. H. (1961). The relation between concentration of food and feeding rate of *Daphnia magna* Straus. *Can. J. Zool.* **39,** 857–868.

Ritz, D. A., and Crisp, D. J. (1970). Seasonal changes in feeding rate in *Balanus balanoides. J. Mar. Biol. Assoc. U.K.* **50,** 223–240.

Roberts, M. H., Jr. (1968). Functional morphology of mouth parts of the hermit crabs, *Pagurus longicarpus* and *Pagurus pollicaris. Chesapeake Sci.* **9,** 9–20.

Roman, M. R. (1977). Feeding of the copepod *Acartia tonsa* on the diatom *Nitzschia closterium* and brown algae (*Fucus vesiculosus*) detritus. *Mar. Biol. (Berlin)* **42,** 149–155.

Roman, M. R. (1978). Ingestion of the blue-green alga *Trichodesmium* by the harpacticoid copepod, *Macrosetella gracilis. Limnol. Oceanogr.* **23,** 1245–1248.

Ross, A., and Newman, W. A. (1969). A coral-eating barnacle. *Pac. Sci.* **23,** 252–256.

Rubenstein, D. I., and Koehl, M. A. R. (1977). The mechanisms of filter feeding: Some theoretical considerations. *Am. Nat.* **111,** 981–994.

Ryther, J. H. (1954). Inhibitory effects of phytoplankton upon feeding of *Daphnia magna* with reference to growth, reproduction and survival. *Ecology* **35,** 522–533.

Sanders, H. L. (1963). The Cephalocarida. Functional morphology, larval development, comparative external anatomy. *Mem. Conn. Acad. Arts Sci.* **15,** 1–80.

Saunders, C. G. (1966). Dietary analysis of caprellids (Amphipoda). *Crustaceana* **10,** 314–316.

Schindler, D. W. (1968). Feeding, assimilation and respiration rates of *Daphnia magna* under various environmental conditions and their relations to production estimates. *J. Anim. Ecol.* **37,** 369–385.

Schindler, J. E. (1971). Food quality and zooplankton nutrition. *J. Anim. Ecol.* **40,** 589–595.

Schoener, T. W. (1971). Theory of feeding strategies. *Annu. Rev. Ecol. Syst.* **2,** 369–404.

Scully, E. P. (1978). Utilization of surface foam as a food source by the hermit crab *Pagurus longicarpus* Say, 1817. *Mar. Behav. Physiol.* **5,** 159–162.

Sheader, M., and Evans, F. (1975). Feeding and gut structure of *Parathemisto gaudichaudi* (Guerin) (Amphipoda, Hyperiidea). *J. Mar. Biol. Assoc. U.K.* **55,** 641–656.

Southward, A. J., and Crisp, D. J. (1965). Activity rhythms of barnacles in relation to respiration and feeding. *J. Mar. Biol. Assoc. U.K.* **45,** 161–185.

Starkweather, P. L. (1978). Diel variation in feeding behavior of *Daphnia pulex*. Influences of food density and nutritional history on mandibular activity. *Limnol. Oceanogr.* **23,** 307–317.

Stone, R. L., and Barnes, H. (1973). The general biology of *Verruca stroemia* (O. F. Müller). I Geographical and regional distribution; cirral activity and feeding. *J. Exp. Mar. Biol. Ecol.* **12,** 167–185.

Strickler, J. R., and Bal, K. (1973). Setae of the first antennae of the copepod *Cyclops scutifer* (Sars): Their structure and importance. *Proc. Natl. Acad. Sci. U.S.A.* **70,** 2656–2659.

Sullivan, B. K., Miller, C. B., Peterson, W. T., and Soeldner, A. H. (1975). A scanning electron microscope study of the mandibular morphology of boreal copepods. *Mar. Biol.* (*Berlin*) **30,** 175–182.

Tezuka, Y. (1971). Feeding of *Daphnia* on planktonic bacteria. *Jpn. J. Ecol.* **21,** 127–134.

Turner, J. T. (1978). Scanning electron microscope investigations of feeding habits and mouthpart structures of three species of copepods of the family Pontellidae. *Bull. Mar. Sci.* **28,** 487–500.

Valiela, I., Babiec, D. F., Atherton, W., Seitzinger, S., and Krebs, C. (1974). Some consequences of sexual dimorphism: Feeding in male and female fiddler crabs, *Uca pugnax* (Smith). *Biol. Bull.* (*Woods Hole, Mass.*) **147,** 652–660.

von Westernhagen, H. (1976). Some aspects of the biology of the hyperiid amphipod *Hyperoche medusarum*. *Helgol. Wiss. Meeresunters.* **28,** 43–50.

White, G. E., Fabris, G., and Hartland-Rowe, R. D. (1969). The method of prey capture by *Branchinecta gigas* Lynch, 1937 (Anostraca). *Crustaceana* **16,** 158–160.

White, H. H. (1979). Effects of dinoflagellate bioluminescence on the ingestion rates of herbivorous zooplankton. *J. Exp. Mar. Biol. Ecol.* **36,** 217–224.

Wickstead, J. H. (1959). A predatory copepod. *J. Anim. Ecol.* **28,** 69–72.

Wickstead, J. H. (1962). Food and feeding in pelagic copepods. *Proc. Zool. Soc. London* **139,** 545–555.

Widdows, J. (1978). Combined effects of body size, food concentration and season on the physiology of *Mytilus edulis*. *J. Mar. Biol. Assoc. U.K.* **58,** 109–124.

Wilson, D. S. (1973). Food size selection among copepods. *Ecology* **54,** 909–914.

Wu, R. S. S., and Levings, C. D. (1978). An energy budget for individual barnacles (*Balanus glandula*). *Mar. Biol.* (*Berlin*) **45,** 225–235.

Zimmerman, S. T. (1973). The transformation of energy by *Lucifer chacei* Crustacea, Decapoda). *Pac. Sci.* **27,** 247–259.

Zipser, E., and Vermeij, G. J. (1978). Crushing behavior of tropical and temporate crabs. *J. Exp. Mar. Biol. Ecol.* **31,** 155–172.

4

Interactions of Chemical and Osmotic Regulation with the Environment

R. GILLES AND A. PEQUEUX

I. INTRODUCTION

A. Delimiting the Subject

Osmotic regulation (i.e., cell volume maintenance) should be considered as including all the processes at work in control of the thermodynamic activity of water in the different biological fluids. Chemical regulation, on the other hand, may include any kind of metabolic or transport process

THE BIOLOGY OF CRUSTACEA, VOL. 8

ISBN 0-12-106408-5

involved in the adjustment of the level of both inorganic and organic constituents of crustaceans. Viewed in this way, the subject of the interactions of chemical and osmotic regulation with the environment is very complex and voluminous. We will thus concentrate in this chapter on ecological aspects of interaction between the chemical regulation implicated in cell volume control and some physical or chemical factors in the environment. This choice was dictated by the fact that other aspects of chemical regulation are extensively reviewed by other authors in this treatise. Moreover, blood osmoregulation is discussed at length in Chapter 2 of Volume 5; the part of the present review dealing with mechanisms of osmoregulation will thus be essentially concerned with the processes at work in intracellular fluid osmolarity control. The reader interested in blood osmolarity control mechanisms is referred to the contribution of Mantel and Farmer (Chapter 2 of Volume 5).

B. The Media and Their Problems

Crustacea have aquatic, semiterrestrial, and terrestrial representatives. With the exception of a very few species which show hypoosmotic blood regulation in seawater (see Chapter 2 of Volume 5 and Section II,A,1 of this chapter), most marine crustaceans have blood osmolarity close to that of the environmental medium. Therefore, the ecological incidence of osmoregulation is essentially important in the other media colonized by representatives of the class. In fact, it can be considered that the capability to ensure proper cell volume maintenance is an essential prerequisite to the establishment of communities in media of fluctuating salinity such as the estuarine or intertidal environment, in brackish and fresh waters, in semiterrestrial biotopes such as the high shore levels, and in terrestrial habitats.

Brackish and intertidal environments are probably among the most exacting and stressful aquatic biotopes; the establishment of crustacean communities in such habitats supposes various highly adapted physiological features (refer to Chapters in Volume 5 and in this volume). Brackish waters enter the category of mixohaline waters, which may range from oligohaline to polyhaline media. Referring to the salinity limit between fresh and brackish water accepted at the Venice Symposium (1959), the salinity may vary from about 0.5 to 30‰ (see Table I).[1] Most of the brackish waters are found in restricted coastal regions such as estuaries or salt marshes and in partially landlocked seas such as the Caspian or the Baltic. In such large sea areas, the salinity, although it may be far lower than in seawater, is not

[1]The symbol ‰ indicates parts per thousand (ppt).

TABLE I

Salinity Ranges and Current Classification of Species and Waters[a]

Water classification	Salinity range (‰)[b]	Ecological classification of species			Number and diversity of species
Limnic	below 0.5	Limnic stenohaline	Limnic euryhaline		+++++++
Oligohaline	0.5–5/8				+++
		Brackish water		Euryhaline	
Mesohaline	5/8–18				+
Polyhaline	18–30		Marine euryhaline		+++
Oceanic	30–40	Marine stenohaline			+++++++

[a] The salinity limits are given according to salt tolerance studies on different crustaceans. They have been selected essentially for the sake of classification; no such straight boundaries can be found in nature.

[b] Parts per thousand.

subjected to change as frequently and as rapidly as it is in estuaries. The physiological adaptations required for settlement in these two types of brackish biotopes are basically the same, however. The fauna found in both habitats may be relatively similar, being mainly dependent on the mean salinity of the medium, provided that the nature of the bottom is the same. Most of the time, however, the ecological picture will be quite different in the various brackish-water ecotones, although the more or less stressful salinity conditions that prevail remain a major ecological parameter.

The inhabitants of the intertidal zone may also be subjected to important osmotic stresses. Salinity in tide pools may become rapidly higher or lower than that of seawater, depending on the climatic conditions. Another important feature of this habitat is related to tidal movements. Many invaders of the intertidal zone which have only limited or no locomotive abilities will have to face the rapid and drastic changes in external medium conditions that are linked to the tidal rhythm. Besides the problem of oxygen availability, species of intertidal fauna, such as barnacles, will have to withstand temperature stress and desiccation. With the exception of the intertidal zones of cold regions, where freezing may become an important physiological problem, temperature stress is essentially related to an osmotic problem; high temperatures will increase the evaporative water loss and the risk of desiccation.

Invaders of semiterrestrial or terrestrial habitats are faced with problems quite similar to those of the barnacles in the intertidal zone, as far as desiccation and temperature stresses are concerned. Besides the problem of oxygen availability, they also must solve the important problem of water availability. This is especially true for terrestrial species, since most semiterrestrial crustaceans are sand-dwellers and/or have good locomotive abilities and can easily return to a more humid and thus more favorable environment. Physiological, morphological, and behavioral adaptations to terrestrial ecotones, as well as their ecological correlates, are described at length in Chapter 6 of this volume; we shall thus not deal any longer with this matter in the present review.

The ability to tolerate osmotic stress is one of the important clues to settlement of a population in the intertidal zone. On the other hand, good osmoregulatory capabilities can be considered an essential clue to establishment and maintenance of communities in fresh and brackish waters as well as in estuaries, coastal lagoons, or mangroves. Since these last biotopes are considered more and more as possible zones for extensive aquaculture in temperate and tropical latitudes, special reference will be made to the various effects of osmotic stress on populations inhabiting these media.

Salinity will affect aquatic organisms both directly and indirectly. Direct effects are related to the organismic and molecular capability of animals at any stage of their life cycle, which ensure proper water balance in their internal fluids, either intracellular or extracellular. Indirect effects are essentially of two types: osmotic shocks may produce modifications of the biocenotic composition of an ecosystem, which will result in changes in the biotic background for the remaining forms; salinity stress may also become a limiting factor when acting together with one or several other stressful conditions. A species, which in the laboratory can easily withstand large fluctuations of salinity, may not survive even much smaller changes in the field if these changes are associated with other factors which may themselves constitute a stress. Examples of such factors are oxygen availability, temperature, presence of pollutants, changes in the physical or chemical nature of the substratum and changes in the nature of the diet. These indirect effects, when viewed in the framework of the survival of a population in a given medium, are obviously of the utmost importance. However, they are difficult to apprehend and quantify at the moment. Indeed, an approach to the problem in field studies is almost impossible because of the large number of parameters that may interact. To date, studies have been largely restricted to the understanding of the different mechanisms of osmoregulation used by animals to ensure proper water balance. However, it is our belief that research exploring the indirect effects of other factors should prove very rewarding in the future.

II. DIRECT EFFECTS OF SALINITY

A. On Adult Organisms

A major problem that all animals have to face is to maintain a cellular volume compatible with the different life-supporting cell activities. Crustaceans, as well as all other animals, have evolved molecular mechanisms which enable them to ensure a proper intracellular water balance in a given environment. Water homeostasis in media of fluctuating salinities, such as brackish waters or the intertidal zone, implies the necessity of having mechanisms of osmoregulation powerful enough to respond to the sometimes extremely rough demands of the medium. The relatively small number of species that succeeds in these ecotones, when compared to the number of species found in freshwater or marine habitats, can be considered as a reflection of the difficulty of the problems encountered.

All species living in such biotopes are euryhaline. They may be good or poor osmoregulators, depending on the salinity range they can withstand. Good osmoregulators will rely on very efficient mechanisms of osmoregulation, whereas poor osmoregulators will take advantage of several peculiarities which enable them, not to cope with, but to escape large modifications in the salinity of the external medium for more or less long periods of time.

Osmoregulation can be effected in two different ways. The first is to keep the osmoconcentration of the extracellular fluids constant regardless of the salinity of the external medium. The second is to maintain the intracellular fluid isosmotic to the extracellular fluid. Both methods tend toward the same goal, which is to avoid water movement (loss or gain) at the cellular level.

1. OSMOREGULATION OF THE EXTRACELLULAR FLUID AND SALT TOLERANCE

As previously stated, it is not the purpose of this contribution to discuss the molecular mechanisms implicated in the control of the osmolarity of the extracellular fluids. The reader interested in these processes is referred to Chapter 2, Volume 5 of this treatise for detailed information. Rather, we shall examine the possible correlations between blood osmoregulatory abilities and the distribution of species in different ecotones with important salinity fluctuations.

In euryhaline species, the extracellular fluid osmolarity can evolve in different ways as a function of the osmolarity of the surrounding medium. This is illustrated in Fig. 1. Species able to maintain blood osmolarity at a more or less constant level independent of the osmolarity of the external

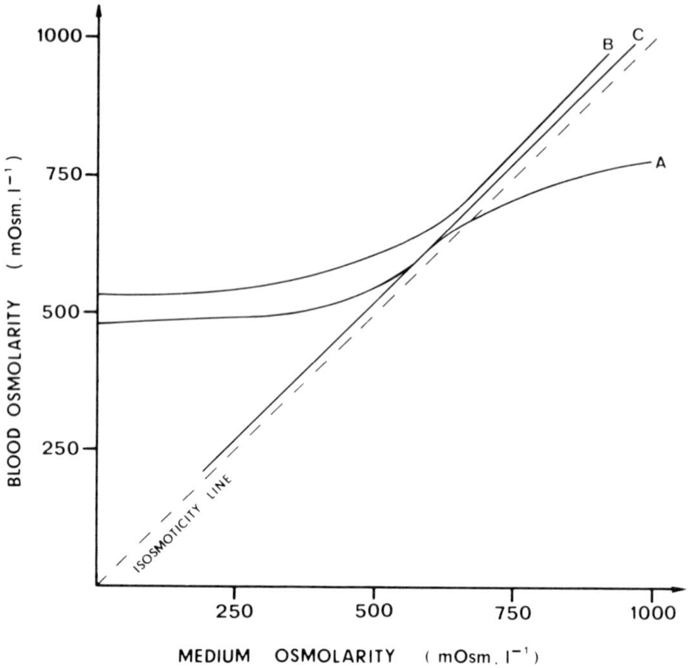

Fig. 1. Schematic representation of different types of blood osmolarity control as a function of the osmolarity of the environmental medium in crustaceans. (A) Hyper-hyporegulators; (B) hyperregulators; (C) osmoconformers.

medium enter the general category of the so-called hyper-hypoosmotic regulators. They are able to hyperregulate their blood osmolarity in dilute media and to hyper- or hyporegulate it in concentrated ones (Fig. 1, curve A). Hyper-hyporegulation represents in crustaceans the most elaborate adaptation to salinity stress; it occurs in only a few species such as *Palaemonetes varians* or *Artemia salina*. These are extremely powerful osmoregulators, which can cope with any fluctuation of the salinity of the external medium that normally occurs in nature. This, however, does not mean that communities of such species will necessarily be successful in any aquatic ecological niche. For instance, *Palaemonetes varians* is often considered a brackish water species (Potts and Parry, 1964).

The second category illustrated in Fig. 1 comprises the hyperosmoregulators (curve B). They can maintain a more or less pronounced hyperosmotic state in dilute media. In more concentrated media, the blood osmolarity strictly follows the osmolarity of the external environment. Most euryhaline crustaceans are hyperosmoregulators; well-known examples are the crabs, *Carcinus maenas* and *Eriocheir sinensis*. Many species of this group are relatively weak osmoregulators. Their natural distribution may in some cases be related to their respective capacities for regulation in hypoosmotic media. The mud shrimp, *Callianassa jamaicense,* which shows limited power for osmoregulation, will be found in river mouths in relatively high salinities (Felder, 1978). *Carcinus maenas* can be found in salinities down to 10‰; *Eriocheir sinensis* is successful in any medium from fresh water to seawater (Gilles, 1974a, 1975a). Freshwater species, such as the crayfish *Astacus astacus,* belong to the same category (Duchâteau and Florkin, 1961). With respect to their salt tolerance or normal habitat, we will find in this group limnic stenohaline and more or less limnic euryhaline species; brackish water species, such as *Gammarus duebeni* (Forsman, 1951); marine euryhaline species, such as *C. maenas;* fully euryhaline species, such as *E. sinensis;* or marine stenohaline species, such as the littoral mole crab, *Emerita talpoida* (Bursey and Bonner, 1977) or *Maja squinado* (Duval, 1925). The power of blood osmoregulation in *M. squinado* is so weak (Fig. 2) that it can be considered a member of the group of osmoconformers (Fig. 1C) (that is to say animals in which blood remains isosmotic to the external medium in all salinities down to the lethal limits). Other examples of osmoconformers are *Hyas araneus* (Schlieper, 1929) and *Libinia emarginata* (Gilles, 1970). These first considerations illustrate the problem of matching laboratory studies of blood osmoregulatory capabilities or of salt tolerance with field studies of ecological niches. If, as already stated, some correlation can be found, there may be large discrepancies between salt tolerance and blood osmoregulatory power, as well as between blood osmoregulatory capabilities or salt tolerance and distribution range. *Eriocheir sinensis* and *C. maenas* show similar blood osmoregulation power. However, *C. maenas* is rarely found in salinities lower than 10‰, while *E. sinensis* is currently caught in freshwater lakes. In the same way, *P. varians,* a strong hyper-hypoosmoregulator, is not commonly found in the sea, although it establishes large communities in brackish waters. *Emerita talpoida* (Bursey and Bonner, 1977), *Pisidia longicornis, Porcellana platycheles* (Smaldon, 1973), and *Upogebia pugettensis* (Thompson and Pritchard, 1969) all show fair blood hyperosmoregulation in dilute media. However, *E. talpoida,* a species limited to seashore habitat, shows, like *P. platycheles,* a salt tolerance between 17 and 51‰, while the thalassinid mud shrimp, *U. pugettensis,* tolerates ranges of 25–125% seawater (SW). On the other hand, *Callianassa major* and *C. islagrande* cannot osmoregulate, but tolerate limited reduction of salinity (Felder, 1978).

It is apparent from the above discussion that species showing strong, good, fair, or even weak power of blood osmoregulation can live or transitorily invade media of fluctuating salinity or brackish waters. Thus, it is obvious that blood osmoregulation is far from being the essential element in

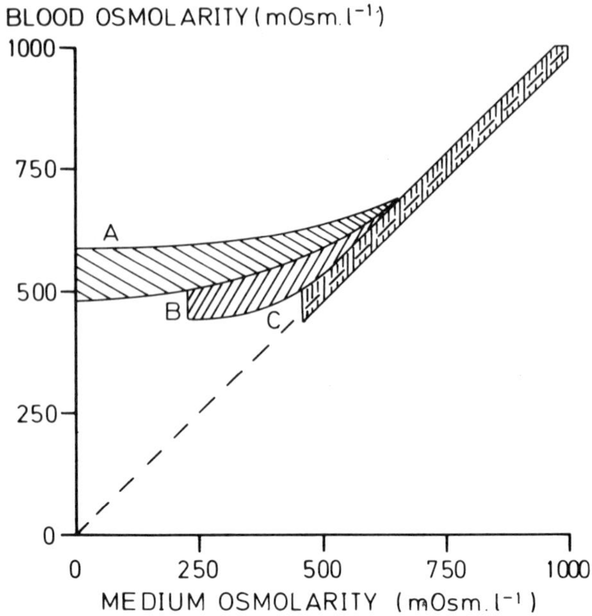

Fig. 2. Schematic representation of blood osmolarity control and salt tolerance in marine crustaceans. In each category, the salt tolerance varies from species to species. The denominations euryhaline, marine euryhaline, and stenohaline are based on natural distribution. (A) Good hyperosmoregulators: euryhaline and marine euryhaline species; (B) fair to weak hyperosmoregulators: marine euryhaline and stenohaline species; (C) osmoconformers: marine euryhaline and stenohaline species. Examples: group A: *Carcinus maenas, Eriocheir sinensis,* and *Callinectes sapidus;* group B: *Callianassa jamaicense, Emerita talpoida, Porcellana platycheles,* and *Pisidia longicornis*; group C: *Maja squinado, Hyas araneus, Libinia emarginata,* and *Callianassa major.*

limiting the distribution of crustaceans in such media. Although correlation can be found between the salt tolerance of some species and the efficiency of their blood osmoregulatory process, this is far from being a general rule. Such lack of correlation indicates that some other physiological mechanisms (such as intracellular fluid osmoregulation) must be taken into consideration to account for salt tolerance. Discrepancies are also found when considering distribution patterns and salt tolerance. In this case, however, distribution patterns cannot exceed the salt range tolerated. On the other hand, the salt range can largely exceed the distribution patterns. Distribution can thus be shifted within the salt range tolerated because of interactions between salinity and other factors. Such factors may be abiotic (chemical and/or physical factors of the environment); they may also be biotic (reproduction, competition, predation). Let us consider these possibilities separately.

2. OSMOREGULATION OF THE INTRACELLULAR FLUID AND SALT TOLERANCE

In a species such as *Eriocheir sinensis,* adaptation from seawater to fresh water results in a decrease in blood osmolarity of about 550 mOsm/liter. If the osmolarity of the intracellular fluid were to remain at its level in the seawater animal, i.e., 1100 mOsm/liter, the cells would have to withstand a swelling pressure of about 12 atm (Gilles, 1974a). This is, of course, a rough calculation made only on the vant'Hoff-Arrhenius equation; nevertheless it shows that the swelling pressure is enormous and that very effective regulatory mechanisms must be at play to avoid the bursting of the cells. As a matter of fact, when *E. sinensis* is passing from a seawater to a freshwater medium, there is first a swelling of the tissue; however, this swelling is regulated rapidly, as shown by the evolution of the water content of the muscle tissue (Fig. 3). After the initial swelling phase, the muscle tissue fluid comes into isosmotic equilibrium with the blood (Fig. 3). This demonstrates the existence of active processes implicated in the regulation of the intracellular fluid osmolarity.

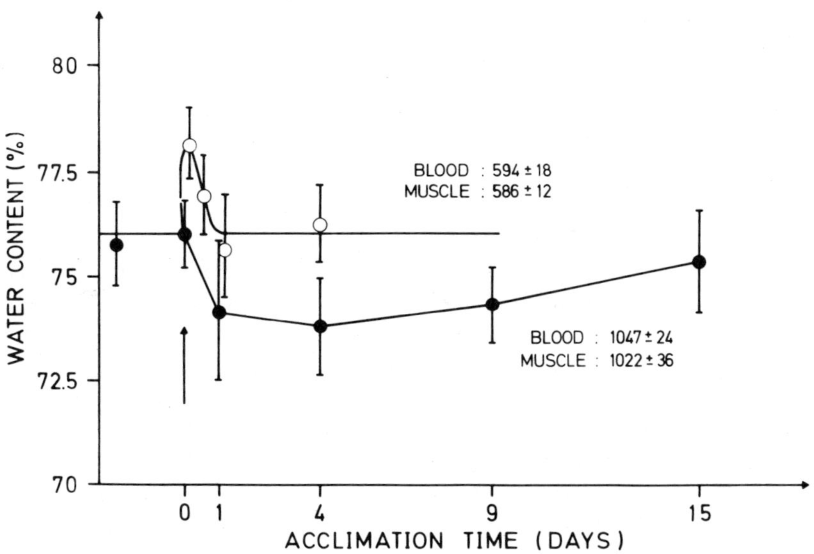

Fig. 3. Change in the water content of the muscle of *Eriocheir sinensis* during acclimation from fresh water to seawater (●) or from seawater to fresh water (○). Acclimation starts at time 0, indicated by an arrow. Control values for seawater and freshwater acclimated animals are not significantly different. The water content is expressed as percent of the tissue wet weight. Blood and tissue fluid osmolarity measured at the end of the acclimation period are given in mOsm per liter or Kg tissue wet weight. (After Gilles, 1977, and unpublished.)

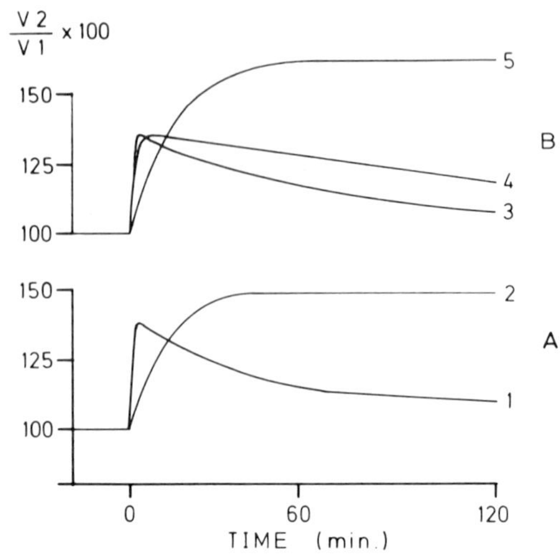

Fig. 4. Effect of an hypoosmotic shock on the volume of isolated muscle fibers (part A) and isolated axons (part B) of some crustaceans. The amplitude $¶_1/¶_2$ of the shock applied is approximately equal to 1.5 for muscle fibers and 2 for axons. (1) *Callinectes sapidus* (Lang and Gainer, 1969); (2) *Homarus gammarus* (Gainer and Grundfest, 1968); (3) *Eriocheir sinensis* (Gilles, 1973); (4) *Carcinus maenas* (Kévers *et al.*, 1979b); (5) *Homarus gammarus* (Gilles, 1973).

It is significant that the duration of the regulatory phase after hypoosmotic shock (adaptation from a concentrated to a diluted medium) is quite different, depending on the species studied. In *Eriocheir sinensis,* complete regulation is achieved in about 24 hr (Fig. 3). In *Carcinus maenas,* control volume is only resumed after some 70 hr (Harris, 1976).

Such differences can be traced to the level of isolated tissues. As shown in Fig. 4, no volume regulation can be observed in isolated nerve cord of the stenohaline lobster, *Homarus gammarus* (= *vulgaris*), after 3 hr of experiments, while volume regulation is almost complete in isolated axons of the fully euryhaline *E. sinensis.* In the marine euryhaline species, *C. maenas,* which shows narrower limits in its distribution pattern than *E. sinensis,* isolated axons are far from completing volume regulation during the 3 hr of experiments. In the same way, isolated muscle fibers of the blue crab, *Callinectes sapidus,* resume their control volume in a few hours (Lang and Gainer, 1969), while during the same time period, no regulation can be observed in muscle fibers of the crayfish, *A. astacus* (Reuben *et al.*, 1964) or the lobster, *H. gammarus* (Gainer and Grundfest, 1968).

This indicates the possibility of correlations between the distribution of these species and the efficiency of their mechanisms of intracellular fluid isomotic regulation. *Homarus gammarus* is a stenohaline species, restricted to seawater. Although *A. astacus* can withstand brackish waters of about 15‰ under laboratory conditions for a couple of months (Duchâteau-Bosson and Florkin, 1961), it is limited in nature to freshwater habitats. In these two species, no volume regulation can be observed, at least on the time scale of the experiments performed. In *C. maenas,* which is found in waters of salinities down to 10‰, complete regulation is achieved in a much slower way than in the excellent euryhaline species, *E. sinensis,* which can live in any medium from fresh water to seawater. Further, the observed differences in volume response efficiency are probably of ecological importance in allowing settlement of species in areas of rapidly fluctuating salinities, such as estuaries. Fast volume control in hypoosmotic conditions, as seen in *E. sinensis* or *C. sapidus,* can indeed confer a significant advantage in such media. It is also worth noticing that volume regulation is, as a rule, more effective in hypoosmotic conditions than in hyperosmotic situations. Such a phenomenon, when considered on a scale of large and relatively fast salinity fluctuations, makes the volume response and the possibility for complete acclimation almost unidirectional. This may be part of a physiological mechanism that has evolutionary significance. Possibly, the marine ancestors of actual limnic species were able to invade oligohaline and then fresh waters, but once the move occurred it became almost irrevocable, because of the ineffectiveness of the volume control mechanisms following application of hyperosmotic conditions.

3. MECHANISMS OF OSMOREGULATION OF THE INTRACELLULAR FLUID

It appears that osmoregulation of the intracellular fluid is a physiological factor at least as important as blood osmoregulation in the ecology of crustaceans. This is especially true in view of the fact that, to our knowledge, no crustaceans have osmoregulatory capabilities allowing full control of blood osmolarity. Therefore, changes in environmental salinity will always induce changes in blood osmolarity, the species coping with them more or less successfully depending on their possibilities of intracellular fluid osmoregulation. Osmoconformers or very weak hyperosmoregulators such as *Libinia emarginata* will even rely entirely on intracellular fluid osmoregulation to invade estuaries with salinities as low as 20‰ (Gilles, 1970). Since the mechanisms at work in intracellular fluid osmoregulation have not been dealt with in other chapters of this treatise, we shall consider them here in more detail. This section will, however, be relatively brief since extensive reviews have been published recently (Gilles, 1975a, 1978, 1979).

Several mechanisms can be involved in cell volume control, which occurs in tissues withstanding osmotic stress. Some are intracellular, while others may be dependent on relations between the intra- and extracellular fluids. As far as this last category of mechanisms in concerned, the plasma membrane could be impermeable to some ions of the external medium in the same way as it is impermeable to large anions of the intracellular fluid. Under such conditions, the cell would be in a "double Donnan equilibrium," which might contribute to the removal of an osmolarity difference between the inside and outside media. Van Slyke *et al.* (1925) and Boyle and Conway (1941) published equations for such a system based on impermeability for Na^+. Such an impermeability for Na^+ cannot, of course, be retained any longer. However, it is possible that some blood macromolecules are implicated in a "double Donnan equilibrium." To our knowledge, this possibility has never been investigated in detail, although some recent studies show modifications in the protein content of the blood serum of various crustaceans submitted to osmotic stresses. We shall come back to this problem later (see Section II,A,3,b). Nevertheless, such a mechanism should play only a minor role in the cellular volume control process. Cell volume regulation can indeed be observed in isolated tissues incubated in artificial salines containing only ions to which the plasma membrane is permeable.

a. Control Mechanisms in Isolated Tissues and Extracts. Application of hypoosmotic conditions induces a biphasic reaction of tissues isolated from euryhaline crustaceans (Fig. 5).

Upon transfer to the hypoosmotic saline, the tissue rapidly swells, and volume is then progressively decreased to resume values close to the control. If, after the phase of regulation, the tissue is placed back in the control saline, the volume decreases to values significantly lower than the control. On the other hand, it can be seen that the maximum swelling achieved is in the range of 25–40%, while it should be of much larger amplitude if the cells were acting as perfect osmometers following the van't Hoff relation. This may be partly due to the fact that possible changes in the extracellular fluid volume of the tissue have not been taken into consideration. However, in recent experiments on isolated axons of *Carcinus maenas,* in which the extracellular space has been taken into account, the maximum swelling estimated to be achieved at the cellular level is 35%, whereas it should have been 70% according to the van't Hoff relation when considering an inactive volume of water of 30% (Kévers *et al.* 1979a).

These first results point out some important facts: (1) A hormonal control is not primarily implicated in volume regulation, since this phenomenon can be demonstrated on isolated tissues incubated in artificial salines. (2) Maximum swelling achieved appears to be lower than what could be ex-

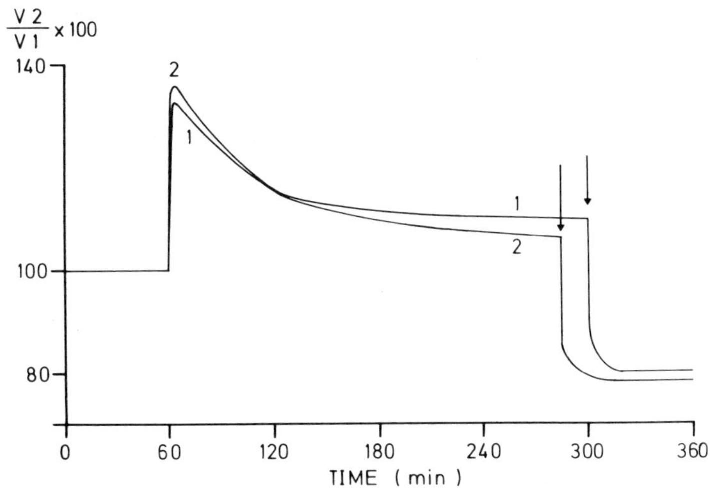

Fig. 5. Effect of hypoosmotic conditions on the volume of isolated muscle fibers of *Callinectes sapidus* (curve 1) or of isolated axons of *Eriocheir sinensis* (curve 2). The tissue is returned to control saline at the time marked by the arrow. The osmotic shock applied ($¶_1/¶_2$) is approximately equal to 1.5 in the case of the muscle fibers (Lang and Gainer, 1969) and to 2 in the case of the isolated axons (Gilles, 1973).

pected if the cells were behaving as osmometers. This indicates that the overall process of volume regulation can be dissociated into two distinct phases. There is first a phase of "volume swelling limitation"; this phase is rapid and avoids the bursting out of the cell by limiting swelling to values compatible with the elasticity of the plasma membrane. This early phase is followed by a much slower phase of "volume readjustment" during which volume progressively resumes values close to the control. (3) The maximum swelling achieved and the time course evolution of volume readjustment are different, depending on the tissue and species studied. This indicates that the effectiveness of the mechanisms at work are different, depending on the tissue and/or the species considered. The possible ecological incidence of such differences is discussed in a previous section (see Section II,A,1). (4) Volume shrinks to values lower than the control when the tissue is placed back in the control saline at the end of the "readjustment phase." This shows that volume regulation is achieved by a decrease in the amount of intracellular osmotic effectors, thus bringing the intracellular fluid close to isosmotic equilibrium with the environmental medium.

This leads us to the questions: What are the osmotic effectors implicated in volume regulation? What are the mechanisms at work in the control of their concentration during both the swelling limitation and volume readjustment phases?

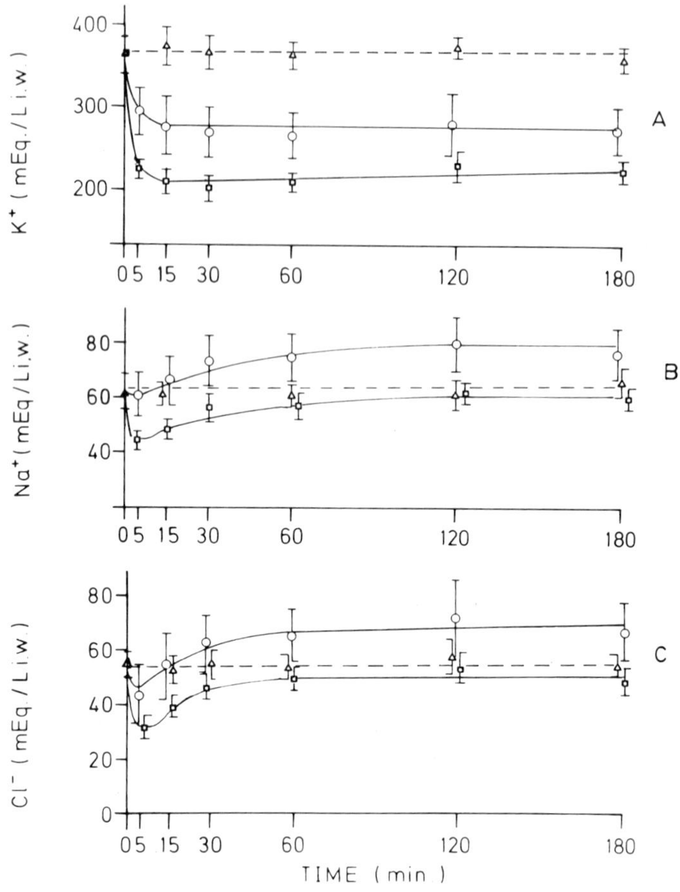

Fig. 6. Effect of an hypoosmotic shock on the intracellular content of K^+ (part A), Na^+ (part B), and Cl^- (part C) of *Carcinus maenas* isolated axons. △, Control conditions; □,○, hypoosmotic conditions; ○, results corrected for the measured changes in intracellular volume. (Compiled from results of Kévers *et al.*, 1979a.)

When isolated axons of *C. maenas* are subjected to hypoosmotic conditions, there is a rapid decrease to new steady-state values of the intracellular content of K^+ (Fig. 6). Transient decreases in the level of Na^+ and Cl^- can also be observed. However, if the results are corrected for the modifications in cellular hydration due to osmotic movements of water, most of the Na^+ and Cl^- decrease can be accounted for by the dilution of the intracellular fluid. As far as K^+ is concerned, the decrease of about half of its intracellular level cannot be explained on the basis of a dilution process.

On the other hand, application of a hypoosmotic stress does not bring about any significant changes in the total amount of proteins, once the results have been corrected for the variations in cellular hydration (Fig. 7). This is at variance with the situation encountered when considering "peptides" and ninhydrin positive substances (NPS), the intracellular amount of which shows an important decrease. It is worth noticing that besides the rapid decrease occurring during the "swelling limitation phase," a further, slower decrease in the level of these organic compounds takes place, while volume readjustment proceeds.

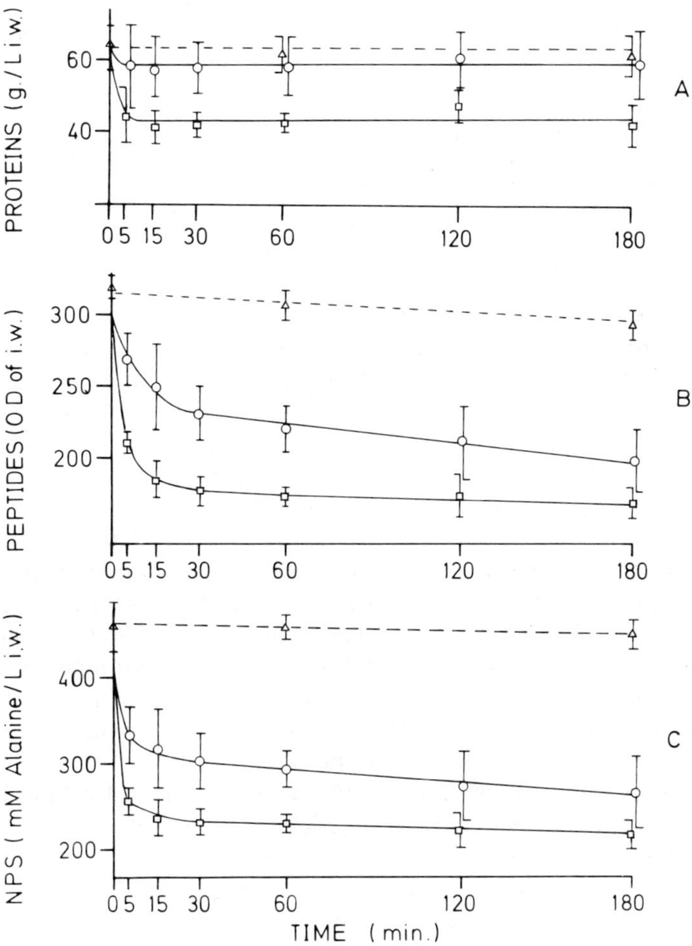

Fig. 7. Effect of a hypoosmotic shock on the intracellular content of proteins (part A), "peptides" (part B), and ninhydrine positive substances (part C) of *Carcinus maenas* isolated axons. △, Control conditions; □,○, hypoosmotic conditions; ○, results corrected for the measured changes in intracellular volume. (Compiled from results of Kévers *et al.*, 1979a)

It appears that the immediate volume regulation response of isolated tissue to hypoosmotic shocks implicates control of the amount of inorganic as well as organic osmotic effectors. K^+, NPS, and "peptides" seem to play a part in the early swelling limitation process; indeed, concomitant with the increase in volume, there is an important decrease in the intracellular amount of these compounds which cannot be accounted for by changes in cellular hydration.

The inorganic ions do not seem to assume a significant role in the volume readjustment phase. If it were the case, a slow decrease in their amount should be recorded after the early modifications related to the swelling phase. Rather, volume readjustment appears to be essentially related to a decrease in NPS and "peptides." The decrease in the amount of these organic compounds could indeed account, at least partly, for the decrease in intracellular fluid osmolarity, which is associated with the outward osmotic movement of water occurring during volume readjustment.

We can now come to our second question: what are the mechanisms implicated in the control of the osmotic effectors level?

i. INORGANIC IONS. As already stated, K^+ does not seem to be implicated in volume readjustment. The fact that its concentration remains steady during this phase also indicates that this ion is leaving the cells during the outward osmotic movement of water in such a way that the solution expelled appears isotonic to the intracellular fluid with respect to K^+. Na^+ and Cl^- do not participate in volume readjustment either; there is even a slight, but significant, increase in the amount of these ions during this process (Fig. 6). Such an increase goes against the expected change in intracellular fluid osmolarity and might account for the fact that volume regulation in *C. maenas* axons is slower than that reported in axons of *C. sapidus* or *E. sinensis* (Fig. 4). As previously stated (see Section II,A,2) this difference appears to have interesting ecological correlates, and a comparative study of the ionic changes occurring in tissues of different euryhaline species might be of interest in this context. However, such a study remains to be done before a more complete picture can be drawn.

Increase in Na^+ concentration could be associated at least partly with a transient increase in permeability, leading to an increase in Na^+ passive influx. This is consistent with results of Kévers *et al.* (1979b), showing that application of hypoosmotic conditions induces a decrease in the ^{22}Na half-renewal-time of the intracellular compartment of *C. maenas* axons. The Na^+ pumping mechanisms have been blocked with both ouabaīn and ethacrynic acid. These results are in agreement with the increase in Na^+ permeability reported to occur in vertebrate tissues and cells withstanding hypoosmotic conditions (Doljanski *et al.*, 1974; Schmidt-Nielsen, 1975; for review, see Hoffmann, 1977; Rorive and Gilles, 1979). It should, however, be pointed out here that further readjustments, probably involving further modifications in transport, must occur *in vivo* and possibly *in vitro* on a longer time scale. As a matter of fact, the intracellular amount of Na^+ (as well as that of K^+) is decreasing, not increasing in *C. maenas* axons after acclimation from seawater to dilute media (Table II). Data on ion transport in tissues following acclimation of crustaceans to different media are unfortunately totally lacking. However, since after acclimation the Na^+ amount is decreased and not increased, it may be considered that somewhere along the line the activity of a Na^+ pump must be increased in order to bring the Na^+ concentration to its new steady-state value. Such an increase could be initiated by the early increase in Na^+ intracellular level due to the recorded increase in permeability, as well as by the decrease in K^+ amount. It is, however, worth noting that the Na^+/K^+ active transport system does not play an important part in volume readjustment observed during short-term, *in vitro* experiments. As a matter of fact, it has been shown repeatedly that ouabaīn, a well known inhibitor of the Na^+/K^+ active transport, remains without significant effect on the volume regulation process. This is true for crustacean tissues (Lang and Gainer, 1969; Gérard, 1975), as well as for the vertebrate tissues and cells studied up to now (for reviews, refer to Hoffmann, 1977; MacKnight and Leaf, 1977; Rorive and Gilles, 1979). This lack of effect of ouabaīn might be related to a possible implication of ouabaīn-insensitive Na^+ or K^+ specific pumps, as postulated in the case of vertebrates tissues. It may be considered, as stressed above, that inorganic

TABLE II

Intracellular Ionic Content of *Carcinus maenas* Axons[a,b]

		A		B	
Ions	SW saline	1	2	SW acclimated	SW/2 acclimated
Na^+	69.3 ± 13.4	64.7 ± 3.9	74.8 ± 6.7	62.2 ± 6.8	26.0 ± 1.2
K^+	365.1 ± 13.7	245.4 ± 1.7	283.7 ± 10.3	366.4 ± 10.6	284.7 ± 23.3
Cl^-	58.8 ± 9.8	59.8 ± 2.7	69.1 ± 5.1	56.6 ± 11.2	29.0 ± 2.8

[a] After C. Kévers, unpublished.

[b] Values are given in mEq/liter of intracellular fluid. Results in column A have been obtained with isolated axons from crabs acclimated to seawater and incubated for 3 hr either in a control seawater saline or in a hypoosmotic saline corresponding to the blood composition of crabs acclimated to twice diluted seawater. Results in subcolumn A2 have been corrected for dilution of the intracellular fluid induced by the osmotic influx of water occurring in the hypoosmotic saline. Results in column B have been obtained on axons from crabs acclimated either to seawater or to twice diluted seawater.

ions do not play an important part in volume readjustment. It indeed appears that, at least in crustacean cells, the osmotic effectors playing a major role in volume readjustment are organic compounds; the mechanisms at work in the control of their concentration are probably ouabaīn insensitive.

If inorganic ions do not play an important part in volume readjustment, K^+ appears to be an important osmotic effector implicated in the swelling limitation process. There is a large decrease in its concentration immediately after application of the hypoosmotic conditions (Fig. 6). As shown in Fig. 8, this decrease can be related to an increase in permeability occurring during the swelling phase.

It appears that the modifications in intracellular ionic content taking place during volume regulation can be accounted for by an increase in plasma membrane permeability occurring during the swelling phase. The fact that an early increase in Na^+ level, concomitant to the decrease in K^+, cannot

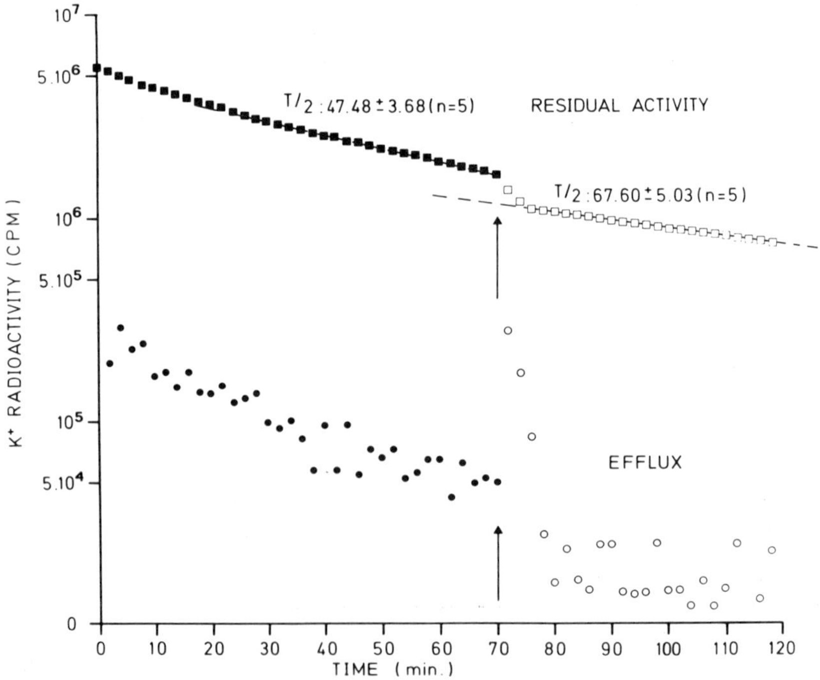

Fig. 8. Effect of hypoosmotic shock on the efflux of ^{43}K from isolated axons of *Carcinus maenas*. The osmotic shock is applied at the time marked by an arrow, in the intracellular compartment of the system. (From C. Kévers, A. Péqueux, and R. Gilles, unpublished.)

be observed may be due to the much lower membrane permeability to Na^+. Moreover, the effect of the increased influx of Na^+ is, during the early swelling phase, counterbalanced by dilution due to osmotic inward movement of water, while dilution increases the effect of increased permeability on the intracellular amount of K^+ (i.e., decrease in concentration).

We may now turn to the question: What are the primary causes of these changes in permeability? The first thing which should be pointed out in this respect is that the changes are not specific for the inorganic ions. Modifications in permeability to organic molecules, such as amino acids, have also been reported in crustacean tissues submitted to hypoosmotic stress (Gérard and Gilles, 1972a; Gérard, 1975; Gilles, 1978). As previously proposed (Gilles, 1974a, 1975a, 1978), such nonselective changes in permeability might be due to the osmotic swelling per se, which would induce modifications in the plasma membrane conformation. This view finds some support in the fact that the increased efflux of alanine observed in *C. sapidus* isolated axons withstanding hypoosmotic stress is completely suppressed when the tissues are dropped in saline (the ionic composition of which is identical to that of the hypoosmotic saline, but which is kept isosmotic to the control saline by addition of sucrose) (Gérard and Gilles, 1972a). As far as the inorganic ions are concerned, the situation appears to be far more complicated. For instance, recent studies from this laboratory show that incubation of *C. maenas* isolated axons in media of various ionic and osmotic concentrations lead to modifications in the K^+ intracellular content which differ depending on both the K^+ and total osmotic concentration of the external medium (Fig. 9). The larger decrease in K^+ axonal amount is observed in hypoosmotic saline with dilute Na^+ and K^+. The decrease is far less important in hypoosmotic saline with K^+ maintained to control level, or in a saline with dilute Na^+ and K^+ but supplemented with sucrose to avoid tissue volume modifications. Only a slight decrease in K^+ content remains in a medium supplemented with sucrose in which K^+ is kept at control level. Thus it seems that the K^+ intracellular level is primarily sensitive not only to the osmotic pressure per se, but also to the amount of that ion in the environmental medium. This would indicate that the reduced level of intracellular K^+ achieved during hypoosmotic stress represents a balance between the necessity to contribute the osmotic equilibrium and to maintain a K^+ gradient across the plasma membrane compatible with bioelectric phenomena. More data are needed to shed some light on this problem. In this respect, data on membrane potential during hypoosmotic adaptation, such as those recently produced by Treherne and his co-workers on various polychaete worms, would be of the utmost interest (Treherne and Pichon, 1978; Benson and Treherne, 1978; Carlson *et al.*, 1978).

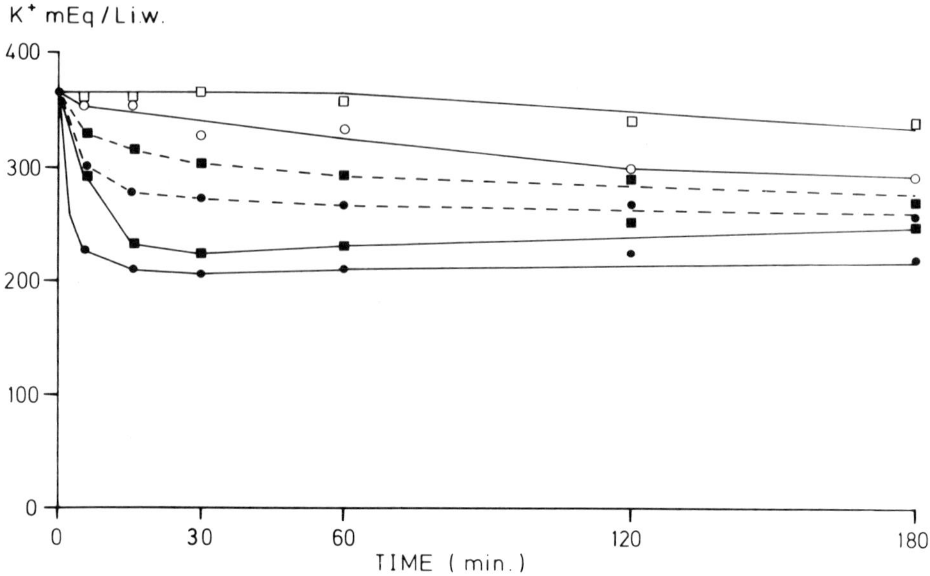

Fig. 9. Effect of various hypoosmotic and isosmotic salines on the K^+ intracellular content of isolated axons of *Carcinus maenas*. Data are given in mEq/liter of intracellular fluid. ●——●: hypoosmotic saline containing half the amount of Na^+ and K^+ of the control saline; ●- - -●: figures from the same experiments corrected for dilution of the intracellular fluid. ■——■: hypoosmotic saline with control K^+ level and half the Na^+ content; ■- - -■: figures from the same experiments corrected for dilution of the intracellular fluid. ○——○: Saline containing half the amount of Na^+ and K^+ but kept isosmotic with control saline by addition of sucrose. □——□: Saline containing control K^+ level and half Na^+ content but kept isosmotic to control saline by addition of sucrose.

ii. ORGANIC OSMOTIC EFFECTORS. As already stated in a previous section of this chapter (see Section II,A,3,a) organic osmotic effectors seem to play a prominent role in crustacean tissue cell volume regulation. Although amino acids are not the only organic solutes implicated in the process, they appear to be quantitatively the most important. Since the early work of Florkin and co-workers on the modifications in amino acid concentration in tissues of invertebrates after acclimation to media of different salinities (Duchâteau and Florkin, 1955; Duchâteau *et al.*, 1959; Duchâteau-Bosson *et al.*, 1961; Florkin, 1962), much work has been devoted to the part taken by these compounds in the process of cell volume maintenance that Florkin (1962) proposed to term "isosmotic intracellular regulation." Most of these early works have been surveyed several times during the last decade (Huggins and Munday, 1968; Florkin and Schoffeniels, 1969; Schoffeniels and Gilles, 1970, 1972; Lange, 1972), and accounts of this literature have also been given in more recent reviews (Gilles, 1974a, 1975a, 1979; Schoffeniels, 1976). To briefly summarize the results of these studies, let us say that (1) amino acids appear to play an important part in cell volume readjustment in all the crustacean species studied so far; (2) the quantitative role of these compounds as well as the nature of the most important amino acids varies from tissue to tissue and in a same tissue from species to species; (3) with the exception of arginine, it is always the so-called nonessential amino acids (alanine, aspartate, glutamate, glycine, proline, and serine) that show the largest variations upon acclimation of euryhaline species from seawater to dilute media. The question that now arises is: What are the mechanisms implicated in the control of the intracellular amino acid pool?

Several experiments show modifications of the NPS content of crustacean isolated tissues following application of an osmotic stress in artificial salines (Schoffeniels, 1960; Lang and Gainer, 1969; this chapter, Fig. 7). This indicates that the mechanisms at work in the adjustment of the amino acid pool are primarily intracellular and not dependent on some hormonal process. The amino acid content must be essentially regulated by modifications in (1) the steady state between synthesis and degradation of these compounds, and/or (2) the equilibrium between them and proteins, and/or (3) the rate of their transport through the plasma membrane.

While clear-cut information about equilibrium between amino acids and proteins is still needed, it is clear that amino acids are released from isolated tissues during volume readjustment following hypoosmotic stress. The efflux of labeled material in relation to different osmotic conditions has been studied on isolated axons of the blue crab, *C. sapidus,* preloaded with [^{14}C]alanine by Gilles and co-workers (Gérard and Gilles, 1972a; Gérard *et al.*, 1974; Gérard, 1975; Gilles, 1978). During hypoosmotic stress, there is an increase in the efflux of radioactive material, while in hyperosmotic conditions, no significant changes can be recorded. If unidirectional fluxes are considered, hypoosmotic shock induces both an increase in efflux and a decrease in influx (Table III); this last result implies an inhibition of the inward active movement of alanine. During hyperosmotic stress, there is an increase in the uptake of the amino acid, while no significant changes can be recorded in the passive efflux. To our knowledge, these are the only available data on fluxes of organic osmotic effectors in crustacean tissues. Changes in efflux of organic material upon application of hypoosmotic shocks have, however, been shown to occur in tissues and cell types of species from other phyla (Jonsson, 1971; Pierce and Greenberg, 1973; Hoffmann and Hendil, 1976). It therefore seems likely that increased efflux of amino acids is a phenomenon of general occurrence in tissues withstanding hypoosmotic conditions. As already stated (see Section II,A,3,a,i) this outward movement is probably controlled by permeability changes related

TABLE III

Effect of Osmotic Stresses on the Unidirectional Fluxes and Permeability Coefficient to Alanine in Isolated Axons of *Callinectes sapidus*

	Efflux	Influx (Pmole/cm²/sec)	Active component of influx	Permeability coefficient (10^{-8} cm/sec)
A. Steady state conditions (SW saline)	0.81		0.79	2.13
Hypoosmotic conditions (75% SW saline)	2.17	0.57	0.52	5.72
B. Steady state conditions (75% SW saline)	0.71		0.66	4.87
Hyperosmotic conditions (SW saline)	0.70	0.95	0.91	4.38

[a] After Gérard (1975) and Gilles (1978).

[b] Experimental group A: animals acclimated to seawater (SW) Experimental group B: animals acclimated to 75% seawater. The alanine concentration in the incubation medium is 1 m*M*. For applied equations refer to Gérard (1975).

to tissue swelling. In the tissues where it has been studied (*C. sapidus* isolated axons: Gérard, 1975; flounder, *Platichthys flesus* erythrocytes: Fugelli and Reiersen, 1978), it appears that the modifications in amino acid influx is Na^+ dependent. The influx mechanism would be quite similar to the one implicated in the so-called "Na-gradient hypothesis" in which the uphill transport of amino acids is coupled to the downhill movement of Na^+ following its concentration gradient (see Schultz and Curran, 1970). In such conditions, it is reasonable to assume that the decrease in Na^+ external content caused by application of hypoosmotic shock induces the decrease in amino acid influx.

Control of amino acid transport is implicated in the concentration adjustment that occurs during osmotic regulation. However, the amount of many so-called nonessential amino acids found in the incubation saline of isolated axons of *E. sinensis* after application of a hypoosmotic stress cannot entirely account for the changes measured in the tissue (Gilles, 1974a). This indicates that another process must be involved in the control of the amino acid intracellular level.

Hypoosmotic conditions induce an increase in O_2 consumption as well as in $^{14}CO_2$ production of isolated axons of *C. sapidus* preloaded with various radioactive amino acids (Gilles, 1972). Such a result is consistent with the idea that osmotic shocks induce changes in the intermediary metabolism and in the related catabolism of amino acids. Control of the release of amino acids from the metabolic pool is also consistent with the modifications in the specific activity of amino acids labeled from radioactive glucose induced by osmotic shock at the level of *C. sapidus* isolated axons (Gilles and Gérard, 1974). Changes in intermediary metabolism and amino acid catabolism also agree with findings showing that osmotic stresses induce modifications in the oxidoreduction level of the pyridine nucleotides of *C. sapidus* isolated muscle fibers (Gilles and Jöbsis, 1972); in the C1:C6 ratio calculated from $^{14}CO_2$ production of isolated axons of *E. sinensis* loaded with [1-^{14}C] glucose or [6-^{14}C] glucose (Gilles, 1974b); or in the distribution of ^{14}C from glutamic acid into several Krebs cycle intermediates of *C. maenas* muscle tissue (Chaplin *et al.*, 1970). Since nitrogen metabolism in relation to salinity is being discussed at length in Chapter 3, Volume 5 of this treatise, we shall no longer deal with the subject. Let us simply point out that there seems to be enough data for us to conclude that changes in patterns of intermediary metabolism and amino acid oxidation are implicated in control of the amino acid pool during volume regulation. On the other hand, there is presently no clear-cut evidence about adjustment of amino acid synthesis from keto precursors. Some indication that such a process might be at work can be found in various results from our laboratory (see Gilles and Gérard, 1974). However, more data are needed to confirm this hypothesis. We come now to the question: What could be the mechanisms triggering the modifications in metabolic activity?

Increase in saline osmolarity by addition of sucrose causes a decrease in the NPS content of *E. sinensis* isolated axons instead of the increase which is recorded when hyperosmotic shock is achieved by increasing the ionic concentration of the incubation saline (Schoffeniels, 1960). Such results could argue that the changes in metabolic activity recorded during osmotic shocks may be associated with modifications in the intracellular ionic concentration. More arguments supporting this view can be found in the results of studies on mollusks, bacteria, or vertebrate cell lines (refer to Gilles, 1979, for detailed discussion). This raises two important questions: (1) What are the mechanisms implicated in the control of the metabolic activity by the inorganic ions? (2) What are the ions essentially involved in these mechanisms.

The ionic composition may control directly or indirectly the level and/or the activity of key enzymes implicated in the metabolic pathways of deamination and oxidation of amino acids. A first possible control process would be related to genetically controlled modifications in the level or the kinetics characteristics of key enzymes. In another view, the ionic concentration could modulate the activity of some key enzymes.

As to the first possibility, Chaplin *et al.* (1965) have not been able to demonstrate a significant difference in glutamate dehydrogenase activity between tissues of *Carcinus maenas* acclimated to 50% seawater or 100%

TABLE IV

Characteristics of Glutamate Dehydrogenase (GDH) and Aspartate Aminotransferase (GOT) Extracted from Tissues of *Eriocheir sinensis* Acclimated to Fresh Water (FW) or to Seawater (SW)[a]

	GOT: FW		GOT: SW		GDH: FW		GDH: SW	
	Specific activity	K_m ($MX10^{-4}$)	Specific activity	K_m ($MX10^{-4}$)	Specific activity	K_m ($MX10^{-5}$)	Specific activity	K_m ($MX10^{-5}$)
Muscle	12.14	9.42 ± 1.53	11.94	7.44 ± 1.25	3.89	4.25 ± 2.67	3.54	3.63 ± 1.14
Gills	3.27	5.28 ± 1.18	3.70	4.79 ± 1.72	2.16	4.46 ± 1.89	2.82	4.12 ± 0.98
Hepatopancreas	1.13	4.41 ± 2.48	1.18	7.81 ± 3.02	—	—	—	—

[a] "Specific activity" is expressed in terms of maximum velocity (V_{max}) given in optical density units per minute per mg of tissue DNA. K_m values are those for α-ketoglutarate.

seawater. In the same way, there is no change in the specific activity and in the K_m for α-ketoglutarate of glutamate dehydrogenase or aspartate amino transferase from tissues of *Eriocheir sinensis* acclimated to fresh water or to seawater (Table IV). These findings indicate that the changes in amino acid concentration are not primarily related to modifications in the genetically controlled level or kinetics characteristics of key enzymes. Such a possibility appears unlikely because of the fast rate at which the changes in metabolic activity can be effected in the experiments performed on isolated tissues. More information would be needed before definite conclusions could be drawn.

As to the possibility of a direct control of enzyme activity by the inorganic composition of the incubation medium, several studies report on the effect of NaCl or KCl on enzymes implicated in:

1. The amino acid metabolism, such as glutamate dehydrogenase, serine hydrolase, alanine, and asparate aminotransferases.
2. The Krebs cycle and its anaplerotic pathways, such as malate hydrolase, malate dehydrogenase (E.C.1.1.1.37 and 1.1.1.40), isocitrate dehydrogenase, oxaloacetate decarboxylase, and succinate dehydrogenase.
3. The fate of reducing equivalents, such as lactic dehydrogenase, 3-glycerophosphate dehydrogenase, or glyoxylate reductase.

The results of these experiments have been presented in a number of papers (see below) and have been discussed in light of metabolic osmotic adjustments in recent review articles (Schoffeniels and Gilles, 1970; Gilles, 1974a, 1975a, 1978, 1979; Schoffeniels, 1976). We shall only briefly summarize in a schematic form (Fig. 10) the main conclusions drawn. Increasing the NaCl or KCl content of the incubation medium leads to a decrease in the activity of the enzymes implicated in the Krebs cycle. This is consistent with the decrease in $^{14}CO_2$ production from labeled amino acids, with the decrease in O_2 consumption of isolated axons of *C. sapidus* submitted to hyperosmotic shock, as well as with the fact that the pyridine nucleotides are oxidized in *C. sapidus* muscle fibers when in hyperosmotic conditions (see this Section, p. 131).

As far as the sequence of amination–deamination of the amino acids is concerned, the activity of serine hydrolyase is decreased by increasing concentrations of NaCl (Gilles, 1969). On the other hand, the activity of alanine and aspartate aminotransferases is not affected (Chaplin *et al.*, 1967; Huggins and Munday, 1968; Gilles, 1969).

Increasing amounts of NaCl or KCl also induce an increase in the activity of reductive amination of glutamate dehydrogenase (Shoffeniels and Gilles, 1963; Chaplin *et al.*, 1965; Gilles, 1969, 1974c). It is noteworthy that under

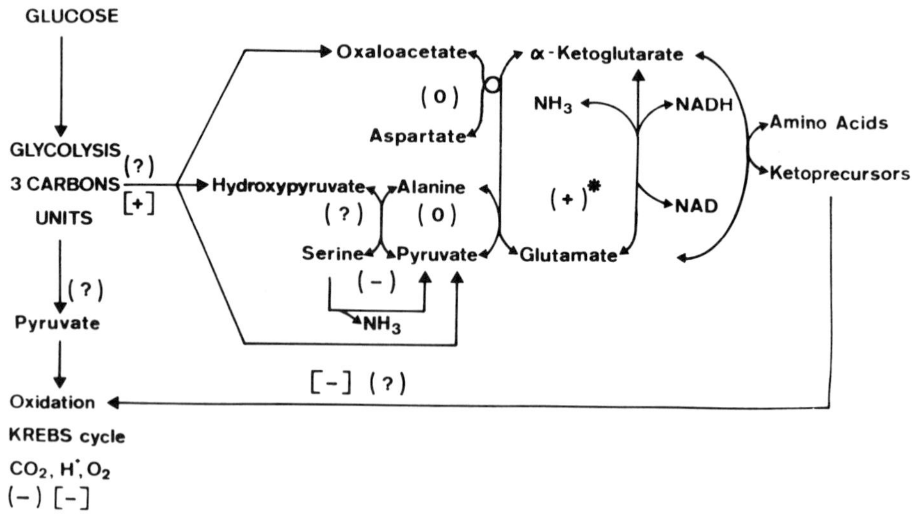

Fig. 10. Relations between glycolysis, oxidative metabolism, and the transamination sequence ruling the equilibrium between amino acids and their keto precursors. Effect of NaCl addition to the incubation medium of enzymes implicated in these pathways (+) activation; (−) inhibition; (○) no effect; (?) unknown; (*) the effect depends on the NADH concentration. Signs between [] refer to effects obtained in the course of studies other than enzymatic, essentially metabolic, studies on isolated tissues. See text for further comments.

the same experimental conditions, oxidative deamination of glutamate by the enzyme is inhibited (Chaplin *et al.*, 1965).

It may be tentatively concluded that under conditions of increased ionic intracellular concentration, glutamate formation from α-ketoglutarate is enhanced. Under the same conditions, the activity of serine hydrolase is decreased, thus leading to reduced deaminating activity; since the aminotransferase activity does not change while the oxidative activity of the Krebs cycle is decreased, it may be assumed that the equilibrium between amino acids and their keto precursors will be displaced, leading to a lower rate of reaction in the direction amino acids keto precursors. Such an effect, when considered together with an unmodified or eventually increased entrance of the amino acids in the metabolic pool, should result in an intracellular accumulation of these compounds. This concept is in agreement with the increase in NPS observed in isolated *E. sinensis* axons withstanding hyperosmotic stress (Schoffeniels, 1960) and with the increase in amino acids and the decrease in ammonia production observed in crustaceans upon acclimation from a dilute to a more concentrated medium (see Section II,A,3,b).

Although the enzymatic data can be fitted with the results obtained at higher levels of biological organization, the situation as far as enzyme activity control is concerned appears far more intricate than described above. It seems that the possible interplay between ions and different other modulators must be taken into consideration before a more complete picture can be produced. An example of such interactions can be found in the effects of NaCl on the reductive amination of α-ketoglutarate by glutamate dehydrogenase at different concentration of α-ketoglutarate and NADH (Gilles, 1974c). Some of the obtained results are presented in Fig. 11. In fact, the effect of NaCl on the formation of glutamate by glutamate dehydrogenase can change from inhibitory to activatory, depending on the amount of substrate and coenzyme. On the other hand, serine is a positive modulator of the enzyme activity, and the activation induced by both NaCl and serine are cumulative at high inhibitory concentrations of NADH.

It is of interest to recall here that the concentrations of NADH at which these effects are observed are in the range of the concentration found in tissues of crustaceans (Thabrew *et al.*, 1971). These effects may be of physiological significance and provide the cell with feedback control mechanisms of the modulation of NaCl on glutamate formation, depending on the

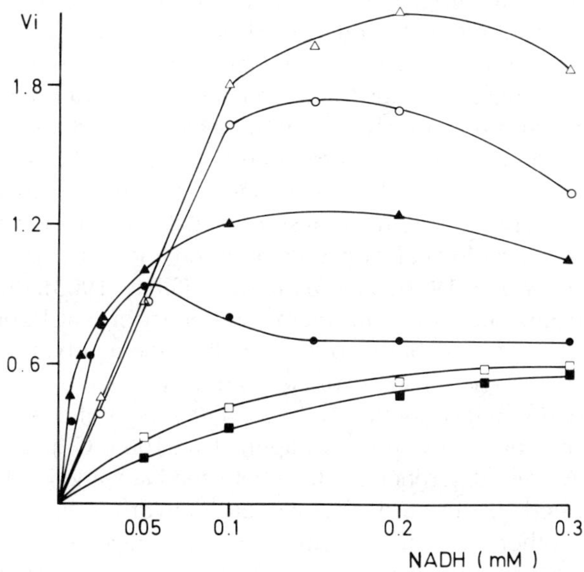

Fig. 11. Modulation by NaCl and serine of the effect of NADH on the reductive amination activity of *Eriocheir sinensis* muscle glutamate dehydrogenase: □,α-ketoglutarate 5 × 10^{-4} mol/liter; ■: α-ketoglutarate 5 × 10^{-4} mol/liter; NaCl 300 mmol/liter; ●: α-ketoglutarate 5 × 10^{-3} mol/liter; ○, α-ketoglutarate 5 × 10^{-3} mol/liter; NaCl 300 mmol/liter; ▲: α-ketoglutarate 5 × 10^{-3} mol/liter; serine 20 mmol/liter; △: α-ketoglutarate 5 × 10^{-3} mol/liter; serine 20 mmol/liter; NaCl 300 mmol/liter. (Modified from Gilles, 1974c.)

cellular level of different amino acids, or on the cellular availability of reducing equivalents and therefore on the energetic status of the cell. Besides, such interactions may be of interest in order to understand species differences observed in the evolution of deamination activity following application of osmotic stresses. Such differences might be related to differences in the ecology of the species considered. For instance, the extremely euryhaline crab, *E. sinensis,* shows only transient changes in ammonia production during acclimation to dilute media (Jeuniaux and Florkin, 1961), while the less euryhaline *C. maenas* shows much longer perturbations in its ammonia output (Needham, 1957).

Differences have also been observed by Schoffeniels and co-workers when considering the effect of NaCl on lactic dehydrogenase and 3-glycerophosphate dehydrogenase extracted from a limmic euryhaline crayfish and from a marine stenohaline lobster (Schoffeniels, 1968). Also, the isozymic pattern of lactic dehydrogenase is modified differently under the influence of salinity in *E. sinensis* and in less euryhaline species (Bollette-Dugaillay and Schoffeniels, 1970; Schoffeniels, 1970, 1976). These enzymes are generally considered important among those controlling the extramitochondrial ratio NADH/NAD, as well as the transfer of reducing equivalents from the cytoplasm to the mitochondria. This, when taken together with the controls that exert NADH on the modulation of glutamate dehydrogenase by ions and with the change in pattern of utilization of 3-C units of glycolysis (Fig. 9), leads to the conclusion that control of the fate of reducing equivalents is probably an important factor in the metabolic adjustment of the amino acid pool occurring during osmotic shocks.

A last question now arises: How can NaCl affect the activity of enzymes implicated in the amino acid metabolism? Since the first demonstrations of a NaCl effect in physiological concentrations on the activity of glutamate dehydrogenase (Snoke, 1956; Schoffeniels and Gilles, 1963), different studies have been devoted to this problem. The effect of NaCl appears to be independent of a variation of the ionic strength of the incubating medium. It must be related to a specific effect of the cationic and/or the anionic species. In fact, it appears that it is essentially the anionic species that is implicated in the modulation of enzyme activity (Chaplin *et al.*, 1965; Gilles, 1969; Gilles *et al.*, 1971). A slight difference in effect between Na^+ and K^+ has, however, been reported in the case of glutamate dehydrogenase (Schoffeniels, 1966). On the other hand, it is worth noting that the series obtained when placing the various anions tested in order of increasing effectiveness is the same for all the enzymatic systems studied up to now, i.e., $A_c^- < Cl^- < B_r^- < NO_3^-$ (Gilles *et al.*, 1971; Gilles, 1979). This order is similar to the one in which these anions disrupt the structure of various macromolecules as determined by physical methods. This, as well as the changes in enzyme thermosensitivity, brought about by addition of NaCl to the incubation medium, argue that the anion effect results from changes in configuration of the enzyme protein rather than from some indirect effect on association ion–substrate, substrate p*K*, or electrostatic shielding.

It appears that the changes in the intracellular ionic content induced by application of osmotic shocks (see Section II,A,3) can in turn induce modifications in metabolic activity contributing to the adjustment of the level of organic osmotic effectors playing a part in cell volume regulation. However, if the present information on the molecular control mechanisms can be fitted into the framework of the metabolic events occurring at the level of isolated tissues or intact animals, it should be kept in mind that the proposed scheme is only tentative. More data as to the intracellular localization of the enzymes and the ionic composition prevailing locally in the cell and on the nature of the effect of salts on the activity of other enzymes (particularly on those implicated in keto precursor formation from 3-carbon units of glycolysis) are obviously needed before a more detailed picture can be produced.

b. Control Mechanisms in the Whole Animal. Some important questions remain to be discussed in the framework of this review. Among other things: (1) Are the mechanisms described at the level of isolated tissues also at work in the whole animal? (2) Could some additional mechanisms also take part in the isosmotic regulation processes of intact organism? (3) Could these additional processes also be related to the ecological fitness of crustaceans to their environment?

As seen in the previous section of this chapter (see Section II,A,3,a), metabolic adjustments and variations in transport activity are the basis of the control of the amino acid pool in isolated tissues submitted to osmotic stresses.

Increased efflux of amino acids from tissues to blood in euryhaline crustacean species submitted to hypoosmotic stress is indicated by several studies showing that an increase in the blood amino acid content takes place concomitantly with the decrease observed at the tissues level (Jeffries, 1966; Clark, 1968; Binns, 1969; Vincent-Marique and Gilles, 1970a,b; Gérard and Gilles, 1972b; Gilles, 1977).

As far as modifications in metabolic activity are concerned, several studies report changes in NH_3 excretion by euryhaline species submitted to osmotic stresses (Needham, 1957; Jeuniaux and Florkin, 1961; Emerson, 1969; Mangum *et al.*, 1976). These results may be indicative of modifications in tissue deamination activity, provided that they are concomitant with parallel changes in ammonia, blood, and tissue level. That it is the case is indicated by the increase in NH_3 level of *Callinectes sapidus* blood (Gérard and Gilles, 1972b; Mangum *et al.*, 1976) or *Eriocheir sinensis* muscle tissue (Vincent-Marique and Gilles, 1970a,b) recorded during acclimation to dilute media.

In the crayfish *Orconectes limosus* acclimated to a concentrated medium, the CO_2 production decreases by 20% (Siebers, 1972). On the other hand, the $^{14}CO_2$ production from injected labeled glutamate increases by 50% in *Carcinus maenas,* acclimating to a dilute medium (Chaplin *et al.* 1970). Such results are in agreement with the modifications in oxidative activity demonstrated on isolated tissues.

These results indicate that the mechanisms, the existence of which has been demonstrated on isolated tissues, can be traced through the more complex physiological interactions occurring in whole organisms. Could other mechanisms be implicated in tissue volume control and/or in the related control of the intracellular free amino acid pool in whole animals? Data concerning these problems are almost totaly lacking in the literature. To our knowledge, there are only two examples of interactions between tissues which might help cell volume control.

An example of such interactions can be found in the reversible modifications of blood protein concentrations that are observed following application of osmotic stresses to whole animals (Siebers *et al.*, 1972; Gilles, 1977; Péqueux *et al.*, 1979). In the euryhaline species studied, acclimation to a dilute medium causes a significant increase in the blood serum protein concentrations. In the case of *E. sinensis* (Gilles, 1977), the variation can be traced to the level of all the protein bands, as revealed by acrylamide gel electrophoresis and particularly at the level of the Cu^{2+} containing proteins. This last result may be of interest, since the variation in the content of these proteins might be part of a physiological adaptation related to the changes in oxygen demand that appear in euryhaline species upon acclimation to media of different salinities. As discussed elsewhere, most euryhaline species show a higher oxygen consumption when acclimated to dilute media; this increase can be related, at least partly, to the activity of the Na^+ transport mechanism at work in the maintenance of the blood hyperosmotic state (see Chapter 2, Volume 5 of this treatise). In such conditions of acclimation to dilute media, the increase in the blood level of Cu^{2+} containing proteins, essentially respiratory pigments, may help the animals in meeting the increased tissue oxygen demand. On the other hand, osmotic stresses induce modifications in the efficiency of the oxygen carrier in various euryhaline species (Mangum, 1976). It is possible that the observed changes in Cu^{2+} blood proteins are also part of an adaptation to modifications in oxygen carrier efficiency. That acclimation to media of different salinities causes such modifications in the species studied remains to be demonstrated. Moreover, the changes in blood protein content observed in euryhaline species could be related to the variations in the amino acid transport that occur at the level of the plasma membranes following application of osmotic stresses. During hypoosmotic shock, part of the amino acids leaking out of the cells could indeed be stored under the form of blood serum proteins. On the contrary, the hyperosmotic stress would induce an increased degradation and/or decreased synthesis of blood serum protein material. The free amino acids left could then be actively transported into the cells. On the other hand, it is interesting to consider that the modifications in blood protein content may also directly help volume regulation at the cell level. The cells are indeed in a double Donnan equilibrium involving nondiffusible compounds on both sides of the plasmatic membrane. The increase in the number of negative, nondiffusible charges in the extracellular fluid observed upon acclimation to a dilute medium may reduce the swelling pressure the tissues have to withstand.

A different picture emerges when considering stenohaline or semiterrestrial species. As recently shown by Péqueux *et al.* (1979), blood protein content decreases in the stenohaline *Libinia emarginata* upon acclimation to dilute media, while in the semiterrestrial *Uca minax* or *Ocypode quadrata,* high blood osmolarity corresponds to high blood protein levels. This last result may be interpreted in the light of the role that blood proteins play in the establishment of the colloid osmotic pressure of invertebrate body fluids (Mangum and Johansen, 1975). In these semiterrestrial species permanently confronted with problems of water availability, changes in blood proteins level might be implicated in an adjustment of the balance of hydrostatic pressure and colloïd osmotic pressure related to water economy through control of water movements at the level of various organs such as gills, pericardial sacs, or excretory system.

The fact that no relation between blood osmolarity and blood proteins similar to those described either for aquatic euryhaline or semiterrestrial species can be seen in the stenohaline *L. emarginata* could be related to the fact that this stenohaline spider crab is an aquatic osmoconformer. It appears that the study of blood proteins in relation to blood osmolarity and water availability may have interesting ecological correlates. However, more results are needed to assess this hypothesis.

Another example of interaction in whole organisms can be found in the participation of NH_3 and CO_2 (arising from tissue catabolic activity) in the inorganic ion transport at the gill level. Accordingly, the active uptake of Na^+ that occurs at the gill level of animals acclimated to dilute media would implicate an exchange with NH_4^+ ions and also possibly with H^+ ions. On the other hand, the active influx of Cl^- appears to be related to a mechanism of Cl^-/HCO_3^- exchange (see Chapter 2, Volume 5). During acclimation to dilute media, there is an increased blood ammonia level and an increased tissue oxidative activity which results in larger CO_2 production (Section II,A,3,a,ii). Both metabolic products can be used at the gill level as counter-ions in the mechanisms of active uptake of Na^+ and Cl^-. The fact that the gill tissue of the euryhaline crabs shows (Na^+/NH_4^+)ATPase activity may be considered an argument favoring this hypothesis (Towle, 1974; Towle *et al.*, 1976; Péqueux and Gilles, 1977). In the same way, part of the amino acids released from the tissues during the adjustment to hypoosmotic conditions could be used at the gill level to provide this tissue with a supplement of ammonia and CO_2 required for Na^+ and Cl^- active transport. Recent experiments performed in our laboratory show that a Na^+/NH_4^+ coupling can only account for a small part of the Na^+ uptake occurring in isolated perfused preparations of the three posterior pairs of gills of the Chinese crab, *Eriocheir sinensis*. This is illustrated in Fig. 12, where it can be seen that the influx of Na^+ is much larger than the efflux of NH_4^+, regardless of the salinity of the external medium. On the other hand, the total influx of Na^+ is not dependent on the addition of NH_4^+ to the perfusion saline (Péqueux and Gilles, 1978). It appears that an important part of the Na^+ active uptake occurring in the gills of this species must be accounted for by another type of transport mechanism. An electrogenic Na^+ transport or a Na^+/H^+ coupling is the most likely process that may be evoked. In this last hypothesis, oxidation of amino acids at the gill level could provide this tissue not only with the ammonia, but also with at least part of the H^+ ions required for Na^+ uptake. Oxidation of amino acids in the gills may also contribute by providing the Na^+ and Cl^- transport mechanisms with part of the energy they need. Further experiments are needed to critically assess the role of amino acids in the gill ions transport processes.

c. *Conclusions.* In summary, it appears that the capability of cell volume maintenance has interesting ecological correlates in crustacean species inhabiting permanent or transitory media of fluctuating salinity. Although some correlation can be drawn between blood osmoregulation and distribution patterns, this is far from being a general rule; other mechanisms implicated in cell volume maintenance (i.e., cell volume regulation by osmoregulation of the intracellular fluid) must be taken into consideration. Detailed analysis of the processes implicated in osmotic regulation of the intracellular fluid indicate that correlations with ecology could be traced to the level of molecular events participating in cell volume regulation. Differences seem to exist between stenohaline, good euryhaline, and poor euryhaline species in (1) the response of blood protein constituents to osmolarity changes; and (2) the effectiveness of the cell volume readjustment response following application of osmotic stresses. Such differences might be related to differences in the control of transport and/or metabolic processes at work in the regulation of the amount of intracellular osmotic effectors.

The relationships suggest that if they satisfactorily explain most of the

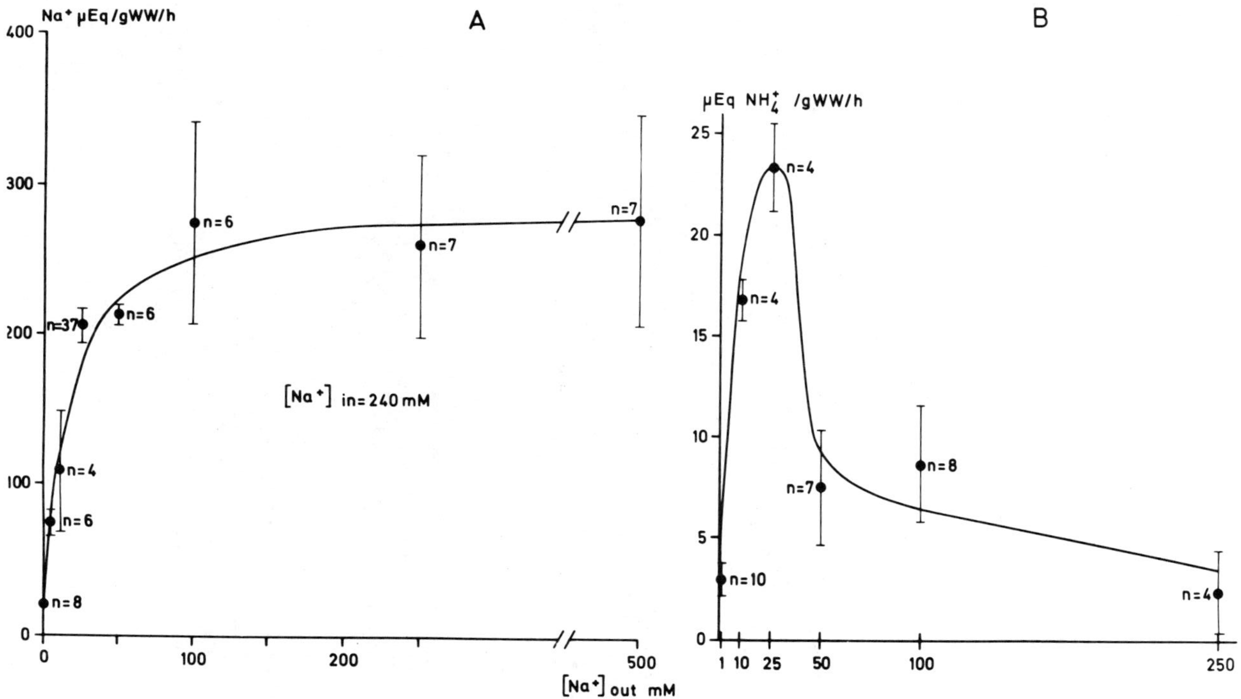

Fig. 12. Relation between external Na^+ concentration and Na^+ influx (part A) or NH_4^+ outflux (part B) in posterior perfused gills isolated from freshwater-acclimated *Eriocheir sinensis*. (Compiled from data of Péqueux and Gilles, 1978.)

experimental data, they by no means exclude other mechanisms. Additional data will be needed before a more detailed picture can be produced.

B. At Different Stages of the Life Cycle

Successful establishment of a population in a given environment is only possible if the invaders can withstand the demands of the medium at any moment during their life cycles. This problem is particularly acute for animals experiencing ecotones with fluctuating salinities. Salinity will indeed affect reproduction and development in areas where it undergoes pronounced changes. Unfortunately, data about these problems are scanty, and it seems difficult at the moment to make general statements. Moreover, it is impossible to bring an analysis of the ecological effect of salinity on growth and development down to the molecular level.

In the great majority of crustaceans, the adults can tolerate more pronounced salinity fluctuations than can the young stages. Reduced reproductive capacities and even complete sterility have been reported to occur as a function of salinity. For instance, the euryhaline amphipod, *Gammarus duebeni,* loses its reproductive potential completely when in fresh water (Hynes, 1954). Eggs of *G. duebeni* will not develop in salinities below 1‰, yet the adults tolerate pure fresh water (Kinne, 1953). The optimal range of salinity for successful development in this species seems to be between 2 and 10‰ (Kinne, 1959). Further, it seems that the average number of eggs produced is higher at 10‰ than in higher (30‰) or lower (2‰) salinities. Eggs of the shore crab, *Carcinus maenas,* will develop normally only in salinities between 28 and 40‰, while adults easily tolerate more dilute media (Broekhuysen, 1936). Similar findings have been reported for several species of *Balanus* (Barnes, 1953). Also, egg development in *Balanus eburneus, B. amphritrite,* and *Chelonibia patula* takes place at a normal rate at 25–40‰, but the portion of eggs hatched decreases in lower (25–15‰) or higher (40–60‰) salinities (Crisp and Costlow, 1963). In the same way, eggs of the tube building amphipod, *Corophium triaenonyx,* can develop in a fairly wide range of salinities (7.5–37.5‰), although the adults are able to tolerate a range between 0.6 and 59.8‰ (Shyamasundari, 1976). However, the satisfactory percentage (70%) of egg development is observed in a slightly smaller range (20–32.5‰) and the maximum percentage only between 27.5 and 30‰.

It appears that salinity will markedly affect both egg production and embryonic development in a variety of crustaceans. To our knowledge, experiments on gamete formation are totally lacking in the literature. Such information would be of interest, especially in view of the generally higher salt tolerance of adult forms and of the migratory behavior of some of them.

Salinity will also affect postembryonic development, and in many cases, the salt tolerance limits of the larval stages will be narrower than that of the adults. For instance, adults of the migrating shrimp, *Macrobrachium vollenhovenii,* can live in fresh water, whereas the juveniles only develop in brackish waters (Ville, 1971). The same holds true for *M. acanthurus* (Hughes and Richard, 1973). Similarly, adults of *Sesarma reticulatum* occupy habitats in low salinity environments in which complete larval development is not possible (Foskett, 1977). Salinity tolerance of larval stages of a variety of crustaceans has been studied by Costlow and co-workers. This work includes studies on *Sesarma cinereum* (Costlow *et al.*, 1960), *Panopeus herbstii* (Costlow *et al.*, 1962), *Rhithropanopeus harrisii* (Costlow *et al.*, 1966; Christiansen and Costlow, 1975), *Callinectes sapidus* (Costlow and Sastry, 1966; Costlow, 1967), and *Menippe mercenaria* (Ong and Costlow, 1970). These studies, as well as others (Robertson, 1968; Sandifer, 1975; Young and Hazlett, 1978), show that salinity (and also the combination salinity–temperature) affects larval development. Few attempts, if any, have been made up to now to (1) explain such effects in terms of molecular events, (2) relate the discrepancy between adult salt tolerance and larval salt tolerance to osmoregulatory capabilities, and (3) relate the salt tolerance of larvae to the distribution patterns of populations.

In most studies, the relationship between larval salt tolerance and species distribution is taken for granted. As far as the effect of salinity on molecular events occurring during development is concerned, the only indication we have been able to find is related to the effect of external salinity on protein turnover described in *Artemia salina* nauplii (Conte *et al.*, 1973). Possibly, such effects might reveal important factors in development, especially in view of the fact that during this period protein synthesis mechanisms are highly solicited.

Several studies deal with the osmoregulatory capabilities of larval stages as compared to those of adult forms. To our knowledge, these studies have been previously restricted to blood osmoregulation. We have been unable to find in the literature any reports dealing with cell volume response and control of intracellular fluid osmolarity. Such data would be of interest, since efficiency of cell volume control appears to be an important parameter in the ecology of euryhaline crustaceans.

It appears that in many crustaceans the maximum osmolarity gradient maintained between blood and external medium by adults in dilute media is much larger than for the larval stages. For instance, blood osmoregulatory power is less in larvae of *Rhithropanopeus harrisii* (Kalber and Costlow, 1966) than in adults (Smith, 1967). This is true for *Sesarma reticulatum* (Foskett, 1977) and, as stated by Foskett (1977), for *Cardisoma guanhumi.* Such differences between blood osmoregulatory capabilities in adults and larvae can account at least partly for the differences observed when studying salt tolerance. Obviously, they are also parameters to take into consideration in the distribution patterns of species. It appears that the capability of blood osmoregulation may be different according to the larval stage considered. In some species, such as *Cardisoma guanhumi,* the ability to osmoregulate seems to increase with postembryonic development (Kalber and Costlow, 1968). However, this is far from being a general rule. Larvae of *Sesarma reticulatum* show no significant changes in blood osmoregularity power during development (Foskett, 1977). In the same way, larvae of *Rhithropanopeus harrisii* osmoregulate similarly during the different larval stages (Kalber and Costlow, 1966).

It is interesting to consider that in some species, larvae can maintain a blood hyperosmotic state, while adults cannot. This is the case of the osmoconformers, *Hepatus epheliticus* and *Libinia emarginata,* in which blood osmoregulatory capability shown by larvae is gradually lost in megalops and settling crab stages (Kalber, 1970). This could confer some advantage to species that are osmoconformers, but live in the vicinity of estuaries where salinity can be subjected to moderate changes. Larvae are indeed unable to avoid salinity changes that adults with good locomotive abilities can escape.

In summary, there appears to be no general trend toward development of adult osmoregulatory patterns through larval life, although in many species, the osmoregulatory capabilities progressively increase during postembryonic development. This can be related to the migrating behavior of the various invaders of estuaries in which larvae cannot osmoregulate as effectively as adults. Other correlations also emerge from the data at hand between osmoregulation in larval stages and the ecology of some species. More results, particularly about cell volume control capabilities of larvae, are needed before a clear picture can be produced.

III. INDIRECT EFFECTS OF SALINITY

A. Stress Relations

Besides osmotic problems, the inhabitants of many biotopes have to face various other stresses which are integral parts of the ecological framework of these media. Changes in temperature and oxygen availability may become important parameters in shallow estuarine waters, marshes, mangroves, and semienclosed lagoonal ponds. These parameters even constitute essential environmental factors in the intertidal zone. The species living in this biotope are subjected to important modifications in temperature and oxygen availability following the tidal rhythm. These media, which occupy quite restricted areas, form the borderline between land and ocean and, as such, are subjected to increasing pollution from both rivers and oceans. Temperature and water availability are also parameters of the utmost importance for terrestrial species.

Temperature–osmoregulation relationships in terrestrial species are discussed in Chapter 6 of this volume. Similarly, problems related to oxygen availability are discussed in Chapters 1 and 2. The present part of this review will be restricted to the presentation of salinity–temperature and salinity–pollution relationships in aquatic media and to a discussion, whenever possible, of the effect of such relationships on animal distribution. At the onset, the difference between synergic effects of and interactions between environmental parameters should be stressed. It is obvious that in any biotope, one stress is better withstood than two or more together. Such synergic effects have to be taken into consideration in the definition of biocenoses; however, they cannot be studied at the molecular level, since there is no reciprocal dependency of one effect on the other. The picture should be quite different when considering interactions. In this case, one factor acts as a positive or negative modulator of the effect of the other. Unfortunately, it is still impossible to analyze the ecological effects of environmental factor interactions at the molecular level. Up to now, studies have been essentially concerned with species survival in different combinations of external parameters. We shall thus limit ourselves to a rapid survey of the available data in the hope that some guidelines for future research might emerge.

1. SALINITY–TEMPERATURE RELATIONSHIPS

Temperature will act on the mechanisms of ion transport implicated in blood osmoregulation as it acts on any chemical interaction. Carrier mediated transports should be enhanced by an increase in temperature and decreased by a temperature drop. If, during acclimation, there is no inhibitory effect of thermocompensation on this process, it might be speculated that osmoregulation of extracellular fluids is more effective at high rather than at low temperatures. Lockwood (1960) showed that the rate of Na^+ uptake increased with temperature in the isopod, *Asellus aquaticus*. Similarly, the rate of adjustment of blood osmolarity increases with temperature in two species of *Crangon* (Spaargaren, 1971). Also, the ability for blood osmoregulation is impaired by low temperature in two species of *Penaeus* (Williams, 1960). This would indicate that in euryhaline species a hypo-osmotic stress is better withstood at high temperature. In such a view, invasion of dilute media would be favored by high environmental temperature.

The picture is far from being that simple. There is no significant difference in blood osmoregulatory capability between 7° and 20°C in *Gammarus duebeni* (Kinne, 1952) or in *Crangon crangon* between 5° and 15°C (Flügel, 1963). Similarly, no general trend of salinity–temperature interaction can be found in salt tolerance studies. While various species can withstand salinity fluctuations better at high temperatures, e.g., the shrimp, *Crangon septempinosa* (Haefner, 1969), high temperature will narrow the salinity tolerance range in others. *Hemigrapsus edwardsi* survives better at low salinities only at low temperatures (Hicks, 1973). If low salinities appear to be better tolerated at high temperatures by the copepod, *Acartia tonsa,* the related species, *A. bifilosa,* will support low salinities better at medium temperatures (Lance, 1963). Different effects of salinity–temperature interactions in related species or in populations of the same species have been described many times (Williams, 1960; Flügel, 1963; Segal and Burbanck, 1963; Spaargaren, 1971, 1972; Biggs and McDermott, 1973; Dorgelo, 1974; Bradley, 1975), and this literature has been the subject of several reviews (see Kinne, 1971; Alderdice, 1972; Dorgelo, 1976; Vernberg and Silverthorn, 1979). It seems that no clear, simple, pattern can emerge from these studies. In his extensive review of salt tolerance in Crustacea, Dorgelo (1976) attempted to classify salinity–temperature interactions based on survival curves. In animals from high salinity regimes, four main types of beneficial salinity–temperature combinations can be distinguished: (1) low salinity–low temperature; (2) high salinity–high temperature; (3) low salinity–high temperature; and (4) high salinity–low temperature. In animals from low salinity regimes, types 1, 2, and 3 are represented.

The picture is still different in species of the intertidal semiterrestrial habitats in which salt tolerance does not appear to be affected by temperature (Sjöberg, 1967; Jansen, 1971; Jones, 1972; Harvey *et al.*, 1973; Dorgelo, 1974). This does not hold true for all species of the intertidal zone, since temperature has been shown to influence salt tolerance in many of them (Todd and Dehnel, 1960; Remmert, 1967; Jansen, 1971; Marsden, 1973; Harvey *et al.*, 1973; Barnes and Klepal, 1974). Thus, there seems to be no simple trend of correlation between distribution patterns of adult forms and salinity–temperature interactions.

Maintenance of a breeding population is an essential factor in the establishment of a community in a given environment. Many papers have been published during the last decade dealing with embryonic and postembryonic development under different temperatures or combinations of salinity and temperature. These studies are concerned with a variety of species such as *Hyalella azteca* (Bovie, 1951), *Gammarus fasciatus* (Clemens, 1950), *G. duebeni, Lagunogammarus salinus,* and *G. zaddachi* (Kinne, 1959, 1960, 1961), *G. duebeni, G. lawrencianus, Eulimnogammarus obtusatus, G. oceanicus* and *G. setosus* (Steele and Steele, 1973), *Corophium triaenonyx* (Shyamasundari, 1976), *Sesarma cinereum* (Costlow *et al.*, 1960), *Rhithropanopeus harrisii* (Costlow *et al.*, 1966; Christiansen and Costlow, 1975), *Menippe mercenaria* (Ong and Costlow, 1970), *Cancer irroratus* (Sastry, 1970), *Panopeus herbstii* (Costlow *et al.*, 1962; Sastry, 1970), *Palaemonetes vulgaris* (Sandifer, 1973), *Scyllarus americanus* (Robertson, 1968), *Clibanarius vittatus* (Young and Hazlett, 1978), *Pagurus samuelis* (Coffin, 1958), and *Balanus eburneus, B. amphitrite,* or *Chelonibia patula* (Crisp and Costlow, 1963).

As pointed out by Costlow (1967), direct salinity–temperature interactions do not appear to affect larvae. However, synergic effects can be demonstrated. For instance, a steady increase in mortality can be seen in larvae of *Callinectes sapidus* if a decrease in salinity is accompanied by a decrease in temperature. Careful analysis indicates that there are no interactions between these factors (Costlow, 1967). From the data at hand it seems that salinity would be a more important parameter than temperature when considering survival, whereas temperature would be more important in affecting growth and development. On the other hand, what has been said about salinity tolerance appears also to hold true for temperature tolerance; i.e., in many species, tolerance is higher in adults than in larvae. Temperature may become a factor of limitation by retarding development of the larval stages. For instance, thermal tolerance of adult *Clibanarius vittatus* cannot account for the geographical limitation of the species to warm waters, but the poor ability of the larvae to develop at low temperatures might be responsible for such distribution (Young and Hazlett, 1978).

The ecological incidence of synergic effects of salinity and temperature seems clear on a broad scale, i.e., large and rapid fluctuations of both factors are generally more harmful than large and rapid fluctuations of one of them. This is the expression of the fact that one stress is better supported than two, and species distribution will obviously be limited accordingly. However, a detailed analysis of the ecological correlates of synergic effects, as well as of interactions between salinity and temperature when these factors are kept largely within the lethal limits, does not seem possible at the moment. Nothing is known about the molecular basis of these interactions. Even speculations on the relations between molecular processes and ecological incidences of salinity–temperature interactions are impossible. It can be considered that the physiological mechanisms of temperature acclimation should be taken into account in the ecological picture. As we have seen (Section I,A), osmoregulation both at the blood and the cell level can be quite fast, and its effectiveness as well as powerfulness as studied in the laboratory can be fitted with species distribution patterns. Thermocompensation mechanisms are generally much slower because they involve synthesis of new molecular species (Hochachka and Somero, 1973; Gilles, 1975b. In nature, and particularly in the aquatic media, temperature fluctuations over areas large enough to prevent escape reactions are generally very slow. The rate at which thermocompensation processes enter into play is probably not (unlike their powerfulness) an essential factor of limitation. It may very well account, at least partly, for the variability of the responses given by different species in survival studies under laboratory conditions. In these studies, temperature stress is applied abruptly most of the time.

2. SALINITY–POLLUTANT RELATIONSHIPS

Beyond all the natural ecological parameters (temperature, salinity, currents, hydrostatic pressure, and oxygen availability) governing life processes in rivers, coastal, and estuarine environments, animal species sometimes have to cope with physiologically active substances and toxics of natural origin. Nevertheless, such "natural" pollution may be considered of minor importance by comparison with the increasing introduction in the environment of pollutants resulting from human activity. Man-introduced pollution indeed covers several aspects, e.g., domestic and industrial wastes, oils and dispersants, heavy metals, pesticides and herbicides. In addition it covers the alteration of water temperature by industry and nuclear reactor cooling systems.

The biological effects of toxic substances on animal species has received increasing attention during the last few years, and a large amount of information is now available on acute toxicity, bioaccumulation, reproduction, teratology, and some physiological functions, e.g., osmoregulation (Waldichuk, 1974; Vernberg and Vernberg, 1974; Bouquegneau and Gilles, 1979). Nevertheless, information on animal–pollutant interactions at the molecular level still remains scanty.

Most of the work devoted to the effects of pollutants are concerned with definition of tolerance limits. In such studies, it is often almost impossible to distinguish between real synergistic effects and interactions between pollutants and mechanisms concerned with blood or cell osmolarity regulation. The following review of the toxicity and biological effects of toxicants will include some cases of so-called "synergistic" actions where it might be possible that the toxicity results from an interaction with osmoregulatory mechanisms.

In consideration of their relative toxicity and biological effects, chemical pollutants have been divided into three groups that we will continue to consider separately: oils and dispersants, pesticides and PCBs, and heavy metals.

a. *Toxicity and Biological Effects of Toxicants*

i. OIL AND DISPERSANTS. Petroleum contamination of aquatic environments can come from a large variety of sources. If a slight, but significant, quantity of hydrocarbons can be transported through the atmosphere into rivers or the sea, the largest recurring sources of oil pollution obviously are refinery wastes, discharge of oil from nearshore shipping operations, discharge of ballast waters by tankers, and, more intermittently, accidental spills. Despite the increasing importance of the problem, information on the effects of oil on aquatic organisms and particularly on crustaceans has remained surprisingly scanty, and refers mostly to mortality indexes with only small attention to physiological consequences resulting from exposure to oil pollution.

Relatively low concentrations of crude oil have indeed been shown to kill the pelagic larvae of the bottom invertebrates in marine nearshore and estuarine waters, such as the copepod *Anomolocera patersoni* and the hermit crab, *Diogenes pugilator,* within 3 days (Mileikousky, 1970). Adult copepods (such as *Acartia clausi, Paracalanus parus, Penilia avirostris, Centropages ponticus,* and *Oithona nana*) also are killed by concentrations of 0.001 ml/liter (Mironov, 1969a,b). Emulsions of crude oil have also been shown to exert a lethal effect on larvae of the lobster, *Homarus americanus,* at concentrations of 100 ppm, and they appear to be sublethal at concentrations as low as 1 ppm (Wells, 1972). McCauley (1965) reported observations on the tolerance of the amphipod, *Gammarus,* to oil in the sediments.

According to Corner *et al.* (1968), barnacles are the most resistant of the intertidal organisms, but that does not mean that they remain insensitive to petroleum contamination. Recently, Morton and Wu (1977) emphasized the susceptibility of both adult barnacles, *Balanus amphitrite amphitrite* and *Balanus variegatus variegatus,* to kerosene and BP 1002. They showed that barnacle nauplii were killed at lower concentrations than the adults (George, 1961).

While the tolerance to crude oil may significantly differ from one species to another and appears to be dependent on the development stage, the refined products, such as fuel oils, have proven to be generally much more toxic (Anderson *et al.*, 1974). Mortalities and survival times measured by Lee *et al.* (1977) on amphipods, *Gammarus mucronatus* and *Ampithoe valida,* show that toxic effects of southern Louisiana crude begin at 2.4 ppm, but with Exxon No. 2 fuel oil Exxon, effects begin at 0.8 ppm. The major chemical components of the water-soluble fractions of fuel oils (i.e., benzene-extracted organics) can be classified into volatile (naphthalene 2-methylnaphthalene) and nonvolatile (*O*-creosol, *O*-toluidine, dimethylphenols). Lee and Nicol (1978) assessed the toxicity of four components of petroleum oils to the marine amphipod, *Elasmopus pectinicrus,* and observed that volatile components were much more toxic than nonvolatile ones (see also Donahue *et al.*, 1977). Recently Laughlin *et al.* (1978) have investigated the effects of long-lasting exposure of the mud crab, *Rhithropanopeus harrisii*

(Gould), after hatching to water-soluble fractions (WSF) of No. 2 fuel oil. Survival, growth, and development rates were monitored over 6 months. These authors reported that the megalopa and crab stages were not particularly sensitive to continued petroleum exposure, in contrast with the zoeal stages severely affected by 20% of the WSF (0.36 ppm total naphthalenes, 1.26 ppm total hydrocarbons). Nevertheless, mean duration of the megalopa and first crab stages were significantly modified, depending on the WSF concentration. The data also indicated that these crabs possess a considerable ability to recover from the effects of chronic sublethal exposure to petroleum hydrocarbons. Berdugo *et al.* (1977) have also observed that exposure of the estuarine copepod, *Eurytemora affinis,* to WSF for periods less than 4 hr results in significant sublethal effects on reproduction during the subsequent life span of the female. This suggests that such short-term exposure in an estuarine situation might have longer-term effects on population dynamics continuing after the hydrocarbon levels in the water have resumed residual concentrations.

From the few results briefly presented above and from the examination of many reports on oil spilled from tanker wrecks, it appears that considerable confusion and often contradictory conclusions exist concerning the effects of oil and derivatives on crustacean larvae and adults.

Few data are available on the synergistic effects of interactions between oil toxicity and salinity stress in crustaceans. Investigating oil toxicity to shrimps, Lytle (1973, cited by Heitz *et al.*, 1974) reported that the 96-hr TLm value was about 35 ppm emulsified oil in the fall, while in the winter, the TLm value dropped to 15 ppm; simultaneously, the salinity of the surrounding water was observed to drop from 15 to 22‰ to a level of 1–4‰. While some seasonal, temperature, or cyclic phenomenon cannot be discarded *a priori,* it seems that the increase in oil toxicity has to be related to a decrease in salinity.

In addition to survival, growth, and development studies during oil exposure, little is known about the sensitivity of various physiological mechanisms, and, to our knowledge, there is a real lack of experimental results on the direct effects of oils and dispersants on osmoregulation in crustaceans.

ii. PESTICIDES AND POLYCHLORINATED BIPHENYLS (PCBs). The widespread use of insecticides, pesticides, herbicides, and polychlorinated biphenyls (PCBs) has elicited increasing interest in their toxicity and in the mechanisms of their accumulation by living organisms to levels largely exceeding those found in the environment. These pollutants are indeed characterized by a physical and chemical stability (Gustafson, 1970) which makes them particularly persistent in contaminated environments. If animals can accumulate these toxicants to high levels, additional accumulation also occurs at successive steps of the food trophic chain.

As a consequence, long-term sublethal effects resulting from recurrent exposure of living organisms to low levels of these toxicants might have the most important impact on aquatic communities and particularly estuarine populations confronted with osmoregulation problems.

Early works of George *et al.* (1957) have described disturbed motor coordination in fiddler crabs, *Uca pugnax,* after their tidal marsh environment was sprayed with strobane, DDT, and BHC.

Since that time, the effects of several pesticides and PCBs on the sensitivity and the behavior of several species of fiddler crabs have been investigated. Sensitivity has been reported to DDT (Odum *et al.,* 1969), chlorinated hydrocarbons absorbed from sludge (Krebs *et al.,* 1974), PCBs (Nimmo *et al.,* 1971), dieldrin (Klein and Lincer, 1974), and the insecticide, Temefos (Ward and Howes, 1974; Ward and Busch, 1976; Ward *et al.,* 1976). These studies show that pesticide exposure results in decreased density of natural populations of fiddler crabs due to direct mortality or disturbance of their behavior which render them more susceptible to predation. In the latter case, maladaptive behavior generally was attributed to malfunctions in their neurophysiology.

The toxicity of methoxychlor, a substitute for DDT, has been investigated by Armstrong *et al.* (1976) in acute and chronic studies using larval, juvenile, and adult stages of the Dungeness crab, *Cancer magister.* Depending on the length of exposure, the insecticide affects the survival of larvae and adults, and the molting mechanisms; it has been demonstrated to be concentrated more rapidly by juvenile crabs than by adults. Concentrations of methoxychlor in individual tissues were found to be highest in the exoskeleton, gill, and hepatopancreas. Similar studies were conducted on the same crab with other insecticides and fungicides by Buchanan and Stewart (1970) and by Caldwell *et al.* (1978).

In addition to survival and development studies illustrated by the few examples above extracted from the abundant literature on the subject, direct information dealing with sensitivity of osmoregulatory mechanisms is now available.

Nimmo and Blackman (1972) have reported a decrease in cation concentrations (Na^+, K^+, Mg^{2+}) in the hepatopancreas of the shrimps, *Penaeus aztecus* and *Penaeus duorarum,* after exposure to sublethal concentrations of DDT in seawater. Methoxychlor, at concentrations which are sublethal in acute tests, has been shown to decrease the tolerance of the crab, *Hemigrapsus nudus,* to reduced salinities (Caldwell, 1974). However, the latter author has failed to demonstrate that the osmotic regulation in *H. nudus* or the osmotic and ionic regulation in *Cancer magister* are significantly impaired by the methoxychlor treatment. In that work, the only evidence for alteration of ionic regulation involves Mg^{2+} transport across the bladder wall.

In a more recent study, Roesijadi *et al.* (1976a,b) have investigated the effects of PCB exposure on hemolymph chloride and osmotic regulation, chloride space, chloride fluxes, and water fluxes of the shrimp, *Palaemonetes pugio.* Exposure to sublethal and 96-hr LC50 levels of Aroclor 1254 did not seriously alter hemolymph chloride and osmotic concentrations, chloride space, or chloride-exchange kinetics in adults; on the contrary, disruption of hemolymph chloride regulation was associated with juveniles sustaining large mortalities. Shrimps exhibited reduced water permeability at 17‰ salinity. At the level of the cellular free amino acid pool, no appreciable effect of Aroclor 1254 has been detected, while changes in the concentrations of individual amino acids, such as glycine, reflect an altered metabolic state.

Organochlorine pollutants appear to alter ion and water transport processes related to ionic and osmoregulation in crustaceans, but seem to exert their toxicity in different ways according to whether adult, larval, or juvenile stages are concerned. More results are needed before a clear picture of the interaction between osmoregulation and pesticides and overall effect on crustacean ecology can be produced.

iii. HEAVY METALS. Among the various sources and kinds of pollutions, pollution by heavy metals, either natural or related to human activity, has been the type most extensively documented and studied, probably because of the disastrous effects and diseases that these are known to produce in man after consumption of contaminated food. Tragedies, such as those of Minamata and Niigata, Japan, have emphasized the importance of the problem of bioaccumulation of mercury by aquatic organisms. The concern of metal pollution has been largely related to its effects on man; less attention has been paid to the toxicity of metals to aquatic organisms, particularly crustaceans.

Low concentrations of inorganic mercury are lethal to many marine larvae, e.g., the larval fiddler crab, *Uca pugilator* (DeCoursey and Vernberg, 1972). Larval stages appear to be much more sensitive to mercury poisoning than adults (Connor, 1972; Vernberg and Vernberg, 1972; Vernberg *et al.*, 1973). It was demonstrated that adult fiddler crabs survived for months in seawater containing 0.18 ppm mercury, while all stage I zoeae died in the same concentration within 48 hr. Furthermore, much lower concentrations of mercury, although sublethal, were shown to significantly affect the metabolism and behavior of the larvae, in addition to decrease in survival.

Shealy and Sandifer (1975) also investigated the effects of several concentrations of mercury [from 0.0 (control) to 0.056 ppm] on survival and development of the larval grass shrimp, *Palaemonetes vulgaris;* they reported reduced survival of the postlarval stages, delayed molting, extended development time, and increased numbers of larval instars and morphological deformities.

Selenium acts as an agent to increase or decrease mercury toxicity in larvae of the crab, *Cancer magister,* as a function of the concentration used (Glickstein, 1978).

Growth inhibition by copper salts have also been reported by Saliba and Krzyz (1976) on the brine shrimp, *Artemia salina*. Copper toxicity has been observed to depend on the particular form of the metal and to increase progressively from the sulphate to the chloride, acetate, and carbonate form.

In a study of the effects of cadmium and salinity, combined with constant and cycling temperature, on the larval development of the two estuarine crab species, *Callinectes sapidus* and *Rhithropanopeus harrisii,* Rosenberg and Costlow (1976) emphasized the importance of salinity stress in enhancing cadmium toxicity. Such interactions between salinity and heavy metals have also been reported to act on the mortality of marine and estuarine species of isopods (*Idotea baltica, I. neglecta, I. emarginata, Eurydice pulchra, Jaera albifrons,* and *J. nordmanni*) (Jones, 1975) and to retard the rate of regeneration and ecdysis in fiddler crabs after multiple autotomy (Weis, 1978). The particular sensitivity of estuarine species to heavy metals and the few available data on their combined action with salinity intuitively support the idea that osmoregulatory processes in crustacean species confronted with dilute media might be affected by heavy metals. Since gills are largely implicated in blood osmoregulation and in exchanges with the environmental medium, several works have concentrated on the effects of heavy metals on that tissue.

In the crayfish, *Orconectes virilis,* Anderson (1978) demonstrated that the concentration of lead in the gills and the exoskeleton increased as environmental concentration of Pb and exposure time increased. Increasing concentrations of Pb were also shown to decrease the ability of *O. virilis* to take up oxygen, while the crayfish was able to acclimate to the metal concentration (0.5–2 ppm) by compensating for decreased gill efficiency through increasing its ventilation volume.

Both respiratory and osmoregulation functions of gills from the two estuarine species, *Carcinus maenas* and *Cancer irroratus,* have been simultaneously investigated by Thurberg (1973) after 48-hr exposure to various concentrations of cadmium (0.12–8 ppm) and copper (0.3–40 ppm). Copper has no effect on gill-tissue oxygen consumption, while cadmium reduces the rate of oxygen consumption in both species. On the other hand, copper-exposed crabs exhibited loss of osmoregulatory power with increasing copper concentration until normally hyperosmotic serum became isosmotic with the surrounding medium (Fig. 13). In contrast, cadmium was

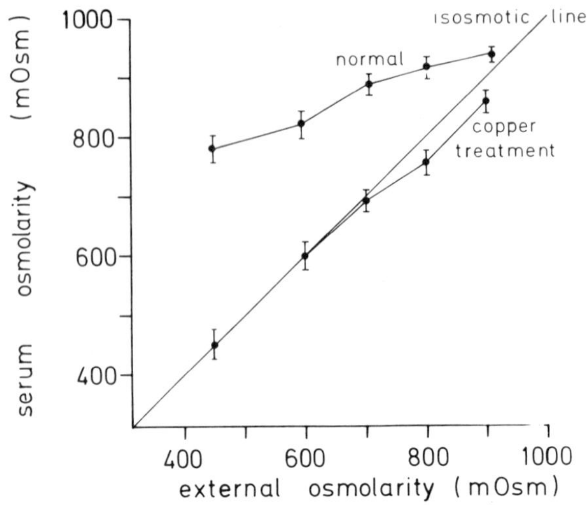

Fig. 13. *Carcinus maenas:* Effect of copper on osmoregulation. Osmoregulatory patterns in normal conditions and after 48-hr exposure to 30 ppm copper. Each point comprises mean serum osmolarity of six crabs. Bars: standard errors. (Compiled after data from Thurberg *et al.*, 1973.)

observed to increase the blood hyperosmotic state of *C. maenas* seen in dilute media.

Impairment of osmoregulatory capability in the sense of severe modifications of the blood osmotic concentration has also been reported for several marine and estuarine isopods after exposure to cadmium, zinc, and mercury (Jones, 1975). Effects of intoxication with mercury salts on osmoregulatory abilities and on the readjustment of ion balance following changes in salinity appear to have received the widest attention.

Long-term intoxications with 50 ppb $HgCl_2$ do not alter the blood Cl^- content of the porcelain crab, *Petrolisthes armatus,* either acclimated or adjusting to different salinities (Roesijadi *et al.*, 1974). At variance with the latter results, Inman and Lockwood (1977) have established that exposure of the amphipod, *Gammarus duebeni,* in 2% seawater to concentrations of 100 ppb and more methylmercuric chloride for periods of 1–4 days (well below 96 hr LC_{50} values) results in a small but significant reduction in hemolymph Na^+ concentration. Treated animals also showed a reduction in the maximal and steady-state values for the active uptake of Na^+, as illustrated in Fig. 14. Prior treatment with sublethal concentrations of CH_3HgCl resulted in a delay of 80 min in the normal time required for the amphipod to increase active Na^+ uptake when transferred from 100 to 2% seawater; this was paralleled by a lower rate of Na^+ loss.

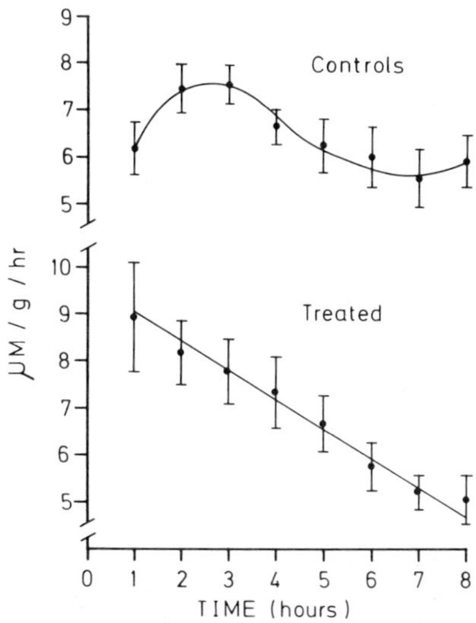

Fig. 14. Active uptake of Na^+ (ordinate) by *Gammarus duebeni* previously treated with methylmercuric chloride when transferred from 2% SW into 2% SW + Na^{22} (± SE $n = 8$). (From Imman and Lockwood, 1977, by permission of Pergamon Press.)

In a recent work, J. M. Bouquegneau and P. Hibaude (unpublished results, personal communication) have studied the toxicity, accumulation, and effects of $HgCl_2$ on osmoregulation of the euryhaline Chinese crab, *Eriocheir sinensis*. One-hundred ppb $HgCl_2$ have been observed to be lethal in freshwater acclimated crabs, while it remains only sublethal in seawater animals. Mercuric chloride appears to be more toxic in animals adapted to fresh water. This could be attributed to the fact that the accumulation rate of the pollutant is higher in fresh water due to the fast water turnover the animal has to maintain in dilute media to ensure proper osmotic balance. In fresh water contaminated with 10 ppm Hg, the gills tissue already contained more than 150 and 400 ppm Hg, respectively, after 1 and 3 hr of intoxication, while in seawater, the mercury content remained lower than 10 ppm. This large mercury accumulation in the gills of freshwater-acclimated crabs seems to affect blood osmoregulation since a large decrease of the Na^+ and Cl^- blood content can occur during intoxication with 10 ppm $HgCl_2$ (Fig. 15).

Several reports show that heavy metals impair blood osmoregulation. This problem seems to be particularly acute when the animals are in dilute media. It appears that in this situation, accumulation at the gill level is faster, thus leading to faster impairment of the mechanisms implicated at that level in the maintenance of blood ionic content. We now come to the question:

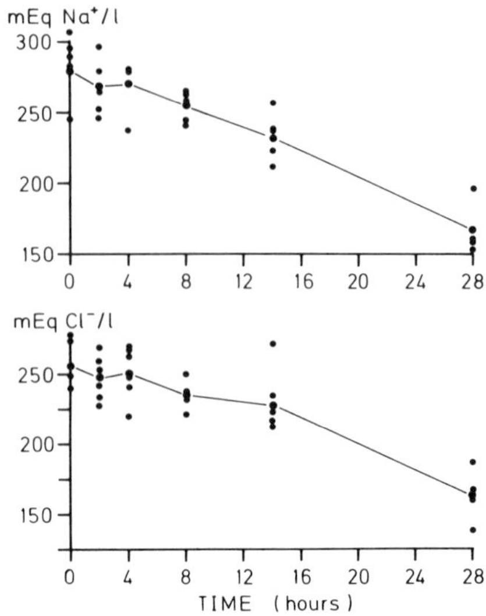

Fig. 15. *Eriocheir sinensis:* changes in the Na^+ and Cl^- content of hemolymph of freshwater-acclimated crabs after exposure to 10 ppm mercury ($HgCl_2$). •, Typical data; ●, mean values. (J. M. Bouquegneau and P. Hibaude, personal communication.)

What are the molecular interactions leading to the effects of the different types of pollutants considered in this review?

b. Mechanisms of Action. Works devoted to the effects of pollutants, other than heavy metals, on the osmoregulatory mechanisms in crustaceans, still remain very scanty, and there is a real lack of information on the effects of oils and dispersants. Nevertheless, their toxicity has been well documented by considering their lethal and sublethal effects at the level of the whole animal. As far as stenohaline marine species are concerned, toxic effects of petroleum derivatives must be due to the disturbance of physiological mechanisms other than osmoregulation. As shown by Blumer *et al.* (1973), oil fractions impede the sensory systems responsible for detection of food in the lobster, *H. americanus*. This results in a reduced ingestion of food, with a consequent reduction in growth and metabolism. Similarly, WSF-induced depression of feeding rate has been demonstrated in the estuarine copepod, *Eurytemora affinis*, by Berdugo *et al.* (1977). Edwards (1978) also concluded, from his study on intoxication of the shrimp, *Crangon crangon*, by water-soluble oil fractions, that these pollutants decrease food intake and/or assimilation efficiency.

Not much is known about other possible mechanisms of action of oils and dispersants. Heitz *et al.* (1974) investigated the effects of *in vivo* oil exposure on the activity of seven enzymes extracted from stomach–hepatopancreas homogenates, but have been unable to detect any significant disturbance. The tissues from which the latter enzymes have been extracted were not in direct contact with the pollutant, which implies that oil first had to enter the test animals in order to exert its toxicity. Enzymes or other macromolecular structures in membranes of tissues more directly exposed to the polluted environment might be more affected. In this respect, studies on the effects of oils and dispersants on gill enzymes and molecular interactions in these tissues would be of the utmost interest. Data about ($Na^+ + K^+$) ATPase would be of interest in view of the importance of this enzyme in blood osmoregulation. Also, it may be reasonably considered that gill coating by oil can act as a barrier modifying the transport processes implicated in osmoregulation. It might further be considered that aromatic hydrocarbons can interact with the lipid components of membranes, inducing modifications in their passive and active properties. Unfortunately, information about such effects is totally lacking in the literature.

As far as chlorinated hydrocarbons, insecticides, PCBs, and herbicides are concerned, their effects on membrane ATPases *in vitro* have been fairly well documented. As an example, Matsumura and Narahashi (1971) and Jowett and Rhead (1978) have found a correlation between the degree of DDT inhibition of ATPases extracted from lobster nerve and the electrophysiological symptoms of DDT poisoning. Those findings support the idea that the principal site of action of pesticides, including DDT, is the nervous system, at the level of which poisons induce disruptions of Na^+ and K^+ transport processes.

By extension and because of the supposed importance of the ($Na^+ + K^+$) ATPase system in hydromineral regulation, it is expected that impairment of osmoregulatory ability by pesticides might be correlated with inhibition of the ATPase activity of the salt transporting system. This correlation has been established by Eisler and Edmunds (1966) in marine fishes. The ($Na^+ + K^+$) ATPase activity of crustacean gills is also severely inhibited by pesticides (Caldwell, 1974; Jowett and Rhead, 1978). However, in the case of methoxychlor, the enzyme inhibition occurs in the crab *Cancer magister* without simultaneous impairment of the capability to osmo- and ionoregulate (Caldwell, 1974). Interpretation of the effects of methoxychlor on the crab *Cancer magister* is confusing. Administration of the pesticide at barely sublethal levels results in an inhibition of the gill ($Na^+ + K^+$) ATPase without concomitant effects on osmotic and ionic regulation. However, despite this latter lack of effect, the crab shows a decreased tolerance to low salinity exposures. This may be an expression of the synergic effects of two completely different stresses acting on completely different mechanisms (as stated in the introduction of this section, one stress is always better supported than two). The decrease in blood osmolarity occurring upon acclimation to dilute media may potentiate the well-known effect of pesticides on the nervous system. More data are needed to assess this hypothesis.

For heavy metals, that their toxicity results mainly from an action on osmoregulatory processes and on cell membrane characteristics is well documented. Passive movements and active transport of ions have been proven to be disturbed by heavy metal exposure in crustaceans, resulting in serious disruption of the osmolarity control of cells and body fluids. By analogy with experimental data collected on fishes (Renfro *et al.*, 1974; Bouquegneau, 1977; Bouquegneau and Gilles, 1979), it is likely that abolishment of active Na^+ transport ability must be related to direct interference with ($Na^+ + K^+$) ATPase activity. Metals are known to combine with enzymes in many ways, among which are sulfhydryl(SH) binding. In the particular case of ($Na^+ + K^+$) ATPases, Skou (1963) has emphasized the importance of SH groups in enzyme activity, and therefore the activity of enzymes can be expected to cease when heavy metals block reactive SH sites. Studies on such interactions in crustaceans tissues are needed. Besides the inhibition of enzymes related to osmoregulation, the inhibition of respiratory enzymes has also been caused by the lethal effects of metals, such as copper in crustaceans (Corner and Sparrow, 1956; Kerkut and Munday, 1962; Hubschmann, 1966, 1967). Except when they act at high concentrations on the surface tissues as protein precipitants, the basic action of heavy metals is thought to be enzyme inhibition, regardless of location and role. Effects on growth might also be related to inhibition of enzymes involved in protein synthesis. However, it is reasonable to think that the particular sensitivity of migrating and estuarine species could be ascribed to the impossibility of synthesizing the enzyme or protein components involved in ion transfer processes and metabolic activity.

Histological changes have also been reported to occur concomitantly in the tissues of crustaceans exposed to heavy metals. According to Bubel (1976), copper, mercury, and cadmium ions induce, at the level of the gills of the isopod, *Jaera nordmanni,* ultrastructural changes including distended microvilli, dilated endoplasmic reticulum, dissociated ribosomes, diffuse cytoplasm, swollen mitochondria, and a basal membrane withdrawn from the basal lamina. There is an increase in the number of hemocytes observed in the hemolymph spaces during heavy metal ion exposure.

Available experimental data do not reveal whether histological changes are the consequence of the alteration of membrane structures and permeability properties, or whether modifications in ion permeability must be accounted for by changes in histological structure.

c. Reaction and Adaptation toward Pollutant Toxicity. Until now, the toxicity and biological effects of various pollutants on crustaceans have been reviewed, and an attempt has been made to understand and explain their mechanisms of action. The question that arises now concerns the means used and developed to remove or render the pollutant harmless, hence enabling the animal to deal with transient contaminations or with slightly polluted areas. This question has been rather well documented in fishes (Bouquegneau and Gilles, 1979). Little information is available on this problem in crustaceans. Several ways of dealing with the toxicant can be considered.

The first and easiest way is by expelling the toxicant back across the body surface or the gills. Excretion of zinc has been shown to occur across the gills in the shore crab, *Carcinus maenas* (Bryan, 1967a).

A second way is by excreting the metal into the lumen of gut (e.g., in the freshwater crayfish, *Austropotamobius pallipes pallipes* for the removal of zinc (Bryan, 1967b) and in the cyprid larva of the barnacle, *Balanus amphitrite niveus,* for the removal of copper (Bernard and Lane, 1961).

The third method of toxicant removal is related to urine excretion. Much evidence has established that crustaceans are able to excrete copper, manganese, cobalt, zinc, and mercury in the urine (Corner, 1959; Bryan, 1964, 1968; Bryan and Ward, 1965).

A fourth method is by storage in a particular tissue. Several organs have been found to be able to accumulate toxicants and more particularly heavy metals. According to Bryan (1964, 1971), the hepatopancreas of the lobster can accumulate a large amount of zinc before gradually releasing it either in the urine or across the body surface. A more striking way of storage consists of accumulating the toxicants in the form of salt granules, which may be considered as the end product of detoxification process. In the shrimp, *Crangon crangon,* excess copper appears to be stored as granules in the cells of the hepatopancreas (Djangmah, 1970). In the barnacle, *Semibalanus balanoides,* an unusually high zinc level has also been localized in small intracellular granules found in parenchyma cells surrounding the midgut (Walker *et al.,* 1975a,b). A qualitative and quantitative analysis of granule-rich pellets has shown that they were composed mainly of zinc phosphate. In the same cells, a second type of granule has been identified, in which copper seems to be complexed with organic matter (Walker, 1977). These structures were demonstrated to be relatively insoluble and inert.

Storage of toxic substances might also be achieved by binding in a nontoxic form involving particular polypeptides called metallothioneins (Suda *et al.*, 1974). These compounds are low-molecular-weight proteins mainly characterized by their high cysteine content and by the virtual absence of aromatic amino acid residues (Weser *et al.*, 1973). It is now well known that they are implicated, at least in vertebrates, in a process of detoxification of mercury salts, as well as in the mechanism of adaptative tolerance. To our knowledge, there are no data on the occurrence of metallothioneins in crustaceans.

Another possible protective agent against toxicity of heavy metals, and more particularly mercury, is the presence of selenium. Reduction in mercury toxicity due to ingested and injected doses of selenium has been rather well documented for mammals, birds, fishes, and humans (Iwata *et al.*, 1973; Stoewsand *et al.*, 1974; Mackay *et al.*, 1975; Rossi *et al.*, 1976). In *Cancer magister* larvae, Glickstein (1978) has also reported that moderate selenium concentrations (10–1000 g.liter^{-1}) tended to decrease mercury toxicity.

d. Conclusions. Information reported in this section has emphasized the importance of various species of pollutants as stressing factors which might interact with salinity on crustacean ecology.

Attention must be drawn to the fact that, depending on the animal studied, on the development stage considered, on the nature of the pollutant, its concentration, the length of the intoxication period, and the synergistic effects of a wide variety of environmental factors, the toxicity and the mechanism of action of toxicants can be totally different.

The mode of action is in most cases far from being known. Examples have been reported of animals showing a decreased tolerance to salinity stress after exposure to toxicants, where the disruption of osmotic and ionic regulation did not seem to be the principal cause of mortality, contrary to what had been established for many other species.

Whatever the real cause of mortality, the synergistic effects and the interactions between pollutants and salinity seem to be detrimental to species normally living in or migrating through media of fluctuating salinity. It is also significant that larval stages are generally much more sensitive to pollutants than adults. The causes of this high sensitivity would merit special attention in view of species distribution.

Intoxication through alimentation has been little studied, and the question which arises is: How do the pollutants accumulated and stored in nontoxic forms (see Section III,A,2,c) behave after ingestion by predators? Until these problems can be solved, the question of the existence of thresholds beneath which a pollutant can be considered nontoxic will remain controversial.

B. Species Relations

Salinity probably affects species interactions in many ways. To date, these problems have not been well investigated. Species–species salinity interactions can be concerned with: (1) direct species–species relations, such as predator–prey or parasite–host relations; and (2) indirect species–species relations, such as utilization of dissolved organic substances in chemical communication or in trophic chains.

1. DIRECT SPECIES–SPECIES RELATIONS

A striking example of direct species–species salinity interaction can be found in sex determination in the euryhaline amphipod, *Gammarus duebeni*. In this species, sex ratio tends to shift in favor of males at high salinity (30o). As shown by Bulnheim (1969), this effect is not direct, but rather is mediated through a salinity-dependent parasite–host relation. Females of *G. duebeni* infested by the microsporidian, *Octosporea effeminans,* will give birth to females almost exclusively, while eggs free from microsporidians will differentiate into females or males according to genetic pattern and photoperiodic conditions. In 30‰ salinity, females will lay eggs free from microsporidians, the percentage of infested eggs varying with the salinity of the environment. Consequently, the sex ratio is shifted in favor of females in low salinities, and more males appear at high salinities.

The main direct species–species interaction in media of fluctuating salinities is probably effected through trophic problems. Such problems can occur as a result of modifications in the nature of the food supply, of competition between species for the same kind of food, or as a result of salinity-induced changes in feeding activity. Various microfaunal and meiofaunal communities are very successful in reproducing and growing in estuarine conditions. They have short generation times, and their production/biomass ratio tends to be considerably higher than for macrofauna (Gerlach, 1971). Moreover, they can be found in extremely dense populations in estuarine habitats; the euryhaline meiofauna, essentially consisting of nematodes and harpacticoid copepods, can contribute up to 58% of total respiration recorded in some areas in the semienclosed lagoon of Arcachon (France) (Lasserre *et al.*, 1976). In this context, a significant part of benthic primary production which is converted to meiofaunal mass might not be passed upward to higher trophic levels. Meiofaunal populations are particularly active in shallow estuarine waters during summer, and summer growth retardation observed in various species of benthophageons macrofauna colonizing these media could therefore be related, at least partly, to a trophic competition between these two types of populations. More data are needed to bring some light on this interesting problem.

2. INDIRECT SPECIES–SPECIES RELATIONS

All aquatic media contain dissolved organic compounds (DOC) from metabolism or degradation of biological material. These substances (ammonia, amino acids, amines, peptides, vitamins, carbohydrates, fatty acids, and more complex biopolymers) could affect several biological processes, such as nutrition (setting-off or on of alimentary behavior), and consequently growth and reproduction. Such effects undoubtedly play a part in the integration of species in their biotopes. Up to now, most studies on the importance of DOC in ecology have been concerned only with the quantitative significance of these compounds as food for marine organisms. Many reports on the uptake of glucose and selected amino acids and fatty acids are available; it seems that uptake activity can result in many cases in a net influx of the studied compounds. Since the early work of Stephens (1962) on the coral, *Fungia scutaria,* uptake of DOC has been shown to occur in many marine invertebrates, such as polychaetes; Stephens (1964), Reish and Stephens (1969), Testerman (1972), Ahearn and Gomme (1975); Pogonophora: Southward and Southward (1972); Mollusca: Bamford and Gingles (1974), Bamford and Campbell (1976); Echinodermata: Epel (1972), Ahearn and Townsley (1975), Stephens and Virkar (1966), Ferguson (1971), Yingst (1976), Stephens *et al.* 1978); Coelenterates: Shick (1975). In several of these species, and particularly in infaunal species living in rich sediments, it can be estimated that absorption of DOC is ecologically meaningful since it can cover a significant part of the energy requirements (Jørgensen, 1976). As far as Crustacea are concerned, Stephens and Schinske (1961) were unable to find any uptake of glycine in *Semibalanus balanoides, Platorchestia platensis, Limulus polyphemus, Palaemonetes vulgaris, Pagurus longicarpus,* and *Uca pugnax*. More recently, Anderson and Stephens (1969) reported that small species such as *Tigriopus californicus, Artemia salina, Limnoria tripunctata,* or *Corophium acherusicum* can take up glucose, acetate, or amino acids. Treatment of the animals with antibiotics reduced the uptake to nonsignificant levels, thus indicating that the uptake measured was caused by microorganisms and not by the animals themselves. Similar results have been obtained in this laboratory when studying incorporation into proteins and oxidation to CO_2 of labeled alanine by the amphipod, *Gammarus insensibilis* (D. Hansoul and R. Gilles, unpublished). It seems that utilization of DOC by adult crustaceans, if it exists, is extremely reduced. It should be pointed out that the situation could be quite different in extremely small species, such as meiobenthic or planktonic copepods, or in eggs and larvae. To our knowledge, the problem of DOC uptake in copepods has never been studied. As far as eggs and larvae are concerned, Pavillon (1978) reported the uptake of eight amino acids, four vitamins, and one fatty acid by eggs at an advanced stage of development and by larvae of *Euphausia krohnii* until the calyptosis I stage. Such an absorption could be of importance for the animal at the larval stage, before the whole digestive system is fully developed. However, no net flux studies or experiments with antibiotics have been achieved, and more data should be obtained to assess the role of DOC absorption in the larval stages of this species.

Interactions between salinity and uptake and incorporation of [^{14}C]leucine have been reported in *Artemia salina* nauplii by Conte *et al.* (1973). At an external concentration of $5 \times 10^{-7} M$, increasing salinity from 0.05 *M* NaCl to 2.5 *M* NaCl decreases the rate of appearance of free-labeled leucine in the internal fluids of the nauplii; similarly, the rate of incorporation of the amino acid into naupliar proteins is decreased. Effects of inorganic ions on amino acid metabolism and transport are well known in crustacean tissues (see Section II,A,3,a,ii; Gilles, 1975, 1978), and the effects of salinity on leucine utilization in *Artemia* nauplii are probably related to such interactions. There are not enough data on the effect of salinity on DOC utilization for a useful discussion of these problems.

It appears that in adult macrobenthic Crustacea, utilization of DOC has no ecological meaning; considerations of interactions between DOC uptake, salinity, and osmoregulation consequently appear meaningless. The situation might be different in planktonic and meiobenthic adult forms or in planktonic larval stages. In these cases, such interactions might reveal important parameters in the establishment of communities. The available data are far too scanty for conclusions to be drawn.

Another interesting aspect of DOC in ecology lies in the role these compounds can play as chemical signals. Such a role has been beautifully exemplified in insects (for reviews, see Butler, 1970; Harborne, 1977). To our knowledge, ecological aspects of chemical communications have not really started in crustaceans. It is our belief that such studies should yield very rewarding results in the future. In this respect, researches dealing with the incidence of interactions between salinity and the responsiveness to ecomones in the establishment and maintenance of communities in media of fluctuating salinity would be of particular interest.

IV. CONCLUSIONS

An attempt has been made in this chapter to review the interactions of chemical and osmotic regulation with the environment, showing clear relevance to ecology. Such organismic or molecular interactions are of the utmost importance in the understanding of the "ecological fitness" of living organisms. The relevance of molecular mechanisms to ecology is too often taken for granted and therefore not studied. When considering the few environmental factors we have been dealing with, it is clear that there is still an enormous gap between the ecological approach and the physiological and biochemical approach. Although environmental physiology and biochemistry are becoming well developed fields of research, their relation with ecology and establishment of biocenoses unfortunately remains largely unexplored.

When considering crustaceans, in particular, the study of salt tolerance and osmoregulation actually constitutes an almost unique example showing that research down to the molecular level can help in the understanding of the underlying causes of species ecological integration.

Careful analysis of osmoregulation processes indeed demonstrates that the capability for volume control has interesting ecological correlates, at least in species inhabiting permanently or transitorily inhabiting media of fluctuating salinity. As far as the mechanisms of osmoregulation of the intracellular fluid are concerned, differences seems to exist between stenohaline, good euryhaline, and poor euryhaline species in (1) the response of blood protein constituents to osmolarity changes; (2) the effectiveness of the cell volume readjustment response following swelling due to decrease in environmental osmolarity; such differences might be related to differences in the control of transport processes at the plasma membrane level and/or in the control of the metabolic processes at work in the regulation of the amount of intracellular osmotic effectors. The relationships suggested, however, are only tentative, and additional data would be needed before a more complete picture could be produced. Moreover, it is clear that species distribution, even in media of fluctuating salinity, cannot be solely related to physiological osmoregulatory processes. As we have seen, other factors must also be taken into consideration, e.g., temperature and pollutants. As far as these two parameters are concerned, only a few studies have been brought to the molecular level. It is clear that pollutants, either acting synergistically with other factors or interacting more directly with osmoregulatory processes, are important limiting factors in the conquest of new biotopes, as well as in the maintenance of populations or survival of individuals in contaminated areas. Further, it appears that pollutants can act on many molecular processes, and that various species have developed different physiological mechanisms to adapt to their toxicity. This renders the experimental approach more complex, but strengthens the need for more information on the adaptative processes at work.

Most temperature studies have dealt with survival and growth. Information as to the effect on ecology of the now well know molecular processes compensating for changes in temperature are totally lacking. Any attempt to explain species distribution in relation with temperature or with the interplay between temperature and other environmental factors on the basis of physiological regulatory processes remains highly speculative.

The molecular bases of crustacean ecology are far from being clearly defined. Further, they cannot be related only to the effect of environmental factors on adult organisms. Data about such effects at all stages of the life cycle need to be integrated before a clear picture can be produced. On the other hand, direct effects of an environmental factor are often concomitant in the field with indirect effects. The many aspects of environmental parameters, either biotic or abiotic, as ecological entities make it difficult to assess the specific causes of an organismic or a community response. Although these causes are difficult to determine, their study is urgently needed for us to develop a clear understanding of what governs animal distribution.

While many marine ecologists have sought to understand crustacean distribution in terms of the environment as a whole, we have tried in this review to demonstrate that the individual and population responses to the environment are determined primarily by physiological adaptative peculiarities. These adaptative features are almost impossible to isolate and study in the framework of field research and should be more easily determined in laboratory studies. It is our belief that the fundamental studies undertaken in this respect should prove very rewarding in the future and should also find a large field of application in ecotone managment.

REFERENCES

Ahearn, G. A., and Gomme, J. (1975). Transport of exogenous D-glucose by the integument of a polychaete worm (*Nereis diversicolor* Müller). *J. Exp. Biol.* **62,** 243–264.

Ahearn, G. A., and Townsley, S. J. (1975). Integumentary amino acid transport and metabolism in the apodons sea cucumber *Chiridota rigida*. *J. Exp. Biol.* **62,** 733–752.

Alderdice, D. F. (1972). Factor combinations. Responses of marine poikilotherms to environmental factors acting in concert. *In* "Marine Ecology" (O. Kinne, ed.), Vol. 1, Part 3, pp. 1659–1722. Wiley-Interscience, New York.

Anderson, J. W., and Stephens, G. C. (1969). Uptake of organic material by aquatic invertebrates. VI. Role of epiflora in apparent uptake of glycine by marine crustaceans. *Mar. Biol. (Berlin)* **4,** 243–249.

Anderson, J. W., Neff, J. M., Cox, B. A., Tatem, H. E., and Hightower, G. N. (1974). Characteristics of dispersions and water-soluble extracts of crude and refined oils and their toxicity to estuarine crustaceans and fish. *Mar. Biol. (Berlin)* **27,** 75–88.

Anderson, R. (1978). The effects of lead on oxygen uptake in the crayfish, *Orconectes virilis* (Hagen). *Bull. Environ. Contam. Toxicol.* **20,** 394–400.

Armstrong, D. A., Buchanan, D. V., Mallon, M. H., Caldwell, R. S., and Millemann, R. E. (1976). Toxicity of the insecticide Methoxychlor to the Dungeness crab *Cancer magister*. *Mar. Biol. (Berlin)* **38,** 239–252.

Bamford, D. R., and Campbell, E. (1976). The effect of environmental factors on the absorption of L-phenylalanine by the gill of *Mytilus edulis*. *Comp. Biochem. Physiol. A* **53A,** 295–299.

Bamford, D. R., and Gingles, R. (1974). Absorption of sugars in the gill of the Japanese oyster *Crassostrea gigas*. *Comp. Biochem. Physiol. A* **49A,** 637–646.

Barnes, H. (1953). The effect of lowered salinity on some barnacle nauplii *J. Anim. Ecol.* **22,** 328–330.

Barnes, H., and Klepal, W. (1974). The general biology of *Verruca stroemia* (O. F. Muller). IV. Effect of salinity and temperature on survival, behaviour and osmotic relations. *J. Exp. Mar. Biol. Ecol.* **14,** 37–46.

Benson, J. A., and Treherne, J. E. (1978). Axonal adaptation to osmotic and ionic stress in an invertebrate osmoconformer (*Mercierella enigmatica* Fauvel). II. Effects of ionic dilution on the resting and action potentials. *J. Exp. Biol.* **76,** 205–219.

Berdugo, V., Harris, R. P., and O'Hara, S. C. (1977). The effect of petroleum hydrocarbons on reproduction of an estuarine planktonic copepod in laboratory culture. *Mar. Pollut. Bull.* **8,** 138–143.

Bernard, F. J., and Lane, C. E. (1961). Absorption and excretion of copper ions during settlement and metamorphosis of the barnacle *Balanus amphitrite niveus*. *Biol. Bull.* (*Woods Hole, Mass.*) **121,** 438–448.

Biggs, D. C., and McDermott, J. J. (1973). Variation in temperature-salinity tolerance between two estuarine populations of *Pagurus longicarpus* Say (Crustacea-Anomura). *Biol. Bull.* (*Woods Hole, Mass.*) **145,** 91–103.

Binns, R. (1969). The physiology of the antennal gland of *Carcinus maenas* (L). IV. The reabsorption of amino acids. *J. Exp. Biol.* **51,** 29–39.

Blumer, M., Hunt, J. M., Atema, J., and Stein, L. (1973). Interaction between marine organisms and oil pollution. *Publ. U.S. Environ. Prot. Agency, Ecol., Res. Ser.* **EPA-R3-73-042,** 1–97.

Bollette-Dugaillay, S., and Schoffeniels, E. (1970). Isoenzymes de la deshydrogenase lactique et stress osmotique. *Arch. Int. Physiol. Biochim.* **78,** 23–27.

Bouquegneau, J. M. (1977). ATPase activity in mercury intoxicated eels. *Experientia* **33,** 941–942.

Bouquegneau, J. M., and Gilles, R. (1979). Osmoregulation and pollution of the aquatic medium. *In* "Mechanisms of Osmoregulation in Animals" (R. Gilles, ed.), pp. 563–580. Wiley-Interscience, New York.

Bouquegneau, J. M., Gerday, C., and Distèche, A. (1975). Fish mercury-binding thionein related to adaptation mechanism. *FEBS Lett.* **55,** 173–177.

Bovie, E. C. (1951). Some effects of temperature on the rates of embryonic, pre-embryonic and adult growth in *Hyalella azteca* Sassure. *Proc. Iowa Acad. Sci.* **59,** 439–444.

Boyle, P. J., and Conway, E. J. (1941). Potassium accumulation in muscle and associated changes. *J. Physiol.* (*London*) **100,** 1–63.

Bradley, B. P. (1975). The anomalous influence of salinity on temperature tolerances of summer and winter populations of the copepod *Eurytemora affinis*. *Biol. Bull.* (*Woods Hole, Mass.*) **148,** 26–34.

Broekhuysen, G. J., Jr. (1936). On development, growth and distribution of *Carcinides maenas* (L.). *Arch. Neerl. Zool.* **2,** 257–399.

Bryan, G. W. (1964). Zinc regulation in the lobster *Homarus vulgaris*. I. Tissue zinc and copper concentrations. *J. Mar. Biol. Assoc. U.K.* **44,** 549–563.

Bryan, G. W. (1967a). The matabolism of zinc and ^{65}Zn in crabs, lobsters and freshwater crayfish. *Radioecol. Conc. Processes, Proc. Int. Symp., 1966* pp. 1005–1016.

Bryan, G. W. (1967b). Zinc regulation in the freshwater crayfish (including some comparative copper analyses). *J. Exp. Biol.* **46,** 281–296.

Bryan, G. W. (1968). Concentrations of zinc and copper in the tissues of decapod crustaceans. *J. Mar. Biol. Assoc. U.K.* **48,** 303–321.

Bryan, G. W. (1971). The effects of heavy metals (other than mercury) on marine and estuarine organisms. *Proc. R. Soc. London, Ser. B* **117** 389–410.

Bryan, G. W., and Ward, E. (1965). The absorption and loss of radioactive and non-radioactive manganese by the lobster *Homarus vulgaris*. *J. Mar. Biol. Assoc. U.K.* **45,** 65–95.

Bubel, A. (1976). Histological and electron microscopical observations on the effects of different salinities and heavy metal ions, on the gills of *Jaera nordmanni* (Rathke) (Crustacea, Isopoda). *Cell Tissue Res.* **167,** 65–95.

Buchanan, D. V., and Stewart, N. E. (1970). Effects of the insecticide Sevin on various stages of the Dungeness crab *Cancer magister*. *J. Fish. Res. Board Can.* **27,** 93–104.

Bulnheim, H. P. (1969). Zur analyse geschlechtsbestimmen der faktoren bei *Gammarus duebeni* (Crustacea: Amphipoda). *Zool. Anz.* **32,** 244–260.

Bursey, C. R., and Bonner, E. E. (1977). Osmotic regulation and salinity tolerance of the mole crab, *Emerita talpoida* (Say) (Crustacea, Anomura). *Comp. Biochem. Physiol. A* **57A,** 207–210.

Butler, C. G. (1970). Chemical communication in insects: Behavioral and ecologic aspects. *In* "Communication by Chemical Signals" (J. W. Johnston, Jr., D. G. Moulton and A. Turk, eds.), pp. 35–78. Appleton, New York.

Caldwell, R. S. (1974). Osmotic and ionic regulation in decapod crustacea exposed to methoxychlor. *In* "Pollution and Physiology of Marine Organisms" (F. J. Vernberg and W. B. Vernberg, eds), pp. 197–223. Academic Press, New York.

Caldwell, R. S., Armstrong, D. A., Buchanan, D. V., Mallon, M. H., and Millemann, R. E. (1978). Toxicity of the fungicide captan to the Dungenes crab *Cancer magister*. *Mar. Biol. (Berlin)* **48,** 11–17.

Carlson, A. D., Pichon, Y., and Treherne, J. E. (1978). Effects of osmotic stress on the electrical activities of the giant axon of a marine osmoconformer, *Sabella penicillus*. *J. Exp. Biol.* **75,** 237–251.

Chaplin, A. E., Huggins, A. K., and Munday, K. A. (1965). Ionic effects on glutamate dehydrogenase activity from beef liver, lobster muscle and crab muscle. *Comp. Biochem. Physiol.* **16,** 49–62.

Chaplin, A. E., Huggins, A. K., and Munday, K. A. (1967). The distribution of L- aminotransferases in *Carcinus maenas*. *Comp. Biochem. Physiol.* **20,** 195–198.

Chaplin, A. E., Huggins, A. K., and Munday, K. A. (1970). The effect of salinity on the metabolism of nitrogen-containing compounds by *Carcinus maenas*. *Int. J. Biochem.* **1,** 385–400.

Christiansen, M. E., and Costlow, J. D., Jr. (1975). The effect of salinity and cyclic temperature on larval development of the mud-crab *Rhithropanopeus harrisii* (Brachyura: Xanthidae) reared in the laboratory. *Mar. Biol. (Berlin)* **32,** 215–222.

Clark, M. E. (1968). A survey of the effect of osmotic dilution on free amino acids of various polychaetes. *Biol. Bull. (Woods Hole, Mass.)* **134,** 252–260.

Clemens, H. P. (1950). Life cycle and ecology of *Gammarus fasciatus* Say. *Contrib. Stone Lab. Ohio Univ.* **12,** 1–63.

Coffin, H. G. (1958). The laboratory culture of *Pagurus samuelis* (Stimpson) (Crustacea, Decapoda). *Walla Walla Collect. Publ.* **22,** 1–5.

Connor, P. M. (1972). Acute toxicity of heavy metals to some marine larvae (1972). *Mar. Pollut. Bull.* **3,** 190–192.

Conte, F. P., Pettersen, G. L., and Ewing, R. D. (1973). Larval Salt gland of *Artemia salina* nauplii. *J. Comp. Physiol.* **82,** 277–289.

Corner, E. D. S. (1959). The poisoning of *Maia squinado* (Herbst) by certain compounds of mercury. *Biochem. Pharmacol.* **2,** 121–132.

Corner, E. D. S., and Sparrow, B. W. (1956). The modes of action of toxic agents. I. Observations of the poisoning of certain crustaceans by copper and mercury. *J. Mar. Biol. Assoc. U.K.* **35,** 531–548.

Corner, E. D. S, Southard, A. J., and Southard, E. C. (1968). Toxicity of oil spill removers (detergents) to marine life: An assessment using the intertidal barnacle *Elminius modestus* *J. Mar. Biol. Assoc. U.K.* **48,** 29–47.

Costlow, J. D., Jr. (1967). The effect of salinity and temperature on survival and metamorphosis of megalops of the blue crab *Callinectes sapidus. Helgo. Wiss. Merresunters.* **15,** 84–97.

Costlow, J. D., Jr., and Sastry, A. N. (1966). Free amino acids in developing stages of two crabs. *Callinectes sapidus* (Rathbun) and *Rhithropanopeus harrisii* (Gould). *Acta Embryol. Morphol. Exp.* **9,** 44–45.

Costlow, J. D., Jr., Bookhout, C. G., and Monroe, R. (1960). The effect of salinity and temperature on larval development of *Sesarma cinereum* (Bosc) reared in the laboratory. *Biol. Bull. (Woods Hole, Mass.)* **118,** 183–202.

Costlow, J. D. Jr., Bookhout, C. G., and Monroe, R. (1962). Salinity-temperature effects on the larval development of the crab, *Panopeus herbstii* Milne-Edwards, reared in the laboratory. *Physiol. Zool.* **35,** 79–93.

Costlow, J. D., Jr., Bookhout, C. G., and Monroe, R. (1966). Studies on the larval development of the crab *Rhithropanopeus harrisii* (Gould) I. The effect of salinity and temperature on larval development. *Physiol. Zool.* **39,** 81–100.

Crisp, D. J., and Costlow, J. D. Jr. (1963). The tolerance of developing cirripide embryos to salinity and temperature. *Oikos* **14,** 22–34.

DeCoursey, P. J., and Vernberg, W. B. (1972). Effect of mercury on survival, metabolism and behavior of larval *Uca pugilator. Oikos* **23,** 241–247.

Djangmah, J. S. (1970). The effects of feeding and starvation on copper in the blood and hepatopancreas and on the blood proteins of *Crangon vulgaris* (Fabricius). *Comp. Biochem. Physiol.* **32,** 709–731.

Doljanski, F., Ben-Sasson, S., Reich, M., and Grover, N. B. (1974). Dynamic osmotic behavior of chick blood lymphocytes. *J. Cell. Physiol.* **84,** 215–224.

Donahue, W. H., Wang, R. T., Welsch, M., and Nicol, J. A. C. (1977). Effects of water-soluble components of petroleum oils and aromatic hydrocarbons on barnacle larvae. *Environ. Pollut.* **13,** 187–202.

Dorgelo, J. (1974). Comparative ecophysiology of gammarids (Crustacea: Amphipoda) from marine, brackish and fresh-water habitats, exposed to the influence of salinity-temperature combinations. I. Effect on survival. *Hydrobiol. Bull.* **8,** 90–108.

Dorgelo, J. (1976). Salt tolerance in crustacea and the influence of temperature upon it. *Biol. Rev. Cambridge Philos. Soc.* **51,** 255–290.

Duchâteau, G., and Florkin, M. (1955). Concentration du milieu extérieur et état stationnaire du pool des acides aminés non protéiques des muscles d'*Eriocheir sinensis,* Milne Edwards. *Arch. Int. Physiol. Biochim.* **63,** 249–251.

Duchâteau, G., Florkin, M., and Jeuniaux, C. (1959). Composante amino-acide des tissus chez les crustacés. I. Composante amino-acide des muscles de *Carcinus maenas* L. lors du passage de l'eau de mer à l'eau saumâtre et au cours de la mue. *Arch. Int. Physiol. Biochim.* **67,** 489–500.

Duchâteau-Bosson, G., and Florkin, M. (1961). Change in intracellular concentration of free amino acids as a factor of euryhalinity in the crayfish *Astacus astacus. Comp. Biochem. Physiol.* **3,** 245–249.

Duchâteau-Bosson, G., Jeuniaux, C., and Florkin, M. (1961). Rôle de la variation de la composante amino-acide intracellulaire dans l'euryhalinité d'*Arenicola marina* L. *Arch. Int. Physiol. Biochim.* **69,** 30–35.

Duval, M. (1925). Recherches physico-chimique et physiologiques sur le milieu intérieur des animaux aquatiques. Modifications sous l'influence du milieu extérieur. *Ann. Inst. Oceanogr. (Paris)* **2,** 232–407.

Edwards, R. R. C. (1978). Effects of water-soluble oil fractions on metabolism, growth and carbon budget of the shrimp *Crangon crangon. Mar. Biol. (Berlin)* **46,** 259–265.

Eisler, R., and Edmunds, P. (1966). Effects of endrin on blood and tissue chemistry of a marine fish. *Trans. Am. Fish. Soc.* **95,** 153–159.

Emerson, D. N. (1969). Influence of salinity on ammonia excretion rates and tissue constituents of euryhaline invertebrates. *Comp. Biochem. Physiol.* **29,** 1115–1133.

Epel, D., (1972). Activation of a Na^+-dependent amino acid transport system upon fertilization of sea urchin eggs. *Exp. Cell. Res.* **72,** 74–89.

Felder, D. L. (1978). Osmotic and ionic regulation in several western atlantic *Callianassidae* (Crustacea, Decapoda, Thalassinidae). *Biol. Bull. (Woods Hole, Mass.)* **154,** 409–429.

Ferguson, J. C. (1971). Uptake and release of free amino acids by starfishes. *Biol. Bull. (Woods Hole, Mass.)* **141,** 122–129.

Florkin, M. (1962). La régulation isosmotique intracellulaire chez les invertébrés marins euryhalins. *Bull. Cl. Sci., Acad. R. Belg.* **68,** 687–693.

Florkin, M., and Schoffeniels, E. (1969). "Molecular Approaches to Ecology." Academic Press, New York.

Flügel, H. (1963). Elektrolytregulation und temperatur bei *Crangon crangon* L. und *Carcinus maenas* L. *Kiel. Meeresforsch.* **19,** 189–195.

Forsman, B. (1951). Studies on *Gammarus duebeni* Lillj. with notes on some rockpool organisms in Sweden. *Zool. Bidr. Uppsala* **29,** 215–237.

Foskett, J. K. (1977). Osmoregulation in the larvae and adults of the grapsid crab *Sesarma reticulatum* Say. *Biol. Bull. (Woods Hole, Mass.)* **153,** 505–526.

Fugelli, K., and Reiersen, L. O. (1978). Volume regulation in flounder erythrocytes. The effect of osmolarity on taurine influx. *Alfred Benzon Symp.* **11,** 418–437.

Gainer, H., and Grundfest, H. (1968). Permeability of alkali metal cations in lobster muscle. A comparison of electrophysiological and osmometric analyses. *J. Gen. Physiol.* **51,** 399–425.

George, J. L., Darsie, R. F., Jr., and Springer, P. F. (1957). Effects on wildlife of aerial applications of Strobane, DDT and BHC to tidal marshes in Delaware. *J. Wildl. Manage.* **21,** 42–53.

George, M. (1961). Oil pollution of marine organisms. *Nature (London)* **192,** 1209.

Gérard, J. F. (1975). Volume regulation and alanine transport. Response of isolated axons of *Callinectes sapidus* Rathbun to hypo-osmotic conditions. *Comp. Biochem. Physiol. A* **51A,** 225–229.

Gérard, J. F., and Gilles, R. (1972a). Modification of the amino-acid efflux during the osmotic adjustment of isolated axons of *Callinectes sapidus. Experientia* **28,** 863–864.

Gérard, J. F., and Gilles, R. (1972b). The free amino-acid pool in *Callinectes sapidus* (Rathbun) tissues and its role in the osmotic intracellular regulation. *J. Exp. Mar. Biol. Ecol.* **10,** 125–136.

Gérard, J. F., Dandrifosse, G., and Gilles, R. (1974). Changes in the extracellular space of *Callinectes sapidus* axons during osmotic shocks. *Comp. Biochem. Physiol. A* **47A,** 315–321.

Gerlach, S. A. (1971). On the importance of marine meiofauna for benthos communities. *Oceanology (Engl. Transl.)* **6,** 176–190.

Gilles, R. (1969). Effect of various salts on the activity of enzymes implicated in amino acid metabolism. *Arch. Int. Physiol. Biochim.* **77,** 441–464.

Gilles, R. (1970). Osmoregulation in the stenohaline crab "Libinia emarginata" Leach. *Arch. Int. Physiol. Biochim.* **78,** 91–99.

Gilles, R. (1972). Amino-acid metabolism and isosmotic intracellular regulation in isolated surviving axons of *Callinectes sapidus. Life Sci.* **11,** Part II, 565–572.

Gilles, R. (1973). Osmotic behaviour of isolated axons of a euryhaline and a stenohaline crustacean. *Experientia* **29,** 1354–1355.

Gilles, R. (1974a). Métabolisme des acides aminés et contrôle du volume cellulaire. *Arch. Int. Physiol. Biochim.* **82,** 423–589.

Gilles, R. (1974b). Glucose metabolism during osmotic stress in isolated axons of *Eriocheir sinensis. Life Sci.* **15,** 1363–1369.

Gilles, R. (1974c). Studies on the effect of NaCl on the activity of *Eriocheir sinensis* glutamate dehydrogenase. *Int. J. Biochem.* **5,** 623–628.

Gilles, R. (1975a). Mechanisms of ion and osmoregulation. *In* "Marine Ecology" (O. Kinne, ed.) Vol. 2, Part 1, pp. 259–347. Wiley-Interscience, New York.

Gilles, R. (1975b). Mechanisms of thermoregulation. *In* "Marine Ecology" (O. Kinne, ed.), Vol. 2, Part 1, pp. 251–258. Wiley-Interscience, New York.

Gilles, R. (1977). Effects of osmotic stresses on the proteins concentration and pattern of *Eriocheir sinensis* blood. *Comp. Biochem. Physiol. A* **56A,** 109–114.

Gilles, R. (1978). Intracellular free amino acids and cell volume regulation during osmotic stresses. *Alfred Benzon Symp.* **11,** 470–491.

Gilles, R. (1979). Intracellular organic osmotic effectors. *In* "Mechanisms of Osmoregulation in Animals" (R. Gilles, ed.), Chapter 4, pp. 111–153. Wiley-Interscience, New York.

Gilles, R., and Gérard, J. F. (1974). Amino-acid metabolism during osmotic stress in isolated axons of *Callinectes sapidus. Life Sci.* **14,** 1221–1229.

Gilles, R., and Jöbsis, F. F. (1972). Isosmotic intracellular regulation and redox changes in the respiratory chain components of *Callinectes sapidus* isolated muscle fibers. *Life Sci.* **11,** Part II, 877–886.

Gilles, R., Hogue, P., and Kearney, E. B. (1971). Effect of various ions on the succinic dehydrogenase activity of *Mytilus californianus. Life Sci.* **10,** Part II, 1421–1427.

Glickstein, N. (1978). Acute toxicity of mercury and selenium to *Crassostrea gigas* embryos and *Cancer magister* larvae. *Mar. Biol. (Berlin)* **49,** 113–117.

Gustafson, C. G. (1970). PCB's-prevalent and persistent. *Environ. Sci. Technol.* **4,** 814–819.

Haefner, P. A. (1969). Temperature and salinity tolerance of the sand shrimp *Crangon septemspinosa* Say. *Physiol. Zool.* **42,** 388–397.

Harborne, J. B. (1977). "Introduction to Ecological Biochemistry." Academic Press, New York.

Harvey, C. E., Jones, M. B., and Naylor, E. (1973). Some factors affecting the distribution of estuarine isopods (Crustacea). *Estuarine Coastal Mar. Sci.* **1,** 113–124.

Heitz, J. R., Lewis, L., Chambers, J., and Yarbrough, J. D. (1974). The acute effects of empire mix crude oil on enzymes in oysters shrimp and mullet. *In* "Pollution and Physiology of Marine Organisms" (F. J. Vernberg and W. B. Vernberg, eds.), pp. 311–328. Academic Press, New York.

Hicks, G. R. F. (1973). Combined effects of temperature and salinity on *Hemigrapsus edwardsi* (Hilgendorf) and *H. crenulatus* (Milne Edwards) from Wellington Harbour, New Zealand. *J. Exp. Mar. Biol. Ecol.* **13,** 1–14.

Hochachka, P. W., and Somero, G. N. (1973). "Strategies of Biochemical Adaptation." Saunders, Philadelphia, Pennsylvania.

Hoffman, E. K. (1977). Control of cell volume. *In* "Transport of Ions and Water in Animals" (B. L. Gupta, R. B. Moreton, L. Oschman, and B. J. Wall, eds.), pp. 285–332. Academic Press, New York).

Hoffmann, E. K., and Hendil, K. B. (1976). The role of amino acids and taurine in isosmotic intracellular regulation in Ehrlich ascites tumour cells. *J. Comp. Physiol.* **108,** 279–286.

Hubschmann, J. H. (1966). Effects of copper on the crayfish *Orconectes rusticus* (Girard). I. Acute toxicity. *Crustaceana* **12,** 33–42.

Hubschmann, J. H. (1967). Effects of copper on the crayfish *Orconectes rusticus* (Girard). II. Mode of toxic action. *Crustaceana* **12,** 141–151.

Huggins, A. K., and Munday, K. A. (1968). Crustacean metabolism. *Adv. Comp. Biochem. Physiol.* **3,** 271–378.

Hughes, D. A., and Richard, J. D. (1973). Some current-directed movements of *Macrobrachium acanthurus* (Wiegmann, 1836) (Decapoda, Palaemonidae) under laboratory conditions. *Ecology* **54,** 927–929.

Hynes, H. B. N. (1954). The ecology of *Gammarus duebeni* Lilljeborg and its occurrence in fresh water in western Britain. *J. Anim. Ecol.* **23,** 38–84.

Inman, C. B. E., and Lockwood, A. P. M. (1977). Some effects of methyl-mercury and lindane and sodium regulation in the amphipod *Gammarus duebeni* during changes in the salinity of its medium. *Comp. Biochem. Physiol.* **58,** 67–75.

Iwata, H., Okamota, H., and Ohsawa, Y. (1973). Effect of selenium on methylmercury poisoning. *Res. Commun. Chem. Pathol. Pharmacol.* **5,** 673–680.

Jansen, K. P. (1971). Ecological studies on intertidal New Zealand *Sphaeromatidae* (Isopoda: Flabellifera). *Mar. Biol. (Berlin)* **11,** 262–285.

Jeffries, H. P. (1966). International condition of a diminishing blue crab population (*Callinectes sapidus*). *Chesapeake Sci.* **7,** 164–170.

Jeuniaux, C., and Florkin, M. (1961). Modification de l'excrétion azotée du crabe chinois au cours de l'adaptation osmotique. *Arch. Int. Physiol. Biochim.* **69,** 385–386.

Jones, M. B. (1972). Effects of salinity on the survival of the *Jaera albifrons* Leach group of species (Crustacea: Isopoda). *J. Exp. Mar. Biol. Ecol.* **9,** 231–238.

Jones, M. B. (1975). Synergistic effects of salinity, temperature and heavy metals on mortality and osmoregulation on marine and estuarine isopods (Crustacea). *Mar. Biol. (Berlin)* **30,** 13–20.

Jonsson, O. (1971). Effects of variations in the extracellular osmolality on the permeability to nonelectrolytes of vascular smooth muscle. *Acta Physiol. Scand.* **81,** 528–539.

Jørgensen, C. B. (1976). August Pütter, August Krogh and modern ideas on the use of dissolved organic matter in aquatic environments. *Biol. Rev. Cambridge Philos. Soc.* **51,** 291–328.

Jowett, P. E., and Rhead, M. M. (1978). *In vitro* changes in the activity of ATPases in the gills of *Carcinus maenas* exposed to various concentrations of P, P'-DDT. *Environ. Pollut.* **17,** 1–6.

Kalber, F. A., Jr. (1970). Osmoregulation in decapod larvae as a consideration in culture techniques. *Helgol. Wiss. Meeresunters.* **20,** 697–706.

Kalber, F. A., Jr., and Costlow, J. D., Jr. (1966). The ontogeny of osmoregulation and its neurosecretory control in the decapod crustacean *Rhithropanopeus harrisii* (Gould). *Am. Zool.* **6,** 221–229.

Kalber, F. A., Jr., and Costlow, J. D., Jr. (1968). Osmoregulation in larvae of the land crab, *Cardisoma guanhumi* Latreille. *Am. Zool.* **8,** 411–416.

Kerkut, G. A., and Munday, K. A. (1962) The effect of copper on the tissue respiration of the crab *Carcinus maenas*. *Cah. Biol. Mar.* **3,** 27–35.

Kévers, C., Péqueux, A., and Gilles, R. (1979a). Effects of hypo- and hyperosmotic shocks on the volume and ions content of *Carcinus maenas* isolated axons. *Comp. Biochem. Physiol.* **64,** 427–431.

Kévers, C., Péqueux, A., and Gilles, R. (1979b). Effects of an hypo-osmotic shock on Na^+, K^+ and Cl^- levels in isolated axons of *Carcinus maenas*. *J. Comp. Physiol.* **129,** 365–371.

Kinne, O. (1952). Zur biologie und physiologie von *Gammarus duebeni* Lillj. V. Untersuchungen über blutkonzentration, Herzfrequenz und Atmung. *Kiel. Meeresforsch.* **9,** 134–150.

Kinne, O. (1953). Zur biologie und physiologie von *Gammarus duebeni* Lillj. I. *Z. Wiss. Zool.* **157,** 427–491.

Kinne, O. (1959). Ecological data on the amphipod *Gammarus duebeni* Lillj. *Veroeff. Inst. Meeresforsch. Bremerhaven* **6,** 177–202.

Kinne, O. (1960). *Gammarus salinus*-einige daten über den Unwelteinfluss auf wachstum, hautungsfolge herzfrequenz und eientwicklungs-dauer. *Crustaceana* **1,** 208–217.

Kinne, O. (1961). Growth, moulting frequency, heart beat, number of eggs and incubation time in *Gammarus zaddachi* exposed to different environments. *Crustaceana* **2,** 26–36.

Kinne, O. (1971). Salinity: Animals-invertebrates. *In* "Marine Ecology" (O. Kinne, ed.), Vol. 1, Part 2, pp. 821–995. Wiley-Interscience, New York.

Klein, M. L., and Lincer, J. L. (1974). Behavioral effects of Dieldrin upon the fiddler crab, *Uca pugilator*. *In* "Pollution and Physiology of Marine Organisms" (F. J. Vernberg and W. G. Vernberg, eds.), pp. 181–196. Academic Press, New York.

Krebs, C. T., Valiela, I., Harvey, G. R., and Teal, J. M. (1974). Reduction in fiels populations of fiddler crabs by uptake of chlorinated hydrocarbons. *Mar. Pollut. Bull.* **5,** 140–142.

Lance, J. (1963). The salinity tolerance of some estuarine planktonic copepods. *Limnol. Oceanogr.* **8,** 440–449.

Lang, M. A., and Gainer, H. (1969). Volume control by muscle fibers of the blue crab. *J. Gen. Physiol.* **53,** 323–341.

Lange, R. (1972). Some recent work on osmotic, ionic and volume regulation in marine animals. *Oceanogr. Mar. Biol.* **10,** 97–136.

Lasserre, P., Renaud-Mornant, J., and Castel, J. (1976). Metabolic activities of meiofaunal communities in a semi-enclosed lagoon. Possibilities of trophic competition between meiofauna and mugilid fish. *Proc. Eur. Symp. Mar. Biol. 10th, 1975* Vol. 2, pp. 393–414.

Laughlin, R. B., Jr., Young, L. G. L., and Neff, J. M. (1978). A long-term study of the effects of water soluble fractions of N°2 Fuel oil on the survival, development rate, and growth of the mud crab *Rhithropanopeus harrisii*. *Mar. Biol.* (*Berlin*) **47,** 87–95.

Lee, W. Y., and Nicol, J. A. C. (1978). Individual and combined toxicity of some petroleum aromatics to the marine Amphipod *Elasmopus pectenicrus*. *Mar. Biol.* (*Berlin*) **48,** 215–222.

Lee, W. Y., Welch, M. F., and Nicol, J. A. C. (1977). Survival of two species of Amphipods in aqueous extracts of petroleum oils. *Mar. Pollut. Bull.* **8,** 92–94.

Lockwood, A. P. M. (1960). Some effects of temperature and concentration of the medium on the ionic regulation of the isopod *Asellus aquaticus* (L.). *J. Exp. Biol.* **37,** 614–630.

Lytle, J. (1973). "Fate and Effect of Oil in the Aquatic Environment. Gulf Coast Region," (4th Quartely Progress Report to Environmental Protection Agency), pp. 20–21. Mississippi State University, State University.

McCauley, R. N. (1965). The biological effects of oil pollution in a river. *Limnol. Oceanogr.* **11,** 475–486.

Mackay, N. J., Kazacos, M. N., William, R. J., and Leedow, M. I. (1975). Selenium and heavy metals in black marlin. *Mar. Pollut. Bull.* **6,** 57–61.

Macknight, A. D. C., and Leaf, A. (1977). Regulation of cellular volume. *Physiol. Rev.* **57,** 510–573.

Mangum, C. P. (1976). The function of respiratory pigments in estuarine animals. *In* "Estuarine Processes" Vol. 1, (M. Wiley, ed.), pp. 356–380. Academic Press, New York.

Mangum, C. P., and Johansen, K. (1975). The colloid osmotic pressures of invertebrate body fluids. *J. Exp. Biol.* **63,** 661–671.

Mangum, C. P., Silverthorn, S. U., Harris, J. L., Towle, D. W., and Krall, A. R. (1976). The relationship between blood pH, ammonia excretion and adaptation to low salinity in the blue crab *Callinectes sapidus*. *J. Exp. Zool.* **195,** 129–136.

Marsden, I. D. (1973). The influence of salinity and temperature on the survival and behaviour of the isopod *Sphaeroma rugicauda* from salt-marsh habitat. *Mar. Biol.* (*Berlin*) **21,** 75–85.

Matsumura, F., and Narahashi, T. (1971). ATPase inhibition and electro-physiological changes caused by DDT and related neuroactive agents in lobster nerve. *Biochem. Pharmacol.* **20,** 825–837.

Mileikousky, S. A. (1970). The influence of pollution on pelagic larvae of bottom invertebrates in marine nearshore and estuarine waters. *Mar. Biol.* (*Berlin*) **6,** 350–356.

Mironov, O. G. (1969a). The effect of oil pollution upon some representatives of the black sea zooplankton. *Zool. Zh.* **48,** 980–984.

Mironov, O. G. (1969b). Viability of larvae of some crustacea in sea water polluted with oil-products. *Zool. Zh.* **48,** 1734–1736.

Morton, B., and Wu, R. S. S. (1977). The toxic effects of hydrocarbons upon the naupilar and adult stages of Balanus (Crustacea: Cirripedia). *Mar. Pollut. Bull.* **8,** 232–236.

Needham, A. E. (1957). Factors affecting nitrogen excretion in *Carcinides maenas* (Pennant). *Physiol. Comp. Oecol.* **4,** 209–239.

Nimmo, D. R., and Blackman, R. R. (1972). Effects of DDT on cations in the hepatopancreas of penaeid shrimp. *Trans. Am. Fish. Soc.* **3,** 547–549.

Nimmo, D. R., Wilson, P. D., Blackman, R. R., and Wilson, A. J., Jr. (1971). Polychlorinated biphenyl absorbed from sediments by fiddler crabs and pink shrimp. *Nature* (*London*) **231,** 50–53.

Odum, W. E., Woodwell, G. M., and Wurster, C. F. (1969). DDT residues absorbed from organic detritus by fiddler crabs. *Science* **164,** 576–577.

Ong, K. S., and Costlow, J. D., Jr. (1970). The effect of salinity and temperature on the larval development of the stone crab, *Menippe mercenaria* (Say), reared in the laboratory. *Chesapeake Sci.* **11,** 16–29.

Pavillon, J. F. (1978). Contribution to studies on the effect of ectarine substances on the development of marine invertebrates. Absorption of labelled (^{14}C or ^{3}H) dissolved organic substances by the eggs and larvae of *Euphausia kroknii*. *In* "Physiology and Behaviour of Marine Organisms" (D. S. McLusky and A. J. Berry, eds.), pp. 149–154. Academic Press, New York.

Péqueux, A., and Gilles, R. (1977). Osmoregulation of the chinese crab *Eriocheir sinensis* as related to the activity of the (Na^+-K^+) ATPase. *Arch. Int. Physiol. Biochim.* **85,** 41–42.

Péqueux, A., and Gilles, R. (1978). Na^+/NH_4^+ co-transport in isolated perfused gills of the chinese crab *Eriocheir sinensis* acclimated to fresh water. *Experientia* **34,** 1593–1594.

Péqueux, A., Vallota, A. C., and Gilles, R. (1979). Blood proteins as related to osmoregulation in Crustacea. *Comp. Biochem. Physiol.* **64,** 433–435.

Pierce, S. K., Jr., and Greenberg, M. J. (1973). The initiation and control of free amino acid regulation of cell volume in salinity-stressed marine bivalves. *J. Exp. Biol.* **59,** 435–440.

Potts, W. T. W., and Parry, G. (1964). "Osmotic and Ionic Regulation in Animals." Pergamon, Oxford.

Reish, D. J., and Stephens, G. C. (1969). Uptake of organic material by aquatic invertebrates V. The influence of age on the uptake of glycine-^{14}C by the polychaete *Neanthes arenaceodentata*. *Mar. Biol.* (*Berlin*) **3,** 352–355.

Remmert, H. (1967). Physiologisch-okologische Experimente an *Ligia oceanica*. *Z. Morphol. Oekol. Tiere* **59,** 33–41.

Renfro, J. L., Schmidt-Nielsen, B., Miller, D., Benos, D., and Allen, J. (1974). Methylmercury and inorganic mercury: Uptake, distribution and effect on osmoregulatory mechanisms in fishes. *In* "Pollution and Physiology of Marine Organisms" (F. J. Vernberg and W. B. Vernberg, eds.), pp. 101–122. Academic Press, New York.

Reuben, J. P., Girardier, L., and Grundfest, H. (1964). Water transfer and cell structure in isolated crayfish muscle fibers. *J. Gen. Physiol.* **47,** 1141–1174.

Robertson, P. B. (1968). The complete larval development of the sand lobster, *Scyllarus americanus* (Smith) (Decapoda, Scyllaridae) in the laboratory, with notes on larvae from the plankton. *Bull. Mar. Sci.* **18,** 294–342.

Roesijadi, G., Petrocelli, S. R., Anderson, J. W., Presley, B. J., and Sims, R. (1974). Survival and chloride ion regulation of the porcelain crab *Petrolisthes armatus. Mar. Biol. (Berlin)* **27,** 213–217.

Roesijadi, G., Anderson, J. W., Petrocelli, S. R., and Giam, C. S. (1976a). Osmoregulation of the grass shrimp *Palaemonetes pugio* exposed to polychlorinated biphenyls (PCBs). I. Effect on chloride and osmotic concentrations and chloride and water-exchange kinetics. *Mar. Biol. (Berlin)* **38,** 343–355.

Roesijadi, G., Anderson, J. W., and Giam, C. S. (1976b). Osmoregulation of the grass shrimp *Palaemonetes pugio* exposed to polychlorinated biphenyls (PCBs). II. Effect on free amino acids of muscle tissue. *Mar. Biol. (Berlin)* **38,** 357–363.

Rorive, G., and Gilles, R. (1979). Intracellular inorganic osmotic effectors. *In* "Mechanisms of Osmoregulation in Animals" (R. Gilles, ed.), pp. 83–109. Wiley-Intersciences, New York.

Rosenberg, R., and Costlow, J. D., Jr. (1976). Synergistic effects of cadmium and salinity combined with constant and cycling temperatures on the larval development of two estuarine crab species. *Mar. Biol. (Berlin)* **38,** 291–303.

Rossi, L. C., Clemente, G. F., and Santaroni, G. (1976). Mercury and selenium distribution in a defined area and in its population *Arch. Environ. Health* **31,** 160–165.

Saliba, L. J., and Krzyz, R. M. (1976). Acclimation and tolerance of *Artemia salina* to copper salts. *Mar. Biol. (Berlin)* **38,** 231–238.

Sandifer, P. A. (1973). Effects of temperature and salinity on larval development of grass shrimp, *Palaemonetes vulgaris* (Decapoda, Caridea). *Fish. Bull.* **71,** 115–123.

Sandifer, P. A. (1975). The role of pelagic larvae in recruitment to populations of adult decapod crustaceans in the York River Estuary and adjacent lower chesapeake Bay, Virginia. *Estuarine Coastal Mar. Sci.* **3,** 269–279.

Sastry, A. N. (1970). Culture of brachyuran crab larvae using a recirculating sea water system in the laboratory. *Helgol. Wiss. Meeresunters.* **20,** 406–416.

Schlieper, C. (1929). Die osmoregulation der Süsswasserkrebse. *Verh. Dtsch. Zool. Ges.* **33,** 214–218.

Schmidt-Nielsen, B. (1975). Comparative physiology of cellular ion and volume regulation. *J. Exp. Zool.* **194,** 207–220.

Schoffeniels, E. (1960). Origine des acides aminés intervenant dans la régulation de la pression osmotique intracellulaire de *Eriocheir sinensis* Milne Edwards. *Arch. Int. Physiol. Biochim.* **68,** 696–698.

Schoffeniels, E. (1966). The activity of the L-glutamic acid dehydrogenase. *Arch. Int. Physiol. Biochim.* **74,** 665–676.

Schoffeniels, E. (1968). The control of intracellular hydrogen transport by inorganic ions. *Arch. Int. Physiol. Biochim.* **76,** 319–343.

Schoffeniels, E. (1970). Isozymic pattern of lactic dehydrogenase isolated from various crustacea of the North Carolinian Coast. *Archs. Int. Physiol. Biochim.* **78,** 135–139.

Schoffeniels, E. (1976). Biochemical approaches to osmoregulatory processes in crustacea. *In* "Perspectives in Experimental Biology" (S. Davies, ed.), pp. 107–123. Pergamon, Oxford.

Schoffeniels, E., and Gilles, R. (1963). Effect of cations on the activity of L-glutamic acid dehydrogenase. *Life Sci.* **2,** 834–839.

Schoffeniels, E., and Gilles, R. (1970). Osmoregulation in aquatic arthropods. *In* "Chemical

Zoology" (M. Florkin and B. Scheer, eds.), Vol. 5, pp. 255–286. Academic Press, New York.

Schoffeniels, E., and Gilles, R. (1972). Ionoregulation and osmoregulation in Mollusca. *In* "Chemical Zoology" (M. Florkin and B. Scheer, eds.), Vol. 7, pp. 393–420. Academic Press, New York.

Schultz, S. G., and Curran, P. F. (1970). Coupled transport of sodium and organic solutes. *Physiol. Rev.* **50,** 637–718.

Segal, E., and Burbanck, W. D. (1963). Effects of salinity and temperature on osmoregulation in two latitudinally separated populations of an estuarine isopod, *Cyathura polita* (Stumpson). *Physiol. Zool.* **36,** 250–263.

Shealy, D. H., Jr., and Sandifer, P. A. (1975). Effects of mercury on survival and development of the larval grass shrimp *Palaemonetes vulgaris. Mar. Biol. (Berlin)* **33,** 7–16.

Shick, J. M. (1975). Uptake and utilization of dissolved glycine by *Aurelia aurita* scyphistomae: Temperature effects on the uptake process; nutritional role of dissolved amino acids. *Biol. Bull. (Woods Hole, Mass.)* **148,** 117–140.

Shyamasundari, K. (1976). Effects of salinity and temperature on the development of eggs in the tube building amphipod *Corophium triaenonyx* Stebbing. *Biol. Bull. (Woods Hole, Mass.)* **150,** 286–293.

Siebers, D. (1972). Mechanismen der intrazellulären isosmotischen regulation der aminosäurekonzentration bei dem Flusskrebs *Orconectes limosus. Z. Vergl. Physiol.* **76,** 97–114.

Siebers, D., Lucu, C., Sperling, K. R., and Eberlein, K. (1972). Kinetics of osmoregulation in the crab *Carcinus maenas. Mar. Biol. (Berlin)* **17,** 291–303.

Sjöberg, B. (1967). On the ecology of the Jaerea albifrons group (Isopoda). *Sarsia* **29,** 321–348.

Skou, J. C. (1963). Studies on the $Na^{+}-K^{+}$ activated ATP hydrolysing enzyme system. The role of −SH groups. *Biochem. Biophys. Res. Commun.* **10,** 79–84.

Smaldon, G. (1973). Osmoregulation in *Pisidia longicornis* (L.) and *Porcellana platycheles* (Pennant) (Decapoda, Anomura) subjected to reduced salinities. *Comp. Biochem. Physiol. A* **44A,** 893–895.

Smith, R. I. (1967). Osmotic regulation and adaptative reduction of water-permeability in a brackish-water crab, *Rhithropanopeus harrisii* (Brachyura, Xanthidae). *Biol. Bull. (Woods Hole, Mass.)* **133,** 643–658.

Snoke, J. E. (1956). Chicken liver glutamic dehydrogenase. *J. Biol. Chem.* **223,** 271–276

Southward, A. J., and Southward, E. C. (1972). Observations on the role of dissolved organic compounds in the nutrition of benthic invertebrates. II. Uptake by other animals living in the same habitat as pogonophores and by some littoral Polychaeta. *Sarsia* **48,** 61–68.

Spaargaren, D. H. (1971). Aspects of osmotic regulation in the shrimp *Crangon crangon* and *Crangon allmanni. Neth. J. Sea. Res.* **5,** 275–333.

Spaargaren, D. H. (1972). Osmoregulation in the prawns *Palaemon serratus* and *Lysmata seticaudata* from Bay of Naples. *Neth. J. Sea Res.* **5,** 416–436.

Steele, D. H., and Steele, U. J. (1973). The biology of *Gammarus* (Crustacea Amphipoda) in the North Western Atlantic. VII. The duration of embryonic development in five species at various temperatures. *Can. J. Zool.* **51,** 995–999.

Stephens, G. C. (1962). Uptake of organic material by aquatic invertebrates I. Uptake of glucose by the solitary coral *Fungia scutaria. Biol. Bull. (Woods Hole, Mass.)* **123,** 648–660.

Stephens, G. C. (1964). Uptake of organic material by aquatic invertebrates. III. Uptake of glycine by brackish-water annelids. *Biol. Bull. (Woods Hole, Mass.)* **126,** 150–162.

Stephens, G. C., and Schinske, R. A. (1961). Uptake of amino acids by marine invertebrates. *Limnol. Oceanogr.* **6,** 175–181.

Stephens, G. C., and Virkar, R. A. (1966). Uptake of organic material by aquatic invertebrates.

IV. The influence of salinity on the uptake of amino acids by the brittle star, *Ophiactis arenosa*. *Biol. Bull.* (*Woods Hole, Mass.*) **131,** 172–185.

Stephens, G. C., Volk, M. J., Wright, S. H., and Backlund, P. S. (1978). Transepidermal accumulation of naturally occurring amino acids in the sand dollar *Dendraster excentricus*. *Biol. Bull.* (*Woods Hole, Mass.*) **154,** 335–347.

Stoewsand, G. S., Bache, C. A., and Lisk, D. J. (1974). Dietary selenium protection of methylmercury intoxication of Japanese quail. *Bull. Environ. Contam. Toxicol.* **11,** 152–156.

Suda, T., Horiuchi, N., Ogata, E., Ezawa, I., Otaki, N., and Kimura, M. (1974). Prevention by metallothionein of Cd-induced inhibition of vitamin D activation reaction in kidney. *FEBS Lett.* **42,** 23–26.

Testerman, J. K. (1972). Accumulation of free fatty acids from sea water by marine invertebrates. *Biol. Bull.* (*Woods Hole, Mass.*) **142,** 160–177.

Thabrew, M. I., Poat, P. C., and Munday, K. A. (1971). Carbohydrate metabolism in *Carcinus maenas* gill tissue. *Comp. Biochem. Physiol. B* **40B,** 531–541.

Thompson, L. C., and Pritchard, A. W. (1969). Osmoregulatory capacities of Callianassa and Upogebia (Crustacea: Thalassinidae). *Biol. Bull.* (*Woods Hole, Mass.*) **136,** 114–129.

Thurberg, F. P., Dawson, M. A., and Collier, R. S. (1973). Effects of copper and cadmium on osmoregulation and oxygen consumption in two species of estuarine crabs. *Mar. Biol.* (*Berlin*) **23,** 171–175.

Todd, M. E., and Dehnel, P. A. (1960). The influence of temperature and salinity on heat tolerance in two grapsoid crabs *Hemigrapsus nudus* and *H. oregonensis*. *Biol. Bull.* (*Woods Hole, Mass.*) **118,** 150–172.

Towle, D. W. (1974). Equivalence of gill $Na^+ + K^+$ ATPase from blue crabs acclimated to high and low salinity. *Am. Zool.* **14,** 1259.

Towle, D. W., Palmer, G. E., and Harris, J. L. (1976). Role of gill $Na^+ + K^+$ dependent ATPase in acclimation of blue crabs (*Callinectes sapidus*) to low salinity. *J. Exp. Zool.* **196,** 315–322.

Treherne, J. E., and Pichon, Y. (1978). Long-term adaptation of Sabella giant axons to hyposomotic stress. *J. Exp. Biol.* **75,** 253–263.

Van Slyke, D. D., Hastings, A. B., Murray, C. D., and Sendroy, J. (1925). Studies of gas and electrolyte equilibria in blood. 8 . The distribution of hydrogen, chloride and bicarbonate ions in oxygenated and reduced blood. *J. Biol. Chem.* **65,** 701–728.

Venice Symposium (1959). Final resolution of the symposium on the classification of brackish waters. *Arch. Oceanogr. Limnol.* **11,** 243–245.

Vernberg, F. J., and Silverthorn, S. U. (1979). Temperature and osmoregulation in aquatic species. *In* "Mechanisms of Osmoregulation in Animals" (R. Gilles, ed.), pp. 537–562. Wiley-Interscience, New York.

Vernberg, F. J., and Vernberg, W. B., eds. (1974). "Pollution and Physiology of Marine Organisms." Academic Press, New York.

Vernberg, W. B., and Vernberg, F. J. (1972). The synergistic effects of temperature, salinity and mercury on survival and metabolism of the adult fiddler crab, *Uca pugilator*. *Fish. Bull.* **70,** 415–420.

Vernberg, W. B., DeCoursey, P. J., and Padgett, W. J. (1973). Synergistic effects of environmental variables on larvae of *Uca pugilator*. *Mar. Biol.* (*Berlin*) **22,** 307–312.

Ville, J. P. (1971). Ecologie des *Macrobrachium* de Côte d'Ivoire. I. Rôle de la température et de la salinité. *Ann. Univ. Abidjan, Ser. E* **4,** 317–324.

Vincent-Marique, C., and Gilles, R. (1970a). Modification of the amino acid pool in blood and muscle of *Eriocheir sinensis* during osmotic stress. *Comp. Biochem. Physiol.* **35,** 479–485.

Vincent-Marique, C., and Gilles, R. (1970b). Changes in the amino-acid concentration in blood and muscle of *Eriocheir sinensis* during hypoosmotic stress. *Life Sci.* **9,** Part II, 509–512.

Waldichuk, M. (1974). Some biological concerns in heavy metals pollution. *In* "Pollution and Physiology of Marine Organisms" (F. J. Vernberg and W. B. Vernberg, eds.), pp. 1–57. Academic Press, New York.

Walker, G. (1977). "Copper" granules in the Barnacle *Balanus balanoides. Mar. Biol.* (*Berlin*) **39,** 343–349.

Walker, G., Rainbow, P. S., Foster, P., and Crisp, D. J. (1975a). Barnacles: Possible indicators of zinc pollution? *Mar. Biol.* (*Berlin*) **30,** 57–65.

Walker, G., Rainbow, P. S., Foster, P., and Holland, D. L. (1975b). Zinc phosphate granules in tissue surrounding the midgut of the barnacle *Balanus balanoides. Mar. Biol.* (*Berlin*) **33,** 161–166.

Ward, D. V., and Busch, D. A. (1976). Effects of Temefos, an organophosphorous insecticide, on survival and escape behaviour of the marsh fiddler crab *Uca pugnax. Oikos* **27,** 331–335.

Ward, D. V., and Howes, B. L. (1974). The effects of Abate, an organophosphorous insecticide, on marsh fiddler crab populations. *Bull. Environ. Contam. Toxicol.* **12,** 694–697.

Ward, D. V., Howes, B. L., and Ludwig, D. F. (1976). Interactive effects of predation pressure and insecticide (Temefos) toxicity on populations of the marsh fiddler crab *Uca pugnax. Mar. Biol.* (*Berlin*) **35,** 119–126.

Weis, J. S. (1978). Interactions of methylmercury, cadmium and salinity on regeneration in the fiddler crabs, *Uca pugilator, Uca pugnax,* and *U. minax. Mar. Biol.* (*Berlin*) **49,** 119–124.

Wells, P. G. (1972). Influence of Venezuelan crude oil on lobster larvae. *Mar. Pollut. Bull.* **3,** 105–106

Weser, U., Donay, F., and Rupp, H. (1973). Cd-induced synthesis of hepatic metallothionein in chicken and rats. *FEBS Lett.* **32,** 171–174.

Williams, A. B. (1960). The influence of temperature on osmotic regulation in two species of estuarine shrimps (Penaeus). *Biol. Bull.* (*Woods Hole, Mass.*) **119,** 560–571.

Yingst, J. Y. (1976). The utilization of organic matter in shallow marine sediments by an epibenthic deposit-feeding holothurian. *J. Exp. Mar. Biol. Ecol.* **23,** 55–69.

Young, A. M., and Hazlett, T. L. (1978). The effect of salinity and temperature on the larval development of *Clibanarus vittatus* (Bosc) (Crustacea: Decapoda: Diogenidae). *J. Exp. Mar. Biol. Ecol.* **34,** 131–141.

5

Ecological Aspects of Reproduction

A. N. SASTRY

I. INTRODUCTION

Crustaceans are a remarkably successful group, both in number of living species and in colonization of different habitats in marine, terrestrial, and freshwater environments. This is reflected in the diversity of life history patterns and the reproductive strategies. The life-span and size of the crustaceans range from relatively small and short-lived species to a larger, long-

THE BIOLOGY OF CRUSTACEA, VOL. 8

ISBN 0-12-106408-5

lived species. Their habitats include benthic environments of the deep oceans, coastal and estuarine regions, and semiterrestrial, terrestrial, and freshwater environments of the ponds and lakes of low to high altitudes. They also live in pelagic environments of temporary ponds and permanent lakes and in the epipelgic to bathypelagic zones of the oceans. Species of crustaceans are found from polar regions to the tropics. The great majority of crustaceans are aquatic and free living. In life styles, crustaceans vary from sessile forms to burrowers and from highly mobile species and permanent swimmers to benthopelagic forms; some have adapted to live as facultative or obligate symbionts and parasites. In food and mode of feeding, they range from filter feeders, detritus feeders, and plant and carion feeders to scavenger and predators. Many crustaceans use more than one method of feeding and various food sources, either as adults or in different stages of the lifecycle. The method of reproduction and the adaptations for reproductive success also vary between species and in different environments. Ecologically, crustaceans play an important role in the food web and in the economy of aquatic ecosystems.

Reproductive phenomena of crustaceans can be examined in terms of molecular, cellular, physiological, neurohormonal, and behavioral adaptations of individuals to their environment. The knowledge gained by elucidating the proximate factors controlling reproduction can provide a better understanding of the evolution of life history and reproductive strategies in different environments. A comprehensive treatment of all interrelated aspects, which can provide a synthetic overview of the reproductive ecology of this enormously diverse group, is not attempted here in any great detail. This chapter is organized to provide a general overview of the adaptation of crustaceans to reproduction in different environments. Specifically, the questions considered concern the influence of internal and external factors and their interactions in the coordination of reproductive activity with the environment, the patterns of breeding with respect to the environment, and the life history and reproductive traits of species in different environments. It should be noted that reproduction can also be considered in terms of population growth and regulation, distribution of species in time and space, and the genetic variability essential for evolution.

II. TYPES OF REPRODUCTION

A. Parthenogenesis

A great majority of crustaceans are diocieous, but parthenogenetic reproduction and hermaphroditism occurs among certain groups without any taxonomic or phylogenetic relationship.

Parthenogenetic reproduction occurs in branchiopods, ostracodes, some harpacticoid copepods (e.g., *Elaphoidella, Epactophanes*), and, among malacostracans, in isopods (Trichoniscidae) and in amphipods of the genus *Rhabodosoma* (Kaestner, 1970). Two types of parthenogenesis occur: geographic parthenogenesis and cyclic parthenogenesis. Geographic parthenogenesis is characterized by the existence of a bisexual form and parthenogenetic form in a single species or closely related species, with the two forms having different geographical distribution.

Cyclic parthenogenesis occurs only in cladocerans. Most of these species are able to reproduce both parthenogenetically and sexually. During some periods of the year they reproduce only females by parthenogenesis, but at another time particular females produce males and resting eggs which cannot develop without fertilization. Many of the species reproducing by parthenogenesis live in standing water, temporary pools, and vernal ponds. The cladocerans have been of particular interest for genetic studies since there are only a few animal groups that reproduce both sexually and asexually (Hebert, 1980; Young, 1979).

There has been continued interest in the factors that initiate the switch in cyclic parthenogenetic populations from asexual to sexual reproduction. It is generally recognized that environmental factors are important in determining this. Overcrowding, scarcity of food, high and low temperatures, and day-length give rise to males, but also females (Banta, 1939; Stross and Hill, 1965). Cladocerans capable of sexual reproduction usually produce resting eggs under crowded or stressful conditions. Experiments with cladocerans (e.g., *Daphnia magna, Daphnia pulex, Daphnia cucullata, Moina brachiata, Chydorus sphaericus*) have indicated that some unknown stimuli in the cultures cause a change in the parthenogenetic females, but this change does not seem to occur when environmental conditions are constant (Kaestner, 1970). Under favorable conditions of food, temperature, and aeration, it has been shown that *Daphnia magna* and *Moina brachiata* exist for several years without sexual reproduction. Experiments with *M. brachiata* have also shown that lipid material of the food may determine the sex of eggs produced (Kaestner, 1970). Populations fed on yeast from which lipid has been removed produce only females. Addition of fatty extract to the yeast cultures results in 30% of eggs as males.

Parthenogenetic reproduction in which each individual reproduces is favorable for a rapid population growth. During adverse conditions, a new sexual generation makes its appearance with the production of resting eggs. These resting eggs are of importance for passive dispersal of these forms. The ability to modify the reproductive cycle particularly allows cladocerans to exploit conditions of the environment provided by temperate freshwater lakes and ponds. Population cycles or pulses are common in branchiopods and especially in cladocerans. Some cladocerans (e.g., *Diaphanosoma, Holopedium, Leptodora, Bythotrephes,* and *Bosmina*) that live in the middle of large lakes and ponds where temperature changes gradually are monocyclic, with only one sexual generation in autumn when temperature decreases. They overwinter as resting eggs. In very large lakes, for example, *Daphnia longispina, Daphnia cucullata,* and *Bosmina longirostris* are acyclic and skip the autumn sexual period. In small ponds where the spring and autumn environments are different, the cladocerans reproduce sexually twice during the year.

The adaptive significance of cyclic parthenogenesis has been related to the doubling of the intrinsic rate of growth, which occurs during parthenogenesis, and the selective importance of such an increase in organisms inhabiting ephemeral environments (Mayr, 1963). Asexual reproduction eliminates the problem of mating encounters, reduced growth rates, and the need to produce males, allowing an asexual clone to grow twice as fast as a sexual population. Gerritsen (1980) suggested that parthenogenesis is adapted for habitats with unpredictable cycles of disturbance and resource abundance. For theoretical considerations of the relative advantages of sexual and asexual reproduction and their significance in evolution and population biology, consult Williams (1975), Maynard-Smith (1978), and discussions provided by Lynch (1980a,b) on the selective pressures in the evolution of cladoceran life histories and by Gerritsen (1980) on population ecology.

B. Hermaphroditism

Hermaphroditism occurs in only a few crustaceans. It is characteristic of Cephalocarida and most free-living and parasitic Cirripedia. Hermaphroditism occurs sporadically among the malacostracans Tanaidacea, Natantia, two groups of parasitic isopods, Cymothoidae and Cryptoniscina, and in certain Oniscoidea (Kaestner, 1970).

Among the Cirripedia, most of the Thoracica (barnacles) are hermaphrodites, but some species are accompanied by complemental males. Complemental males are attached to the females and, in some species, have reduced plates; in others, they are saclike and lack most of the external features except for a pair of antennae and gonads. Certain genera, *Scalpellum, Ibla, Trypetesa,* and *Alcippe* which are usually found on tropical shores or in relatively deep waters, may be either hermaphrodites accompanied by complementals, or both separate sexes. The suborder Balanomorpha are all hermaphroditic, but a few are accompanied by complemental males. In some Ascothoracica that bore into calcareous substrates, the females are accompanied by dwarf males [e.g., (*Alcippe*=*Trypetesa*) *lampas*]. The Poecilasmatidae associated with the benthic organisms, the oceanic

Lepadidae, and the deep-sea Verrucomorpha are all hermaphrodites. The suborder Rhizocephala, which is wholly parasitic primarily on decapods, is also hermaphroditic. In Cirripedia, the ovaries and testes are separate, but self-fertilization rarely occurs, except in Rhizocephala.

Among the malacostracans, hermaphroditism is usually protandrous where the testes develop, function, and then regress prior to the development of ovaries (e.g., *Pandalus*). Protandrous hermaphroditism has been reported in species from the families Hippolytidae, Campylonotidae, and Pandalidae (Carlisle, 1959a,b; Yaldwyn, 1966). They occur in a variety of diverse habitats ranging from deep-water arctic (e.g., *Pandalus borealis*) to the shallow-water temperate regions (e.g., *Campylonotus rathbunae*) (Yaldwyn, 1966) to the antarctic (*Chorismus antarcticus*) (Maxwell, 1977). In *Calocaris macandreae,* the thalassinid distributed from boreal waters in the Eastern Atlantic to the Mediterranean, the testis and ovary grow approximately at an equal rate, until the testis reaches its maximum size and production. Then the testes degenerate and ovaries continue to increase in size, completely filling the space of the degenerated testes and finally reaching maturity (Buchanan, 1963). The coiled vasa deferentia filled with spermatophores is retained, however, throughout the whole life of an individual.

In the order Tanaidacea, *Heterotanais oerstedi* is considered to be a potential hermaphrodite (Kaestner, 1970). In this species, primary males develop from a second neuter stage after a molt, and secondary males develop from females upon molting after brooding. This molt is always accompanied by a reduction of the ovary. In addition, there are also some gonochoristic females. There is considerable lability in the sex determination of this species. If a male lives in an aquarium with a female, it does not change sex. Larvae raised with an adult male develop into a female and vice versa. Larvae raised in isolation become either gonochoristic females or females that later become males. The antarctic species, *Apseudes hermaphroditicus* (Tanaidacea), is a simultaneous hermaphrodite with both mature ova and sperm developing in the gonad (Lang, 1953).

Some parasitic isopods, Cymothoidae and Cryptoniscina, are protandric hermaphrodites. In the oniscoid *Rhyscotoides legrandi,* spermatogenesis continues after activation of the ovary.

Genetic mechanisms, the environmental factors influencing the sex determination (Charinaux-Cotton, 1960, 1965, 1975), and hormonal control of sex differentiation (Charinaux-Cotton, 1965; Adiyodi and Adiyodi, 1970) are discussed in other chapters of this series. Generally, hermaphroditism seems to occur in species that are sessile, parasitic, or adapted to live in specialized environments. The relative advantage of gonochoroism and simultaneous or consecutive hermaphroditism is a subject of considerable theoretical significance to population biology, ecology, and evolution, and is still a matter of controversy. For a detailed discussion of sex and evolution, and sex change theory, Williams (1966, 1975), Ghiselin (1974), Maynard-Smith (1978), and Charnov (1979a,b, 1981) should be consulted.

III. REPRODUCTIVE CYCLES

Reproductive cycles include a series of events within individuals of the populations (Fig. 1). In individuals completing the juvenile instars, these events include proliferation of gonial cells (activation of gametogenesis), differentiation and growth of gametes to maturation (gamete production), reproductive behaviors associated with mating, release of gametes by transfer of sperm during copulation by males and ovulation and oviposition by females, and incubation of embryos until hatching to release larvae or juveniles. This cycle is basic to all the sexually reproducing crustaceans, but the time relationship between certain events and duration of these events varies for different species. Some species produce gametes prior to the pubertal instar (e.g., *Macropodia rostrata*), while others do so only after completing the pubertal instar (Hartnoll, 1963). The time relationship between mating and ovulation varies for different species. In some species, ovulation occurs immediately after copulation, while others do not ovulate for several days or even months after mating. Variation is also found in the method of brood care, the length of time the brood is incubated, and the stage of development at which young are released. Generally, crustaceans brood their em-

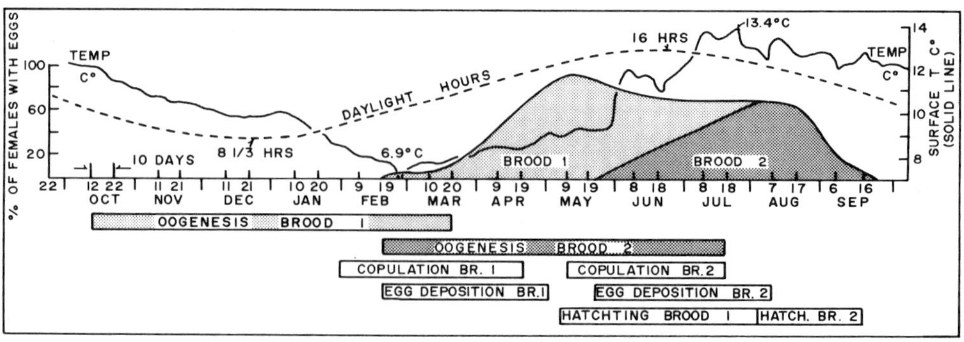

Fig. 1. The reproductive cycle of a brachyuran crab (*Hemigrapsus oregonensis*) showing the typical sequence of reproductive events in relation to the environment. (Modified from Knudsen, 1964.)

bryos by attaching fertilized eggs to certain appendages and retaining them in a brood chamber or an ovisac. Some species release fertilized eggs and attach them to some solid object on the bottom [e.g., fish lice, *Argulus foliaceus;* the freshwater anaspidacean, *Anaspides tasmaniae;* bathynellacean, *Bathynella natans* (Kaestner, 1970)]. Only a few crustaceans shed their eggs into the environment without brood care [e.g., some species of copepods (Marshall and Orr, 1972), some euphausiids (Mauchline and Fisher, 1969), and penaeids except *Lucifer* (Omori, 1974)].

In some species, gametes are produced more or less continuously, and the production of successive broods is interrupted only by the length of the incubation period. In others, gonads may go through a quiescent or spent period, the gonads may regress while the relict gametes are resorbed, or the adult may die. In some species, successive ovulations occur during the same intermolt period, while in others, molting separates ovulation. All these variations in chronology of reproductive events produce different temporal reproductive patterns for different species.

Despite these variations, the basic events of the reproductive cycle take a certain length of time within each individual of a population in a given environment. The interaction between reproductively capable individuals and the environment may be manifested in the timing of movements and/or migrations to suitable locations for mating or breeding, and of behaviors associated with mating and the occurrence of ovigerous females within the year. The temporal pattern of the reproductive cycle of a population relative to the given environment may have been selected for ensuring reproductive success of the species. Hence, the reproductive cycle may be considered a phenotypic expression for successful reproduction of the species in the particular environment. A complex physiological and behavioral integration of the reproductive events with respect to the environment may be involved. Rarely have studies been conducted to elucidate all the proximate factors controlling the temporal pattern of reproduction for a species with respect to environment. In this regard, the boreoarctic barnacle, *Semibalanus balanoides,* has received some attention (Barnes, 1963, 1975; Crisp and Patel, 1969).

A. Gametogenesis and Gonadal Cycles

Studies reporting gametogenesis from either histological sections or gross observations of the gonads show that the sequence of stages is the same for all species, even though the duration of each stage and the amount of synchrony may vary for different species (Ryan, 1967a,b; Haefner, 1976, 1977; Mauchline and Fisher, 1969).

1. SPERMATOGENESIS AND TESTICULAR CYCLES

Basically, primary germ cells develop into secondary gonial cells which multiply by mitosis before they enter into meiosis. After they become spermatocytes, the first and second maturation division occurs, leading to the formation of spermatids. Normally, four spermatids are formed from a single spermatocyte. Finally, spermatozoa are formed. Detailed observations from histological or gross observations of testes have documented spermatogenesis in copepods (Marshall and Orr, 1972), homarid lobsters (Aiken and Waddy, 1980), palinurid lobsters (Heydorn, 1969b), brachyuran crabs (Ryan, 1967a,b; Kon and Honma, 1970a), and in other groups.

The duration for development of spermatozoa from the gonial cells, and the characteristic of either seasonal or continuous production of spermatozoa by an individual of a population, may vary for different species, but precise information is unavailable. In the functionally hermaphroditic boreoarctic barnacles, *Semibalanus balanoides,* there is a well-defined annual cycle of reproduction, with the testis beginning development in late summer and reaching maturity by late autumn when copulation and fertilization occurs (Barnes *et al.,* 1963). Seasonality in spermatogenesis has been also observed in the euphausiids, *Meganyctiphanes norvegica* and *Thysanoessa raschii,* from the Clyde Sea area of England (Mauchline and Fisher, 1969). Spermatozoa are found in November, and all the males have fully formed spermatophores by the beginning of March, when mating takes place. In the Chinese mitten crab, *Eriocheir sinensis,* there is a seasonal testicular activity with an abundance of spermatozoa in testes prior to the breeding season in the autumn and early winter after downstream migration (Hoestlandt, 1948). As the males return to fresh water after mating, phagocytic activity occurs in the testis. A relative increase of spermatozoa during the summer, compared to that in winter, has been observed in the crab, *Cyclograpsus punctatus* (Broekhuysen, 1941).

While seasonal testicular activity seems to occur for some species with a restricted breeding season, others seem to have continuous activity throughout the year. In the crab *Chionoecetes opilio,* the testicular activity is continuous, even though the population has a restricted breeding season (Kon and Honma, 1970b). In the tropical crab, *Portunus sanguinolentus,* sperm production is not correlated with season or molt cycle (Ryan, 1967a). However, after evacuation of spermatophores, several days are required before the male is ready for another copulation. This would indicate that although the population as a whole might be capable of mating throughout the year, individuals within the population go through cycles of testicular activity. Similarly, no seasonality in spermatogenesis occurs in *Homarus americanus* (Aiken and Waddy, 1980) or in palinurid lobsters (Heydorn, 1969a). In contrast, high sperm counts were recorded for *Jasus lalandii* during the summer months when mating occurs (Heydorn, 1969b). Information of the spermatogenic cycle in short-lived species, such as copepods, is scanty (Marshall and Orr, 1972; Corkett and McLaren, 1978). Many of these species are believed to die off after copulation.

Although an increase in testicular size or volume may occur between the inactive or quiescent stage of the testes and the mature condition, quantitative data have rarely been gathered to adequatly describe temporal patterns. The testicular cycles determined by the gonad index method for three populations of tropical decapods, *Uca lactea annulipes, Portunus pelagicus,* and *Metapenaeus affinis,* showed peaks in activity, but there were no pronounced seasonal cycles over the year (Pillay and Nair, 1971). Although there is a continuous testicular activity throughout the year, peaks were observed prior to frequent copulatory activity among members of the populations. In many other species also, peaks in testicular activity have been reported prior to mating regardless of whether a species has a restricted or a continuous breeding season (Heydorn, 1965; Mauchline and Fisher, 1969). In a number of species, peak gonadal activity of both sexes is not necessarily synchronous, but the males will usually mature earlier than the females.

2. OOGENESIS AND THE OVARIAN CYCLES

In contrast to spermatogenic cycles, oogenic cycles have been studied more frequently. The primary oogonia originate from the germinal zone, and as new oogonia are formed, the others are displaced into the lumen of the ovary where they become secondary oogonia. They enter the prophase of the first meiotic division, transform into oocytes, and begin to grow. Meiosis proceeds to pachytene stage, then is arrested until the completion of the cytoplasmic development of the oocytes. At this phase, the nucleus swells into a germinal vesicle and the oocytes enter into regular but slow growth. This may last up to months or years when growth is mostly due to synthesis of new cytoplasm. Little or no yolk is formed in this phase. Following the previtellogenic growth phase, the oocytes enter upon their third growth phase and increase in size within a short time (i.e., days to weeks) to reach their final size. The vitellogenic oocytes are usually found the farthest from the germinal zone of the ovary and in the central part of the lumen. The exact time and location of maturation of the oocytes is not known, but it may occur during ovulation or prior to fertilization (Raven, 1961). Thus, growth of oocytes from newly transformed oogonia essentially follows a sigmoid growth curve. In crustaceans, the multiplication and differentiation of oogonia in the germinal zone and migration of oocytes into the lumen for growth may allow continuous production of oocytes in a succession and permit successive ovulations by individuals. Mauchline (1968) determined the nuclear diameter as a percentage of the diameter of growing oocytes in the euphausiids, *Meganyctiphanes norvegica* and *Thysanoessa raschii,* and found essentially the same features in oocyte development as described above. In the first phase, the nucleus is very large relative to the diameter of the oocytes. The nucleus and size of the oocyte increase during this phase, but the rate of increase in size of the nucleus is much less compared to that of the oocyte as a whole. In the second phase, the nucleus and oocyte size increase about the same rate. Next, the size of cytoplasmic regions increases faster and the yolk globules begin to appear. Finally, in the fourth phase, the oocytes grow fully and become mature. Environmental factors may differentially affect each of the above stages and different species. Interspecific comparison of the environmental influences on reproductive pattern can be complicated considering the number of stages and their differential responses to environmental factors.

In a number of crustaceans, the ovary goes through changes in color and size during oocyte development. Developmental stages have been assigned by different authors based on the occyte size, color, or ovary size to determine the reproductive condition (Heydorn, 1965; Conover, 1967; Marshall and Orr, 1972; Haefner, 1977; Aiken and Waddy, 1980). Yolk deposition in decapods occurs in two phases, termed primary and secondary vitellogenesis (Demeusy, 1965; Adiyodi and Adiyodi, 1970). In *Homarus americanus,* the primary vitellogenesis may extend over several months, resulting in slow growth of the oocytes (Aiken and Waddy, 1980). In this species, regular cycles of deposition and resorption may occur during the primary vitellogenesis, and there may be a resting stage between primary and secondary vitellogenesis during winter months. This tendency toward secondary vitellogenesis seems to occur in the spring months, leading to oviposition. In the crab, *Carcinus maenas,* there appears to be a critical oocyte size up to which spontaneous resorption is possible, but after which the oocytes are committed to irreversible development (Demeusy, 1965).

The duration of the oogenic cycle within an individual is not known precisely for most crustaceans, but may take as little as a few days to a couple of years. In *Calanus hyperboreus,* early oocytes take about 4 days to grow to maturity (Conover, 1967). In the freshwater calanoid copepod, *Diaptomus leptopus,* the oocytes take about 4 days to grow to maturity, and ovulation is followed by a nongravid phase for 1 day (Watras and Haney, 1980). Similarly, the ovary of *Calanus finmarchicus* reverts to a nongravid phase between ovulations (Marshall and Orr, 1972).

In the euphausiids, *Meganyctiphanes norvegica, Thysanoëssa raschii, Thysanoëssa inermis,* and *Euphausia superba,* oogonia released from the germinal zone take up to 4 months to grow to a full size (Mauchline and

Fisher, 1969). In *M. norvegica,* the oocytes are released continuously, but initially the increase in volume of the ovary is irregular. The oocytes grow very little in December and January, and most of the increase in ovarian volume at this time results from the release of new oocytes. As the oocytes grow from mid-January to mid-March, ovarian volume reaches an asymptotic value prior to ovulation. After ovulation, the spent ovaries do not become mature again until the following year.

Ovarian development of the boreoarctic barnacles, *Semibalanus balanoides* and *Balanus balanus,* begins in late summer, and the oocytes develop to maturity prior to copulation and ovulation in late autumn (Barnes *et al.,* 1963). In contrast to the restricted period of ovarian development in the above species, some warm water barnacles go through several cycles of ova production each year (Crisp and Patel, 1969; Fhyn and Costlow, 1977; Hines, 1979).

In the tanner crab, *Chionoecetes opilio,* the ovaries mature from October to June, followed by molting between July and September when mating and ovulation occur (Kon and Honma, 1970a). In this species, the ovarian cycle basically has an annual cycle. In contrast, *Portunus sanguinolentus* ovulates within an intermolt period at intervals of 60, 40, and 40 days in each mature instar (Ryan, 1967b). Prior to each ovulation, oocyte growth requires about 4 weeks from the time of release from the germinal zone. At each ovulation, all the fully grown oocytes are discharged, then immediately following newly released oocytes migrate into the lumen of the ovary for growth to next ovulation. This cycle is repeated again for a third ovulation, which is followed by a molt after a 40-day interval.

Duration of the ovarian cycle in the American lobster *Homarus americanus* is about 2 years (Aiken and Waddy, 1980). It can also take 2 years for oocyte growth in the thalassinid, *Calocaris macandreae* (Buchanan, 1963). These animals initially mature when 5 years old and incubate the eggs for about 8–9 months. After releasing the larvae, the adults molt immediately. A second batch of oocytes, which take about 2 years for maturation, are developed while the females are incubating the eggs. Hence, the females carry no eggs on the abdomen during the seventh year. A second ovulation occurs at the end of the seventh year, and possibly a third ovulation occurs at the end of the ninth year, when high mortalities occur in the population. Although the oogenic cycle in the individuals of *H. americanus* and *C. macandreae* takes 2 years, the populations of both species breed annually.

Growth of oocytes to maturation and ovulation results in growth and regression of the ovary, i.e., the ovarian cycle. Quantitative studies describing the ovarian cycles that can provide information on the period and amplitude of gonadal growth are relatively few. However, Pillay and Nair (1971) determined the temporal patterns of ovarian cycles by the gonad index method (gonad weight/body weight × 100) for three populations of tropical decapods. The reproductive period of *Uca lactea annulipes* extends for several months from July to April, during which time three gonad index peaks occur for the population. The gonad index peaks are followed by the occurrence of a highest number of ovigerous animals. After February, the gonad index values decline steadily to reach an annual minimum in May. Gonads of some individuals of the population redevelop even while the animals are ovigerous, but ovarian growth among members is out of phase. The rate of gonadal recovery varies at different times during the breeding season. The interval between successive ovulations within an individual is about 15–20 days. Thus, the population has an annual reproductive cycle with a breeding season extending for several months, but members of the population go through several oogenic-incubation cycles within the breeding period.

The other two species, *Portunus pelagicus* and *Metapenaeus affinis,* also exhibit more than one gonad index peak during the annual cycle, with each peak followed by the occurrence of a high number of ovigerous animals. A population of *Portunus pelagicus* from the southeast coast of India breeds throughout the year, but it also had distinct peaks of gonad index in certain months of the year (Rahaman, 1967). These tropical populations seem to reproduce for several months in the year, but they exhibit peak reproductive activity only in certain months, with the individuals of the populations going through more than one oogenic-incubation cycle during the year.

Pillay and Nair (1971) have reported that the amplitude of gonad index for females is higher compared to that of males. The females appear to divert a greater amount of energy for the production of ova compared to the males for production of sperm.

The process of oogenesis is basically the same for most crustaceans, but the period and frequency of oogenic cycles vary greatly. The period ranges anywhere from only a few days for some species (e.g., copepods) to as long as 2 years for other species (e.g., lobsters). Species differences in the frequency of oogenic cycles, i.e., the time interval between successive cycles among individuals, vary anywhere from a few days to weeks or months. The duration and frequency of oogenic cycles appear to be characteristic for each species in a given environment. Populations of species may thus vary in duration and frequency of the oogenic cycles in different parts of their distributional range relative to the variations in the environment. The reproductive traits—the rate of oocyte growth, the amplitude of ovarian growth, the frequency of oogenic cycles, the timing within the year, and whether an individual goes through one or more oogenic cycles during one intermolt period—appear characteristic to species in their particular environments. As will be seen later, all these aspects will have important consequences to the reproductive ecology of species.

B. Factors Influencing the Reproductive Cycle

A reproductive cycle may be viewed as an integrated response of the individuals of a population to the environment, both in a functional and a temporal sense. Temporal patterns of the reproductive cycle may result through a complex coordination of a number of endogenous factors with respect to the interacting exogenous factors at a given time and over a time period by members of a population. Crustaceans living in the deep sea, and possibly those living in caves, experience no seasonally related variations in their physical environment, except possibly for seasonal variations in the food supply (Vinogradov, 1962; Mauchline, 1972; Marshall, 1979). On the other hand, crustaceans living in all the other marine, terrestrial, and freshwater environments, and especially those occurring at the higher latitudes, experience pronounced seasonal variations in the environment. Here, a multitude of environmental factors, which interact with each other and with the populations, varies continuously, both in intensity and in rate of change over the year. The intensity and rate of change of the interacting environmental factors vary with the latitude, but may be modified by the prevailing patterns of water circulation in the oceans. Hence, the populations of species occurring in different latitudes or in the same latitudes (e.g., eastern and western Atlantic coasts) may also vary in their patterns of reproductive cycle as a response to the particular characteristics of their environment. The amplitude of change in certain physical environmental parameters is greater in the mid-latitudes relative to that in polar and tropical regions, but the smaller amplitude changes in these regions may be just as important to the populations as those for populations in the mid-latitudes. Regardless of the amplitude of environmental change, the response of a population to a particular environment can be observed as changes in metabolism, growth, and reproductive condition, and also as changes in movements and/or migration and behaviors. The response of different species to a given environment in a given geographical location may also vary based on their own genetically determined tolerances and requirements.

Other aspects of organism–environment interactions, in terms of physiology and behavior, are the time and space scales of change in the environment, on one hand, and life-span of species on the other. Short-time scale and low amplitude changes in the environment may be as important to species with a short generation time as are long-time scale and high amplitude changes occurring over the year for species with a long generation time. As yet, we have only a limited understanding of the significance of time and space scales of environmental variation for the life history and reproductive strategies of species. Nevertheless, populations experiencing variable environments may have to adjust their life history processes to the environment, and timing and temporal partitioning of certain events may be more important. Timing of life history processes relative to changes in the environment requires the organisms' perceptions of environmental change, the transmittal of the signals of environmental input, and the coordination of the appropriate response processes through receptor–neuroendocrine mechanisms to ultimately achieve coordination with the environment. On the other hand, in uniform environments such as deep sea, where timing may not be as important, abiotic environmental cues are doubtless of less importance. Rather the biotic interactions, such as competition for resources, avoidance of predators, and the behaviors for locating food and mates, may become more important to the life history and reproductive strategies of species.

We have yet to learn how populations interact with the environment, how organisms integrate their functions and behaviors to the immediate environment, and what the ultimate adaptive value of these respones is to reproductive success. Much of the available information on many of these aspects is still incomplete, and hence, the endogenous and exogenous factors influencing the reproductive cycles are considered separately in this chapter. It should be recognized, however, that organisms perform functional processes and exhibit behaviors as integrated entities in the context of their total environment, composed of physical and chemical factors, biotic interactions such as competition and predation, and social structure.

1. GAMETE DEVELOPMENT

Both internal factors and external factors influence gamete production within an individual. Among the internal factors, age of the individual, biochemical composition, metabolism, molting, and neuroendocrines appear to be important. Of the external factors, temperature, salinity, daylength, food availability, and chemical factors may interact with the organism, singly or collectively, to influence gamete development. Among these, food availability and all the internal factors may be important to all the species regardless of whether they live in a predictable or unpredictable environment. In any given environment, the relative importance of one external factor over the others may vary, however. The interactive effects of external and internal factors may also vary with respect to different stages of the gonadal cycle, but detailed information is scanty. In many studies, the influence of external factors has been assessed by examining whether or not individuals subjected to different environmental conditions lay eggs or egg masses. Information on the possible interactive effects of environmental factors during the course of gamete development is unavailable.

a. Internal Factors

i. AGE. Generally, gametogenesis occurs in individuals of a population after completion of the juvenile instars (e.g., on molting from copepodite stage V to the adult copepod, or following the pubertal molt in brachyuran crabs), but exceptions, with activation and development of gametes in the earlier stages, are known to occur. Certain morphological changes occur at the end of the juvenile instar, which represents the onset of sexual maturity. These characteristics include structures for copulation as well as other secondary sexual characters. In decapods, the changes following the pubertal molt (Perez, 1929; Hartnoll, 1963) may include changes in color and length of pereopods in the males, and morphological changes in the abdomen and accessory reproductive structures in the females. The nature of the pubertal molt varies from species to species; in some species it is the terminal instar (e.g., oxyrhynchous crabs; *Callinectes sapidus*), while in others it may be followed by one, two, or several instars. The American lobster, *Homarus americanus,* continues to grow throughout its life, even though it molts less frequently as it gets older (Passano, 1960; Aiken and Waddy, 1980). In many brachyuran crabs, the correlation between morphological changes associated with the pubertal molt and the onset of gametogenesis has been recognized (Hartnoll, 1963). This correlation appears to be stricter for females than for males. In males, spermatogenesis may occur before pubertal molt [e.g., *Corystes cassivelaunus, Macropodia rostrata* (Hartnoll, 1963); *Eriocheir sinensis* (Hoestlandt, 1948)]. Hartnoll suggests that accelerated spermatogenesis and spermatophore formation in the final prepubertal molt may be induced by the same hormonal changes that precipitate the pubertal molt at the next ecdysis. Major behavioral changes also occur following the pubertal molt, indicating the beginning of sexual activity. In females, the secondary phase of oogenesis generally occurs after the pubertal molt, but it occurs in some species during the prepubertal molt, at which time the ovaries ripen. Generally, however, mating does not seem to occur prior to pubertal molt.

There is a general tendency for the two sexes to differ in size when reaching maturity. The two sexes may attain sexual maturity over a wide size range in a given species; variation in size at sexual maturity may also occur between geographically separated populations (e.g., some species of homarid and panulirid lobsters: Aiken and Waddy, 1980). Generally, the individuals of populations from the warmer parts of the distributional range mature at a smaller size than those from the colder parts of the range. Size at sexual maturity varies both inter- and intraspecifically for euphausiids, with the species or populations from the colder arctic and subarctic waters maturing at an older age compared to those from warmer waters of the lower latitudes (Mauchline and Fisher, 1969). However, environmentally dependent variations in size at sexual maturity do not always seem to hold for all crustaceans (Aiken and Waddy, 1980). It has been suggested that size variation at onset of sexual maturity between populations may be caused by environmental factors (Templeman, 1936), genetic factors, density-dependent factors (in lobsters) (Chittleborough, 1976), selection pressures from fishing (Charnov, 1981), or predation pressure, as in planktonic cladocerans (Lynch, 1980a,b). The size at which onset of sexual maturity occurs is an important life history trait with consequences to population growth (Cole, 1954), and selection for minimum size for onset of sexual maturity is possibly a response to both density-dependent and density-independent factors.

ii. BIOCHEMICAL CHANGES. Studies on biochemical changes in the various body components of crustaceans relative to gonadal cycles are few compared to studies on biochemical changes associated with molting. Our knowledge of the rates of turnover of any metabolic pools with respect to any other materials for reproduction is still scanty (Adiyodi and Adiyodi, 1970). However, seasonal changes in biochemical composition have been determined for the boreoarctic barnacles *Semibalanus balanoides* and *Balanus balanus* (Barnes *et al.*, 1963) and for the warm-water barnacle *Chthamalus stellatus* (Barnes, 1972). In *B. balanoides,* all the components in the body (e.g., proteins, lipid, and carbohydrates), with the exception of DNA, increase during the spring months when food is most abundant in the environment, and they decrease in autumn, presumably due to partial transfer from storage organs to the newly developing ovary. From June to October, there is only a slight increase in body weight, and changes in biochemical composition of the body occur as a result of redistribution of the proportion of body components. The same general pattern of changes in biochemical constituents are seen for *B. balanus,* but they are somewhat modified by the temporal pattern of the reproductive cycle of this population compared to that of *B. balanoides*. Warm-water species, *C. stellatus,* at the northern limits of its distribution in Scotland, does not have synchronous breeding like the boreoarctic barnacles, and the individuals of the population are found with and without egg masses. Biochemical changes occur as in the other two species, but they are not as clear cut within a population. Pillay and Nair (1972) report that transfer of organic materials from the tissues to the gonad is not evident for the tropical population of the barnacle, *Balanus amphitrite communis*. When populations breed asynchronously or continuously, it would seem that separate analysis of individuals in different stages of the reproductive cycle in samples collected at different time intervals may provide better understanding of the biochemical changes associated with the reproductive cycle than analyses conducted for pooled population samples. Such detailed data has been rarely gathered.

Seasonal changes in biochemical composition (glycogen, total carbohydrates, neutral fats, protein and nonprotein nitrogen) of the ovary throughout the year, and in both the male and female hepatopancreas[1] throughout the molt cycle, have been determined for the crab, *Carcinus maenas* (Heath and Barnes, 1970). No clear changes in either size or composition of the hepatopancreas have been observed, even though this organ has been generally considered the main storage organ in decapods. During the period of ovarian development between September and October, the hepatopancreas of the female does not seem to lose weight, suggesting that it may not store nutrients for reproductive purposes. This species does not store as much material before molting, and molts without accumulating an appreciable amount of reserves. Heath and Barnes (1970) suggested that *C. maenas* may depend on immediate food intake rather than stored reserves for growth and reproduction. However, the changes in ovary determined relative to its increase in weight rather than time of year showed that nitrogen fraction decreases (even though the absolute amount increases) and that the lipid fraction increases both in relative and absolute amounts. The increase in the lipid fraction with the increase in ovarian size may occur because of an increase in lipoprotein content of the ova (Giese, 1966; Holland, 1978). In contrast, glycogen content in the developing ovary decreases relative to that of the spent ovary, where it may probably be used for *in situ* synthesis of lipids and nucleic acids rather than being transported to other organs, since there is no associated increase in glycogen of the hepatopancreas.

Pillay and Nair (1973) have also determined the biochemical constituents of the gonad, hepatopancreas, and muscle in relation to the reproductive cycle of three tropical decapods, *Uca lactea annulipes, Portunus pelagicus,* and *Metapenaeus affinis.* In all three species, lipids fluctuated greatly in relation to the reproductive cycle, with the mature ovary containing more lipid than the immature or spent ovary. The hepatopancreas is apparently a storage organ and contains much lipid and glycogen. There is an indication that at least a part of the lipids from the hepatopancreas are translocated to the ovary. In contrast, changes in biochemical constituents of the testes are not pronounced. On an annual basis, the ovary appears to be richer in lipid while the testes are richer in protein content. Species-specific differences in biochemical constituents of these three species are evident. While lipid is

[1]Also known as midgut gland(s), or digestive gland.

the most important constituent in each of the above three species, the carnivorous *Portunus pelagicus* has a higher protein content in the ovary compared to the other two species, which feed on decaying organic material or detritus.

A decline in lipid content of an Indian lobster between October and January and evidence of storage between February and April has been reported (George and Patel, 1956). Seasonal changes in biochemical composition have also been observed in various body tissues of *Callinectes sapidus* (Thompson and Farragut, 1966). In *C. sapidus,* blood glucose levels increase just prior to ovulation and decline in ovigerous animals (Dean and Vernberg, 1965). *Neomysis integer* breeds throughout the year, with a major brood in March and a decline of egg production during the winter months (Raymont *et al.,* 1966). A significant decrease in lipid content occurs in this species during the months of November, January, and March in relation to breeding, food supply, and metabolic rate. Lipid levels decrease in March with the utilization of lipid reserves for egg development, and in November and January with an increase in respiration rate. An absence of decrease in lipids at other times may be offset by the increase in food supply (phytoplankton) from April onward into the summer. Linford (1965) showed that eggs of *N. integer* contain large quantities of lipid material. The principal protein component of mature ovaries and eggs of six species of decapods were found to be lipoprotein (Wallace *et al.,* 1967). In the antarctic caridean, *Chorismus antarcticus,* females deposit large amounts of lipid in the ovary during the period of egg production in summer (Clarke, 1977). Lipid composition of the ovary and eggs is identical, with over 95% consisting of approximately equal amounts of triglycerides and phospholipid. Free sterol and other lipids are only present in small amounts, and the carotenoids are present in small amounts similar to those found in other decapods (Herring, 1973; Clarke, 1977).

Complete seasonal studies on changes in biochemical constituents relative to the gonadal cycles are of interest for both basic studies and practical use. Studies using isotopic tracers will enable a more detailed account of the partitioning of various materials during development and growth of the gonads. Studies designed to examine the quantitative and qualitative changes in biochemical composition of the body organs relative to the molt cycle and reproductive cycle over the annual cycle should reveal many aspects of biochemical transformation and translocation with respect to molting and reproduction.

iii. METABOLIC ACTIVITY. Metabolic activity in terms of oxygen uptake, either in relation to seasonal variation in temperature or with acclimation, has been studied extensively for a number of crustaceans. Much of this work has centered around the ability of animals to regulate metabolism under the stress of temperature or other environmental factors (Vernberg and Vernberg, 1970). Relatively little attention has been given to the effects of internal state of the animals. Clearly, the internal state of an organism varies with the reproductive cycle and/or season, and it may effect the metabolic rate and possibly give rise to rate compensation. Oxygen uptake of an animal is a measure of its overall aerobic energy demands under the particular conditions at which the measurements are made. Endogenously, oxygen uptake at any given time is a function of the relative biochemical composition of the tissues and the substrates utilized for metabolism. As discussed earlier, biochemical composition of the tissues varies relative to the reproductive cycle and with the season relative to food availability and temperature in animals experiencing fluctuating environments.

Barnes *et al.* (1963) measured the seasonal oxygen uptake rate of two barnacles, *Semibalanus balanoides* and *Balanus balanus,* and examined its relation to changes in the biochemical composition and reproductive condition. In *S. balanoides,* a decline in oxygen uptake rate during late summer and autumn may be associated with a reduced storage of reserves and with an increased amount of semen. After sperms are released with copulation in October, the oxygen demand increases slightly, but the values remain low in the winter. This reduced oxygen uptake rate has been attributed to reproductive anecdysis,[2] low levels of food supply, and uptake of water into body tissues. The oxygen uptake rises again in late winter as the molting frequency increases and tissues return to the normal condition. In spring and summer, the opposing factors of active deposition of energy reserves and increased quantity of inactive material seem to result in a slight rise in oxygen uptake at the time of diatom increase in March. Peak activity is reached with resumption of vigorous growth. Somewhat less activity is associated with the period when ovaries are initiated and material is being transferred to the ovaries. In late summer, as spermatozoa develop and increase in proportion of DNA, the overall oxygen consumption rate declines to the winter level. Similar changes in oxygen uptake rate associated with the seasonal changes in body composition have been observed for *B. balanus* in relation to the pattern of its annual cycle. The authors concluded that differences in weight-specific oxygen uptake rate are due to differences in biochemical composition of the organs of the animals rather than to differences in oxygen requirements or acclimation to different seasonal temperatures.

Barnes and Barnes (1969) also examined the question of metabolic regulation for the barnacles *Semibalanus balanoides, Balanus balanus,* and *Chthamalus stellatus* in relation to the annual reproductive cycle. In *S.*

[2]A short period immediately following egg deposition, when moulting frequency is much reduced due to the debilitated condition of the barnacle.

balanoides, marked metabolic rate compensation is observed between 10°–15°C for most of the months investigated, a time period when reserves are accumulated and utilized for gonad production. Compensation of metabolic rate was also observed in October, when gonad production has been largely completed.

A similar independence of metabolic rate over a limited thermal range has also been observed for a number of brachyuran crabs and among populations from different geographical locations (Vernberg and Vernberg, 1970). Comparisons of species from different habitats having contrasting life history patterns may help elucidate the processes involved in variation in oxygen consumption relative to reproductive condition and other internal factors.

iv. HORMONAL FACTORS. Much of our understanding of hormonal control of molting, metabolism, and reproduction, and their temporal interrelationships is based on the work done with decapods (Passano, 1960; Adiyodi and Adiyodi, 1970; Aiken and Waddy, 1980). The extent to which this information is applicable to other groups of crustaceans remains unclear (Steel, 1980). Hormonal control of molting and reproduction in decapods is reviewed in detail in other chapters in this series. The information presented in here is mostly derived from the reviews by Adiyodi and Adiyodi (1970) and Aiken and Waddy (1980).

Six hormones are known to be involved in decapod reproduction (Fig. 2) (Adiyodi and Adiyodi, 1970), and a seventh has been implicated on the basis of work done with the amphipod *Orchestia* (Junera *et al.,* 1977). Two of these hormones control molting: molt inhibiting hormone (MIH) is involved with reproduction by suppressing the molting, and molting hormones (MH) seem to occur at significant levels in developing ovaries of decapods (Lachaise and Hoffmann, 1977). Two other hormones are primarily involved in control of reproduction. Gonad inhibiting hormone (GIH) prevents precocious maturation of the ovaries (Panouse, 1943, 1947; Rangneker and Deshmukh, 1968). GIH seems to be present in both immature and mature decapods of both sexes and appears to be cyclic in occurrence. In females, GIH appears to regulate vitellogenesis (Charniaux-Cotton and Kleinholz, 1964), affecting primary vitellogenesis more than secondary vitellogenesis. A gonad stimulating hormone (GSH) is also produced by the central nervous system (Otsu, 1963; Gomez and Nayar, 1965) and appears to play a dual role promoting vitellogenesis while suppressing entry into premolt. Two other hormones are involved in sexual differentiation. Androgenic hormone (AH) and the androgenic gland have been reviewed in detail by Charniaux-Cotton (1962) and are discussed elsewhere in this treatise. AH determines the masculinity, and is responsible for the development and maintenance of secondary sexual characters and testes and vasa defer-

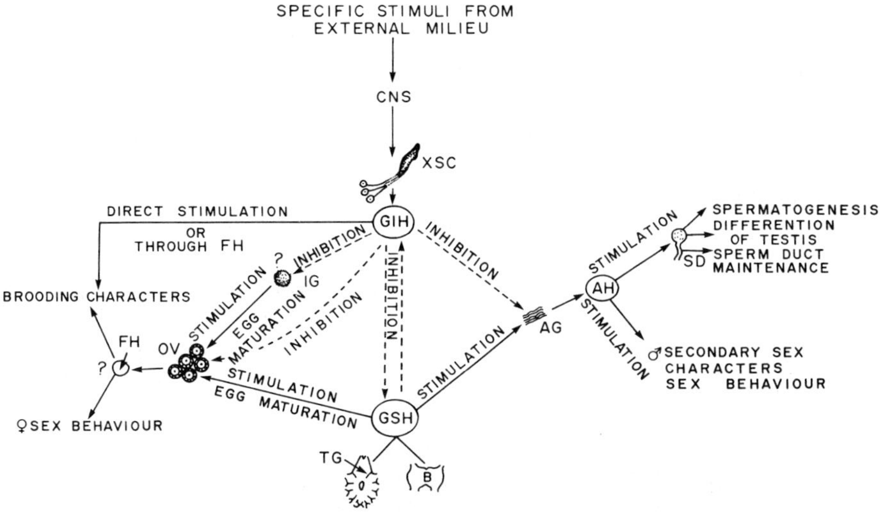

Fig. 2. Diagram showing the hypothesized relationships between gonad inhibiting and gonad stimulating principles and sex hormones in reproduction of the decapod crustaceans. AH, androgenic hormone; B, brain; CNS, central nervous system; FH, female hormone; GIH, gonad inhibiting hormone; GSH, gonad stimulating hormone; IG, intermediate gland; Ov, ovary; SD, sperm duct; TG, thoracic ganglion; XSC, x-organ–sinus gland complex. (From Adiyodi and Adiyodi, 1970.)

entia. The existence of female hormone (FH) is speculative, especially in decapods.

After reaching puberty, the decapods generally develop gametes during their intermolt (C_4) period (Adiyodi and Adiyodi, 1970). The postmolt (A,B,C_1-C_3) is concerned with the formation of new exoskeleton and tissue growth, and the premolt (D) is involved in preparation for the next molt in adults (Passano, 1960). According to Adiyodi and Adiyodi (1970), after completion of C_3 stage of the intermolt cycle, the excess metabolites are diverted primarily to the hepatopancreas for storage. Molting and reproduction are interrelated, but are temporally separated in the decapods. Both molting and reproduction require mobilization of organic reserves from storage sites to the epidermis and gonads. This cyclic mobilization of reserves is coordinated hormonally. In decapods, the eyestalk factors seem to regulate the deposition and subsequent mobilization of organic reserves utilized for reproduction (Adiyodi, 1969; Yamamoto, 1960; Miyawaki and Tanoue, 1962). Free sugars in hepatopancreas of the *Parathelphusa hydrodromus* fluctuate during the ovarian cycle; these same sugars are found during early vitellogenesis, apparently being utilized in the formation of lipoprotein (Adiyodi and Adiyodi, 1970). The eyestalk contains substances that raise blood sugar levels in crabs (Telford, 1968), and a blood glucose increase just prior to ovulation and a decline in ovigerous animals has been observed (Dean and Vernberg, 1965). The hepatopancreas also stores carotoproteins, which are mobilized to the ovary during vitellogenesis (Goodwin, 1951). Adiyodi (1968a,b) reported that a female specific protein (FSP) is formed almost exclusively during vitellogenesis. Its formation is apparently influenced by the reproductive hormones. In crabs, FSP accumulates during the postmolt and reaches a maximum in hemolymph before onset of vitellogenesis. Eyestalk ablation, which eliminates GIH, results in precocious decreases of FSP in hemolymph, suggesting that GIH prevents uptake of FSP by the ovary.

In *Cancer magister,* only two lipoproteins are present in the hemolymph of males and sexually immature females (Allen, 1972). An additional lipoprotein appears in the hemolymph of females underoing vitellogenesis. Lipid concentration in the hemolymph more than doubles with ovarian maturation. *In vitro* labeling experiments performed with *C. magister* in early phases of ovarian maturation showed that the ovarian triglycerides incorporated more label than did any other lipid class of any organ except the hepatopancreas. The flux of hepatopancreas lipids to the triglyceride pool of the ovary is also apparently very rapid.

The concept of antagonism between molting and reproduction resulting in one process occurring only at the expense of the other has been suggested from work done with a number of decapods (Passano, 1960; Bliss, 1966; Cheung, 1969; Adiyodi and Adiyodi, 1970; Sochasky, 1973; Kleinholz, 1976; Aiken and Waddy, 1976). In decapods, eyestalk ablation accelerates molting or reproduction, whichever is dominant at the time. In juveniles, the emphasis is on somatic growth, but in the intermolt stage, female eyestalk ablation accelerates ovarian growth and suppresses premolt. Adiyodi and Adiyodi (1970) suggested that gonad development initiates a shift from a state of synergism between MIH and GIH in the postmolt to antagonism in the intermolt, possibly to allow for a temporal separation of the demands for organic reserves for molting and for ovarian development. In *Homarus americanus,* vitellogenesis suppresses the onset of premolt in adult females (Aiken and Waddy, 1980). In *Carcinus maenas* and *Parathelphusa hydrodromus,* vitellogenic growth of the oocytes takes place in two phases. The first phase of vitellogenesis occurs during premolt, but the second phase seems strictly opposed to somatic growth and hardly takes place during premolt (Demeusy, 1965; Adiyodi and Adiyodi, 1970). Seasonal differences in timing of molting and ovarian development have also been reported for some species (Carlisle, 1953; Cheung, 1969) with the diversion of energy for somatic growth or reproduction varying with season and environmental factors such as temperature, food availability, and day-length.

In the terrestrial isopod, *Oniscus asellus,* molting and reproduction appear to be controlled by separate environmental cues and probably different hormones (Steel, 1980). Field populations of *O. asellus* molt throughout the year, but females incubate a brood of eggs only during part of the season, which suggests separate environmental regulation of molting and reproduction. Specific sensory cues may coordinate molting and reproduction, and maintain the temporal relationship between them. Steel indicates that there is cytological evidence showing separate groups of neurosecretory cells in the brain of *O. asellus;* one group changes with the molt cycle and another group with vitellogenesis. It is suggested that more than one hormone is involved, one for initiating vitellogenesis, the release of which is possibly regulated by day-length, and another, possibly ecdysone, for subsequent uptake of vitellogenin by oocytes. The regulation of vitellogenesis appears to be different in isopods compared to decapods, where the entire vitellogenesis occurs later during intermolt period.

Although complex interaction between hormonal factors, molting, and reproduction is indicated, it is still unclear how these processes are regulated within an individual and coordinated with the environment. Significant changes occur in organic reserves in relation to the molt cycle, and the hormonally controlled cyclic biochemical changes are reflected in the hemolyph of decapods (Passano, 1960; Florkin, 1960). Detailed information on the qualitative and quantitative changes in organic reserves, pools, and fluxes relative to reproduction and on the role of hormonal factors in transformation and translocation of nutrients is unavailable for most crustaceans. Thus far, no endocrinology studies have been conducted to determine the mechanisms in regulation of the reproductive cycle in relation to the environment.

b. Environmental Factors. A number of environmental factors such as temperature, day-length, food availability, and social conditions may influence reproduction. Influence of different environmental factors or their combination on all phases of the reproductive cycle has not been determined for any crustacean species. It is possible that environmental factors may differentially affect the sequence of stages in gametogenesis as well as the other phases of the reproductive cycle. Generally, reproduction occurs within the narrowest environmental range of conditions, compared to those suitable for survival, or even growth of a species. Work with a variety of invertebrates shows that temperature (both level and rate of change), day-length (intensity and periodicity), and food availability (quality and quantity) are the major exogenous determinants of the reproductive cycle. Additionally, many other factors such as salinity for those species inhabiting coastal and estuarine regions, humidity for terrestrial species, chemical factors, or biotic factors such as parasitism also influence reproduction. There is also some evidence to suggest that factor interaction takes place so that changes in a critical level of one factor may be accompanied by changes of the others. It is therefore pertinent to ask to what extent the reproductive cycle in individuals is under the control of environmental factors, to what extent endogenous factors are involved, and what the relation of the latter is to environmental stimuli. Many environmental parameters act as cues; this information, which is channeled through the nervous system, ultimately induces appropriate organismic responses via endocrines to influence growth and reproduction. Our understanding of the complex functional integration of reproduction to the environment is still scanty.

i. LIGHT AND TEMPERATURE. For populations inhabiting the higher latitudes, the seasonal changes in the environment may result in alternating periods of favorable and unfavorable conditions for reproduction. These seasonal changes are accompanied by changes in day-length, temperature, and food availability, and the crustaceans may utilize this information either singly or together to coordinate molting and reproduction with an appropriate season.

Increased day-length has been shown to accelerate normal cyclic increase and decrease in ovarian size of the crayfish *Orconectes virilis* (Stephens, 1952). Complete darkness interrupts this cyclic activity, and the oocytes develop to a mature condition and remain in that state. Suko (1958) found that response of *Procambarus clarkii* to darkness depends on the stage of the ovary when the animal is placed in darkness. Lowe (1961) found that *Cambarellus shufeldti* shows varied responses to lack of light depending on the time of year. Increased photoperiod induces a more rapid cycling of maturation and resorption of oocytes. Decreased day-length tends to stabilize the ovary in a mature condition at one time of the year, while at another time it tends to allow increased proliferation of oocytes without maturation. Low temperatures tend to slow maturation of oocytes, whereas elevated temperatures cause rapid maturation followed by disintegration. Aiken (1969a,b) showed that both temperature and photoperiod influences the endocrine regulation of the seasonally dependent molting and reproduction in *O. virilis*. Breeding condition has been induced in the grass shrimp, *Palaemonetes pugio,* out of season at elevated temperatures and with a combination of increasing temperatures and photoperiod (Little, 1968).

In the boreoarctic amphipod, *Pontoporeia affinis,* a glacial relict, constant illumination inhibits the normal development of gonads (Sergestråle, 1970). Decreased illumination in late summer triggers maturation. The amphipod, *Gammarus setosus,* is distributed in the Western Atlantic from Ellensmere Island in the north to the Gulf of Maine in the south, and it produces a large well-timed brood of young each year over its entire range of distribution (Steel *et al.,* 1977). Duration of the various phases of the reproductive cycle varies inversely with temperature, but temperature by itself does not seem to initiate reproductive activities (Steele and Steele, 1970). Photoperiod accelerates or slows the reproductive cycle of this species, but it does not seem to completely control the cycle. Under all light regimes tested, the resting stage followed the production of clutches and brood production continued under all conditions, including O L : 24 D. Timing of brood production is in partially intrinsic, but actual timing may be determined by the environmental conditions so that the young are released at an optimum time of the year. In the field or under ambient day-length conditions, the population produces a single brood. However, under experimental conditions the animals can produce an asynchronous second brood.

Both temperature and day-length influence molting and reproduction of the terrestrial isopod, *Oniscus asellus,* but they have differential effects on the two processes (McQueen and Steel, 1980; Steel, 1980). Molting is strongly temperature dependent, but is unaffected by day-length. Reproduction is induced by long day-length (16L : 8D) and suppressed by short days (8L : 16D). Reproduction of *O. asellus* appears to be more closely correlated with day-length in the field than with other environmental variables. This serves to confine reproduction to the period from April through September.

Barnes (1963) reported that light of even moderate intensity inhibits gonad development at both high and low temperatures in the boreoarctic barnacle, *Semibalanus balanoides*. For this species, there is a critical temperature (10°C) above which the animals do not reproduce, but this limiting temperature can be raised slightly by total darkness. Breeding is inhibited by constant light, and a period of 4–6 weeks with less than 12 hr of light per day and below critical temperature is necessary for maturation of the gametes. These results indicate that light may act to delay the date of fertilization in this species. Crisp and Patel (1969) indicated that prolonged illumination is unlikely to prevent breeding indefinitely, and that the threshold at which light delays does not remain constant but falls during the conditioning period at 10°C and below. By keeping animals in constant light at 16°–20°C, Tighe-Ford (1967) has obtained delay, both in the appearance of egg masses and in the formation of an elastic sac secreted by the oviducal gland prior to oviposition. When returned to a dark environment, the oviducal gland enlarged, formed a sac, and breeding took place in 4 weeks. Studies concerning ovarian development in relation to lunar phase in *Uca tangeri* (von Hagen, 1962, 1970) and *U. musica terpsichores* (Zuker, 1976) revealed that ovaries are most highly developed around full moon.

ii. SALINITY. The influences of salinity on reproduction is less well studied compared to temperature. Diurnal and annual salinity variations are not as predictable in occurrence as are temperature and day-length. Hence, these variations cannot serve as coordinators for breeding, migration, and other reproductive activities. Generally, salinity effects are less obvious when compared to temperature effects. Female amphipods, *Gammarus dubeni,* produce an average of 18 eggs at 20‰, 27 eggs at 10‰, and 17 eggs at 30‰ salinity per oviposition (Kinne, 1964). *Gammarus dubeni* lays more eggs in brackish water than in fresh water. In the amphipods, *Bathyporeia pelagica,* a population that has migrated into the estuary produces a larger brood than a population living on the open beach (Fish and Fish, 1978).

Kuhl (1966) found that *S. balanoides* grows below 5‰ salinity in Elbe estuary, but fails to reproduce. Barnes and Barnes (1962) indicate that penetration of *S. balanoides* into the Baltic is limited by low salinity combined with high temperatures, and they suggested that these two environmental factors may initially act through their effect on fecundity. Considerable salinity gradient or regular salinity fluctuations exceeding the range suitable for reproduction may differentially affect phases of the reproductive cycle to cause reproductive failure. However, migrations into or out of estuaries are a regular feature associated with the reproduction of some species (e.g., *Callinectes sapidus,* palaemonid shrimps). Downstream migration to higher salinity waters for mating and oviposition is well known for the Chinese mitten crab, *Eriocheir sinensis.* With the exception of *Metapenaeus bennettae,* most Penaeinae that are unable to migrate to waters of suitable salinity do not mature (Wickins, 1976).

iii. FOOD. Crustaceans occupy different trophic levels of the food web, and the temporal patterns of quality and quantity of available food also vary for different species. However, the interrelationships between trophic position of species, food availability, food intake, assimilation efficiency, and reproduction have not been examined systematically. In recent years, theories of feeding strategies on one hand (Schoener, 1971) and reproductive strategies (Stearns, 1976, 1977) on the other have been well developed, but a synthesis of the two has not yet been made.

Changes in gonadal weight or volume measured over a time interval or over the reproductive cycle provides an integrated measure of the rate at which nutrients accumulate in the gonads. However, the relationship between environmental food availability and gamete production is not always evident for many species. Oogenesis involves an intense biochemical synthesis with the mobilization of lipids and proteins for the development of ova. Mobilization of nutrients may be controlled by certain proximate factors, which determine the time that gametes are produced within an individual, both in relation to its own life history and with respect to the environment. Our knowledge of the proximate controls of nutrient mobilization to gonads is still incomplete. The ability of an organism inhabiting a given environment to divert nutrients for gamete production is important for the reproductive success of the individuals, and it also has implications for the distributional patterns of species (e.g., expatriate or fringe populations; seasonal succession of short-lived species).

The relationship between environmental food availability, nutrient storage, and gamete production may vary for different species. In some species, gamete production is coincident with environmental food availability, while in others nutrients are stored when food is abundant and utilized later for gamete production. For many species, it is not known which of these strategies are employed. In all cases, however, gamete production may not occur unless a minimal amount of nutrients are available to the gonads. The nutrient status of an individual, coupled with the current stage of gametogenic cycle, may be important in determining whether or not that individual produces gametes in a given environment. In spite of the central importance of the question of nutrient allocation for reproductive success, both in a proximate and ultimate sense, there still remain many unresolved problems in this area.

The spatial, temporal, qualitative, and quantitative feeding patterns of some copepods (Mullin, 1967; Marshall, 1973; Heinle *et al.*, 1977; Hayward, 1980) and a few decapods (Hartnoll, 1963; Dearborn, 1967; Hughes and Elner, 1979; Elner and Hughes, 1978; Hill, 1979; Wolcott, 1978) have been examined, but the relationship to reproduction is not yet well established.

Planktonic copepods experience temporal changes in food availability associated with seasonal cycles of temperature in the higher latitudes. Some of these planktonic copepods adapt to temporal changes in the food abundance by storage of energy reserves, primarily as lipids and arrested stages of development in their life cycles (Marshall and Orr, 1972; Corkett and McLaren, 1978). Storage of lipids has been related to reproduction and growth (Heinrich, 1962) in the copepods. In *Calanus finmarchicus,* a primary consumer, breeding and development is coincident with spring phytoplankton bloom, and only the molt from copepodite stage V to the adult occurs outside the spring bloom. In others, (e.g., *Calanus cristatus* and *Calanus plumchrus*), breeding and growth of juveniles may occur outside the annual period of phytoplankton abundance; however, these species do not feed as adults, but maintain and reproduce on stored lipids as their major energy source. In some other species, which reproduce throughout the year for extended periods in coastal and estuarine waters of the tropical, temperate, and boreal regions, the brood size is affected by food supply. Ikeda (1974) found that tropical copepods do not have lipid droplets as an energy store. Generally, storage of lipid seems to occur mainly in boreal or deep-water forms, and least so in tropical species.

The copepod, *Pseudocalanus minutus,* stores lipid in oil sacs for short-term storage of energy to buffer reproduction (Corkett and McLaren, 1978).

This species produces a relatively constant volume of eggs in successive clutches, but at reduced food levels the interval between clutches is prolonged while a constancy of clutch volume is being maintained. Females that produce clutches at longer time intervals have reduced oil sacs; apparently the stored lipid is utilized to buffer reproductive output from the immediate food shortage. In *P. minutus,* although egg production is interrupted during the period of starvation, it does not affect later egg production or longevity of the individuals when food supply is resumed. Thus, egg production is prevented below a certain critical food level, but above this level, egg production occurs at a temperature-dependent rate.

In female *Calanus finmarchicus,* egg production is always higher in fed than in starved animals, and starvation for about 4 weeks prevents egg production; but resumption of feeding restores egg laying (Marshall and Orr, 1952, 1972). Food quality also affects egg production in this species. Conover (1967) has indicated that *Calanus hyperboreus* may require abundant food to initiate breeding, but once initiated, the presence or absence of food has little affect on the egg laying. However, viability of eggs from starved animals is low, and the eggs tend to be more buoyant than those from fed animals. Heinle *et al.* (1977) found that starvation of the estuarine harpacticoid copepod *Scottolana canadensis,* shortly after terminal molt, completely prevents reproduction, even if food supply is resumed. In the predatory copepod *Pareuchaeta norvegica,* females resorb oocytes and discontinue the breeding cycle when feeding conditions are inadequate (Hopkins, 1977).

The factors regulating the annual pattern of reproduction in the boreoarctic barnacles, *Semibalanus balanoides* and *Balanus balanus,* have been examined in an extensive series of investigations (Barnes *et al.,* 1963; Barnes, 1963; Barnes and Barnes, 1967, 1968; Patel and Crisp, 1960; Crisp and Patel, 1969). These two species have an annual cycle adapted to the boreoarctic food and temperature conditions. In *S. balanoides,* gonads are developed during summer and early autumn, while copulation occurs in late autumn. The embryos are incubated in the winter and the nauplii are released in spring coincident with the diatom increase (Barnes, 1957). In *B. balanus,* ovarian development continues throughout winter with copulation in mid-winter. There is only a brief retention of embryos before naupliar release in spring. Both species produce only a single brood annually. The annual cycle of these two species centers around the assimilation and storage of nutrients during the spring months and later utilization of the stored reserves for gametic production. Proteins, lipids, and carbohydrates are stored and presumably utilized during gonad production (Barnes *et al.,* 1963). Nutrient storage by these species seems to ensure nutrient availability for gamete production, which occurs when food abundance in the environment is low, and also reduces demands on the immediately ingested food during the period of gonad development.

Even when temperature, food supply, and light conditions are experimentally maintained at their optimal levels, it is not possible to advance the date of egg mass production by more than a few weeks. Barnes *et al.* (1963) suggest that failure to shorten the breeding time of these species is due to biochemical rate limitations during oogenesis. Barnes and Barnes (1967) studied how early nutrient storage must precede gonad development, as well as what the effects of starvation would be at different times during oogenesis in *Semibalanus balanoides*. They found that ovarian development is dependent on an adequate food supply, with interruption of food supply leading to regression of the ovarian tissue. This effect is most obvious in the early stages. Renewal of feeding, however, even after starvation for 5 months, resulted in renewed development of the ovary. Once initiated, ovarian development in the summer is dependent on food supply and possibly on other environmental factors prevailing at the time. In later stages of ovarian development, starvation leads to little or no regression, and development recovers with renewed feeding. According to these authors, the rate of ovarian development is under metabolic control, which in turn is dependent on food supply. Although the rate of gonad development cannot be accelerated by environmental manipulations, the time at which egg masses are deposited and hence the time at which nauplii are released can be delayed by the food supply to animals. In this way, egg mass formation may be delayed from November to mid-January. A prolonged starvation of barnacles into the autumn causes ovarian development time to extend into the period when new ovarian tissue is normally initiated. Under such conditions, initiation of the new tissue takes place and the ovary of the previous year regresses. Fhyn and Costlow (1977) maintained *Balanus amphitrite* without food for 10–40 days. They found that immature oocytes larger than 20 μm disintegrate, but smaller oocytes and oogonia remain normal and the diameter of fully grown oocytes remains unaltered. In contrast to the boreoarctic species, the warm water barnacles, *Chthamalus stellatus, Balanus perforatus,* and *Balanus amphitrite,* breed continuously. These species develop gonads and breed rapidly when temperature is sufficiently high and food is available (Patel and Crisp, 1960). It is not known whether any storage precedes gonad development in these species. It seems, however, that an appreciable storage is unlikely.

Similarly, in certain decapods, the oocyte growth is arrested under unfavorable conditions, and under certain conditions, the developing ova may even be resorbed. Resorption of lipovitellin previously accumulated in the oocytes of *Homarus americanus* occurs at varying degrees (Aiken and Waddy, 1980). For example, if resorption occurs late in secondary vitellogenesis, the quantity of lipovitellin reclaimed from the oocytes turns the hemolymph dark green and causes elevated FSP levels. Cycles of vitellogenesis and resorption also occur routinely for the crayfish, *Cambarus* (Stephens, 1952) and *Nephrops andamanicus* (Berry, 1969). In palinurid lobsters, *Jasus lalandii,* oviposition does not occur in the absence of fertilization, but rather the ripe ovary will be broken down and resorbed (Heydorn, 1969a,b). In *Homarus,* resorption appears to be associated with unfavorable laboratory holding conditions near the expected time of oviposition and results in regression of the ovary to a spentlike condition (Aiken and Waddy, 1980). Resorption is suggested to be favored by low levels of GSH or levels of high GIH in the circulation (Adiyodi and Adiyodi, 1970).

iv. PARASITES. Some species of crustaceans are parasitic and have highly modified reproductive cycles. Alternately, a number of other crustaceans serve as parasite hosts. Reproduction of the parasitic forms (Yamaguti, 1963), or the effects of parasitism on reproduction of free-living crustaceans (Kaestner, 1970) are not considered in detail here. A few examples will illustrate the potential effects of parasitism on reproduction. *Cryptothir balani*[3] inhibits ovarian development in *Balanus* by inhibiting the breakdown of the nutrient reserves which provide for gonad production (Crisp, 1960). The copepod, *Sphaeronella longipes,* feeds ectoparasitically within the marsupium of the host amphipod, *Ampelisca tenuicornis* (Sheader, 1977). Host females with parasites do not produce eggs, and their ovaries are reduced to an immature state. Similarly, *Sphaeronellopsis monothrix,* parasitic on the ostracode *Parasterope pollex,* suppresses egg development in the host (Bowman and Kornicker, 1967). In the ostracode *Candona* sp., parasitized by cestode *Hymanolapsis,* the cysticercareae attach to the ovaries of the host and produce atrophy of the reproductive organs and marked hypertrophy of the carapace. Parasitic infestation of crabs by the rhizocephalan, *Sacculina* prevents males from developing testes, and in many, the form of the abdomen is feminized. For example, males of *Pachygrapsus marmoratus* infected by the *Sacculina* are feminized (Kaestner, 1970).

2. ENERGETICS OF EGG PRODUCTION

Gamete production is an energy demanding process. An organism must allocate a certain amount of the assimilated energy to the gonads in the face of competing demands by the interrelated processes of maintenance and somatic growth. In order to determine the amount of energy allocated to gamete production, one has to determine the flow of energy and fluxes to various tissues over the period of gonadal cycle and/or over the adult lifetime (Pianka, 1976; Calow, 1979). The energy content of mature gonads or

[3]This parasite has been assigned also to *Hemioniscus* by some European workers.

the number of eggs oviposited is an integrated measure of the energy allocated to the gonads at a given point in time. Few studies have measured the energy specifically apportioned to gamete production, per se. A large number of studies have determined the number of eggs oviposited by females, which provides a rough indication of the reproductive effort for a species.

a. *Energy Allocation.* In *Daphnia pulex,* a major portion of the stored energy goes into the production of young (Richman, 1958). The net efficiency of production of offspring increased from 52.1% at a low food concentration to 70.4% at the highest tested. For *Daphnia,* respiration is independent of food consumed, so the energy available for growth and reproduction is directly related to the amount of food consumed. Daborn (1975) examined the energy relationships of the giant fairy shrimp *Branchinecta gigas,* a predaceous species, from an aestival lake in Canada. Males and females of this species assimilated 67 and 94%, respectively, of the ingested food and invested 23 and 35% of the assimilated energy into somatic growth. Molting losses were estimated as 5 and 6%, excretion at 14%, and maintenance at 58% in males and 47% in females, respectively. Females invested 11.3% of the assimilated energy at the expense of somatic growth to egg production, while that invested by the males for sperm production was negligible. For example, growth was temporarily slowed at the time when eggs were about to be released into the egg sac, then the growth resumed after the first clutch was fully developed. Maintenance was the major energy expenditure, such that when food was in short supply, the expenditure for reproduction was curtailed.

The predatory copepod, *Pareuchaeta norvegica,* has a half year life cycle, with the population producing egg sacs in January–February and July– August on the West coast of Norway (Båmstedt, 1979). Females born from the summer generation produce fewer eggs per sac during the next winter than are produced by a winter generation in the summer. Winter breeders transfer about 25% of their body weight to egg sac production while the summer breeders transfer about 37% of their body weight. This reflects the differences in feeding conditions in the respective seasons. It is suggested that high efficiencies in energy transformation occur for energy storage prior to egg laying. Environmental differences appear to influence the energy budget and reproductive output of *P. norvegica* summer and winter generations.

Assimilation efficiencies may depend on the feeding mode of a species (Pandian, 1975; Conover, 1978). The extent to which assimilation efficiencies influence a species strategy is unknown. Intraspecific variation may also occur in the assimilation efficiencies. Prus (1971) reported higher absorption efficiencies in the nonovigerous and ovigerous females (41%) than in males (33%) of the freshwater isopod, *Asellus aquaticus.*

Wu and Levings (1978) constructed an annual energy budget for individual adult barnacles, *Balanus glandula,* for the first year after settlement. A very large portion of the energy is lost for respiration, 67.4%; energy allocated to the remaining processes was 12.3% for egg production, 6.6% for shell production, 3.9% for body tissue, and 2.3% for molting. This species invests a large amount of energy in reproduction. It has an opportunistic type of life history, with crowding of *B. glandula* reducing the amount of energy channeled to reproduction (Wu *et al.*, 1977). More energy is channeled to body tissue and building taller shells than to reproduction in crowded individuals.

A number of factors, including quality and quantity of food, assimilation efficiency, metabolic efficiency or economy, competing demands for somatic growth, age, and environmental factors, influence the amount of energy invested by an individual for gamete production. Many of these characteristics may vary from one species to another, or even between populations over different parts of the distributional range.

b. Reproductive Output. Energy allocated to gamete production may be packaged into a large number of small ova, or a small number of large ova, and these characteristics vary from one species to another. The number of eggs produced by individuals of a species may be characteristic of that species; however, they seem to vary within limits depending upon age, size, and environmental factors, especially temperature and food supply. Reproductive output of crustaceans has been generally determined by counting the number of fertilized eggs ovulated and carried by the female as an egg mass, in a ovisac or a brood chamber. The term fecundity has been often used to represent reproductive performance, although it more accurately denotes the rate at which eggs are produced (Barnes and Barnes, 1968).

i. EGG SIZE. The size of the eggs produced by a species is under genetic control, but it is also in part phenotypically determined (Raven, 1961). Egg size has important consequences to the rate of development and to the size of larvae or juveniles at hatching. Larger eggs produce larger offsprings, better adapting them for feeding and providing greater competitive ability. Generally, species from the colder polar regions or mesopelagic, bathypelagic, and deep-sea benthic environments and some terrestrial and freshwater decapods seem to produce larger eggs compared to those from warm water regions or from the epipelagic zone (Bliss, 1968; Mauchline and Fisher, 1969; White, 1970; Knox, 1970; Omori, 1974; Marshall, 1979). In the meso- and bathypelagic regions, the nonmigratory species of copepods have larger yolky eggs, which allow development through naupliar stages, with feeding beginning in copepodite stage I when the mandibles are fully formed (Marshall, 1979). The penaeid and carid shrimp from the lower mesopelagic and bathypelagic zone produce larger but fewer eggs relative to those from epipelagic zone, with the larvae surviving on internal yolk during their migration to shallower waters where there is a rich food supply (Omori, 1974).

Among the brachyuran crabs, there is considerable interspecific variation in the size of eggs. In addition, egg size may increase during embryonic development, which complicates comparisons (Bourdon, 1962; Wear, 1974). However, species with lecithotrophic larval development and direct development tend to produce larger eggs, while smaller eggs are produced by those with planktotrophic larval development.

A similar pattern of interspecific variation in egg size is also common among other groups of Crustacea (Allen, 1966a; Mauchline and Fisher, 1969). Patel and Crisp (1960) measured the dimensions of fertilized eggs for four species of barnacles produced by adults fed and held at different temperatures during oogenesis. In all four species, the egg volume increased inversely with the holding temperature. Crisp (1960) also found a significant difference in the volume of fertilized eggs of the boreoarctic barnacle, *Semibalanus balanoides,* collected from different latitudes. The size of eggs decreased from colder to warmer waters over a north–south gradient. In barnacles, the size of developing ova does not appear to depend to any significant extent on the nutrition of the parent, for animals that are poorly fed produce fewer, but not necessarily smaller, eggs (Patel and Crisp, 1960). Corkett and McLaren (1978) indicated that the mean number of eggs in an ovisac of the copepod *Pseudocalanus minutus* is a function of female cephalothorax length. In addition, the egg number–length relationship seen in one region cannot be used for predicting egg numbers for populations from other areas, since the mean diameter of the eggs may differ between localities.

ii. NUMBER OF EGGS. The number of eggs and the rate at which eggs are produced by females may be characteristic of a species and may have significance for the life history strategy and ecology. The number of eggs produced influences the dynamics of populations (Edmondson, 1968; Taylor and Saltkin, 1981). Many studies report the number of eggs carried by the females as fecundity, although these data actually represent the number of eggs oviposited at a given time, i.e., the clutch size (Barnes and Barnes, 1968; Corkett and McLaren, 1978). To estimate fecundity, information on the number of eggs in a clutch and the rate of clutch production over the breeding season and/or over the lifetime of an individual is needed. Such information is generally unavailable for most species.

Numerous studies report the number of eggs per female for different species of copepods, barnacles, cumaceans, isopods, mysids, euphausiids, and decapods (Jensen, 1958; Allen, 1966a; Bliss, 1968; Mauchline and Fisher,

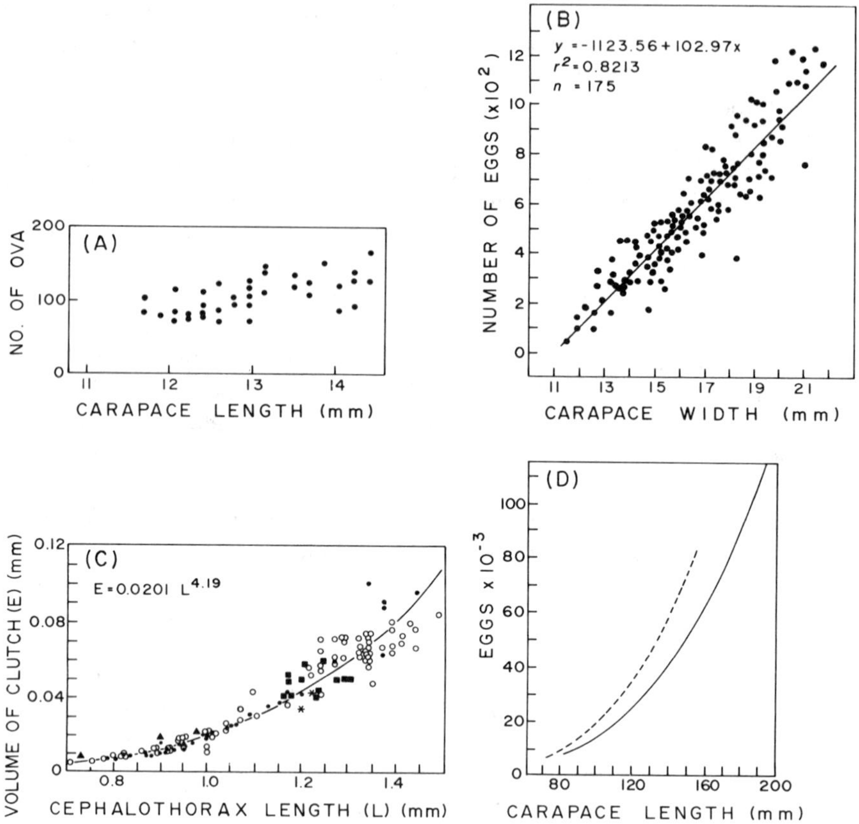

Fig. 3. Relations between number eggs or clutch volume and female body size among crustaceans. (A) Thallassinid, *Calocaris macandreae,* showing no obvious relationship (Buchanan, 1963); (B) the brachyuran crab, *Heterozius rotundifrons,* showing a linear relationship (Jones, 1978); (C) the copepods, *Pseudocalanus minutus,* showing an exponential relationship (Modified from McLaren, 1965); (D) the homarid lobster, *Homarus americanus,* showing a logarithmic relationship (Aiken and Waddy, 1980).

1969; Aiken and Waddy, 1980; Corey, 1981). Generally, the number of eggs produced per female is related to the cube of body length, and this varies from one species to the other (Fig. 3). For some species, however, this relationship is linear (Jensen, 1958; Sheader and Chia, 1970; Fish and Fish, 1978; Jones, 1978). In still others, there is no clear relationship between body length and the number of eggs (Buchanan, 1963). For example, there is much variability in clutch size among females of the same size class in the copepod, *Pareuchaeta norvegica* (Hopkins, 1977). The overall relationship between egg number and size of the adult females is relatively weak for 17 species of cumaceans (Corey, 1981). The data summarized by Allen (1966a) for a number of sublittoral carideans off Northumberland also show that considerable variation prevails in the number of eggs for a given size female of each species.

Most estimates of the numbers of eggs reported for the American lobster *Homarus americanus* show a logarithmic relationship to body length, although a linear relationship has been also reported by some authors (Fig. 3). These discrepancies may be explained by the differences in methodologies, geographical variations of the habitats of populations examined, individual variations in egg output, or time of the year that eggs are counted, since eggs may be lost between oviposition and hatching (Aiken and Waddy, 1980). The number of eggs per a given size female also varies for the Norway lobster, *Nephrops norvegicus* (Thomas, 1964), and the crab, *Heterozius rotundifrons* (Jones, 1978), from one geographical area to the other. However, no geographical variation has been seen in the number of eggs produced by *Panulirus longipes cygnus* (Morgan, 1972). Interspecific variation has been observed in the number of eggs for comparable size females of *Homarus americanus* and *Homarus gammarus* (Hepper and Gough, 1978), and also among different species of palinurid lobsters. Chittleborough (1976) suggested that the number of eggs produced by *P. longipes cygnus* is regulated by the density-dependent factors, so that a high population density results in a smaller size at maturity with fewer eggs per clutch and also fewer clutches per year.

In order to make inter- and intraspecific comparisons of reproductive performance or fecundity, considerable information is needed. Estimates of the food intake, energy investment for gamete production, size of eggs, size of clutches, and rate of clutch production standardized for the growth rate and for the breeding season or for the lifetime of individuals are needed to validly compare reproductive effort between species. Because of the difficulties in obtaining information on these parameters, values for reproductive performance have been rarely obtained for crustaceans.

iii. EFFICIENCY OF EGG PRODUCTION. Barnes and Barnes (1968) made an attempt to compare egg numbers, "metabolic efficiency of egg production" (estimates of consumption have not been determined), and fecundity, and also local and geographical variation in those parameters for a number of common cirripedes. They report that the weight of egg tissue produced per unit time per standard increase in body weight (i.e., efficiency of egg production) is relatively constant for many operculate barnacles, even when they are of different adult or egg sizes. These authors suggest that for comparison or reproductive efficiency, the number of eggs, their size, and the time required for production must all be taken into account. Barnes and Barnes compared the relative brood volume for a variety of barnacles. For each species they calculated a percent brood volume relative to the size of barnacle, using *N* as the number of eggs produced per brood per 50 μg dry body weight and *V* as the volume of egg. From these calculations they found that barnacles fall into three categories: the boreoarctic species have very large broods (*NV* = 1500–3500), temperate and subtropical species have intermediate but much smaller broods (*NV* = 100–500), and a few others with very small broods (*NV* = 30–60). To make interspecific comparisons, of "metabolic efficiency of egg production," (the product of *NVB*), they calculated the number of broods per year (*B*) times the *NV* value of a species. From these calculations, they found the *NVB* values are roughly the same for boreoarctic species producing only one brood and warm water species producing many broods, which suggests that the "metabolic efficiency of egg production" is relatively constant. However, the reproductive effort determined by the *NVB* for *Chthamalus fissus, Balanus glandula,* and *Tetraclita squamosa* from Moro Bay, California, showed that *B. glandula* has the largest reproductive output of these species (Hines, 1978). The *NVB* values, or estimates of reproductive effort, did not support Barnes and Barnes (1968) hypothesis of constant relative reproductive output for different barnacle species. Fecundity is also an important component of fitness, which is dependent on lifetime reproductive performance. The costs associated with reproduction will have important consequences to the life-cycle patterns of species (Calow, 1979).

C. Reproductive Behaviors

Pairing of sexes for sperm transfer by copulation and subsequent fertilization of ova and oviposition is an important phase in the temporal sequence of events in the reproductive cycle. With the exception of parasites and barnacles, crustaceans are capable of limited or extensive mobility. These movements and/or migrations to locate mates, and the reciprocal recognition of sexually receptive females and sexually capable males, forms an integral part of the reproductive process. Vertical migration in the pelagic environment or horizontal movements in the benthic environment serve not only for feeding, avoidance of predators, and locating the preferred environment for survival, but also for locating mates for breeding purposes, a behavior that is common to many species.

Behaviors associated with sex recognition and communication between sexes and social activities related to the courtship may precede copulation, sperm transfer, and oviposition. Components of these behaviors may be temporally separated for varying lengths of time from attainment of sexual maturity to the actual ovulation of fertilized eggs by the females. Furthermore, these activities may take place at a site different from the habitat that the organisms prefer during nonreproductive phase of the life cycle. Reproductive behavioral ecology then forms an integral component of the overall reproductive strategy of a species, adapting it to the environment. Successful reproduction depends on intricate temporal coordination of all or some of these activities, perhaps via sensory, neuroendocrine events to produce young in environments suitable for their survival.

Detailed discussion of animal movement, communication systems, social behaviors, and activity and courtship behaviors can be found in other chapters. These topics are considered here briefly to relate them to the overall sequence of events of reproductive cycles.

1. SENSORY INPUTS

All animal movements, such as vertical or horizontal distribution, social activities, and reproductive behaviors, require that individual animals monitor and interpret their total environment. A variety of sense organs, eyes, statocysts, antennal and mouthpart receptors, and integumental sensila (Waterman, 1961; Laverack, 1968; Mauchline, 1977) may be employed to monitor the environment. As yet, there is insufficient information for assessing the importance and role of different sensory systems in providing information for behaviors associated with reproduction. Knowledge of the structure and function of crustacean sense organs has been reviewed (Laverack, 1968; Mackie, 1975). Descriptions are also available on sense organs on the antennae, mouthparts, walking legs (Laverack, 1968; Waterman, 1961; Mauchline, 1977), hair peg organs (which may detect water currents) (Laverack, 1968), hydrodynamic receptors (Vedel and Clarac, 1976), sensilla on various appendages, and mechanoreceptors and chemoreceptors on mouthparts and thoracic legs (Laverack, 1964; Shelton and Laverack, 1968).

Crustaceans are capable of receiving many types of stimuli (i.e., visual, chemical, and acoustic), and the information provided may be utilized for reproductive purposes. A great deal of inter- and intraspecific variation can be found in terms of how crustaceans perceive their environment. Mauchline (1977) has recognized three broad categories of information pelagic crustaceans may obtain from their environment: (1) information on general characteristics of the environment, i.e., temperature, salinity, pressure intensity, color of light, and planes of polarization; (2) information on events close to the animal such as touch, taste, and movements of the animals relative to water movement; and (3) events taking place at a distance perceived through pressure and wave receptors, chemoreceptors detecting pheromones, and mechanoreceptors detecting water currents. It is not known how the sensory capabilities and information input from the environment are integrated and utilized to perform the specific movements, which result in social and behavioral activities, including those associated with reproduction.

2. MIGRATION, AGGREGATIONS, AND SWARMING

Many crustaceans undergo "temporally organized and spatially oriented" movements and/or migrations that may serve some biological function (Herrnkind, 1980). Migration patterns have evolved and hence must have adaptive value for a species, whether it involves exploiting resources, partitioning of the environment for different ontogenic stages, finding mates, caring for the young, or avoiding predators (Allen, 1966b; Herrnkind, 1980). All of these activities either directly or indirectly affect the reproductive success of a species. In the ostracode, *Philomedes,* the female becomes planktonic for mating, and the swarm ascends 50 m to mate with planktonic males (Kaestner, 1970). After mating, the females descend, and the adult males die in a short time. Vertical migration has evolved in copepods, euphausiids, and penaeid shrimp and is believed to have significance as a life history strategy for reproductive success. A variety of hypotheses have been advanced to explain the adaptive value of this phenomenon (Longhurst, 1976; Marshall, 1979; and others). Many species of isopods may undergo vertical or horizontal migrations in association with reproduction (Naylor, 1955). A number of freshwater prawns migrate horizontally into saline waters, where vitellogenesis and ovulation takes place (Wickins, 1976). A number of brachyuran crabs migrate for reproductive purposes. Smaller individuals of *Callinectes sapidus* migrate in to estuaries for mating, and the females return to high salinity waters for ovulation. The Alaskan king crab, *Paralithodes camtschatica,* migrates to shallower waters in the spring to congregate in pairs and breed (Gray and Powell, 1966). The Chinese mitten crab, *Eriocheir sinensis,* migrates from fresh water to the ocean when mature, and after a single mating and ovulation, the females die (Kaestner, 1970).

Aggregation of cirripedes obviously facilitates copulation and cross fertilization. Individual barnacles of *Pollicipes polymerus* separated by a distance of more than 8 inches fail to copulate and carry developing embryos (Hilgard, 1960). On the European coast, mature males and preadult females of the crab *Maja squinado* aggregate into heaps during summer months and mate in August (Carlisle, 1957).

Swarming of certain crustaceans occurs in relation to reproduction (Clutter, 1969; Omori, 1974). The deep-water mysids, *Gnathophausia gigas* (1500–300 m) and *Eucopia* (2000 m), which live near the bottom, are benthic by day and pelagic by night (Kaestner, 1970). Almost all bathypelagic and bottom dwelling mysids, even those that burrow into sand, rise at night. The ascent is most distinct during spawning periods. Nocturnal swarming occurs for cumaceans (e.g., *Diastylis rathkei*) and appears to be related to molting and mating (Valentin and Anger, 1977).

While migrations of pelagic crustaceans are commonly diurnal and are not primarily related to reproduction, the migration of benthic species seems to be seasonal in nature, often involving inshore and offshore movements. Molting and reproduction are also linked to these migrations. Our knowledge of the mechanisms and factors controlling such movements has been reviewed (Enright, 1975; Herrnkind, 1980). However, migrations of most animals result from an interaction between internal physiological events and certain external stimuli. These migrations may have evolved to place populations of species in environments suitable for particular stages of their life cycle, e.g., for mating and ovulation, release of young, or feeding. Studies on migrations and/or movements of crustaceans in relation to their reproductive strategies would be of interest.

3. SEX RECOGNITION AND COURTSHIP

Recognition of the opposite sex and mechanisms to bring together sexes is essential for successful reproduction of crustaceans. Pairing of sexes, mating, and sperm transfer precede ovulation by the females. Crustaceans typically exhibit sexual dimorphism, but this varies from small differences in certain species to pronounced differences as in some parasitic forms. Various factors may enable males to recognize a receptive female, such as morphological differences or identifying behaviors, which may include olfactory signals or other communication systems.

Crustaceans have evolved a variety of behaviors associated with the reproductive activity which fit into the overall temporal pattern of the reproductive cycle. An intricate coordination between the internal state and the associated neuroendocrinological conditions may produce complex behaviors in individuals of both sexes of a species to find mates, fertilize the eggs, and to produce young at a season favorable for survival of the offspring. Social communication through visual, chemical, acoustic, and tactile signals may allow an individual to advertise its internal state and elicit appropriate behavioral responses from the opposite sex. The nature of the communication system, and the combination of signals employed, varies from one species to the other; these may also vary among individuals of a population in relation to the reproductive condition. Details of reproductive behavioral ecology can be found in other chapters.

a. *Sex Recognition.* Pheromones, or chemical attractants, are probably important for attraction between sexes among aquatic crustaceans. The current status of this field has been reviewed (Mackie, 1975). Dunham (1978) concluded that pheromones have not been conclusively demonstrated in crustaceans. There is some suggestion of chemical attraction between sexes of the copepod *Eurytemora affinis* (Katona, 1973) and *Pseudocalanus minutus* (Griffiths and Frost, 1976). Autoradiographic studies have provided some evidence that aesthetes on the antennules of *Calanus pacificus* are sites of chemoreception and may function to perceive chemicals released by females (Griffiths and Frost, 1976). Both mechanical and chemical stimuli may allow recognition of potential mates and promote contact between sexes in copepods (Stickler and Bal, 1973; Blades and Youngbluth, 1980). in *Pareuchaeta norvegica,* both chemical and physical stimuli may be involved in mate selection (Hopkins, 1977). Male gammarids may also recognize females in reproductive condition by olfactory stimulation. Dahl *et al.* (1970) observed behavioral reactions of male *Gammarus duebeni* to ripe females, and also found that tritium fed to females appeared in molecules at receptor sites on the male antenna. In some other amphipods, males recognize females simply by their passiveness. In the mysid, *Metamysidopsis elongata,* males search for recently molted females. This attraction is species-specific (Clutter, 1969).

In some decapods, sexually receptive females may also be recognized by chemical substances which they release into water (Eales, 1974; Ryan, 1960; Atema and Engström, 1971). In others, e.g., *Callinectes sapidus,* the males are not stimulated by females, but the receptive females respond to male odor. This may be reinforced by visual stimuli of either or both sexes (Teytaud, 1971). The final recognition of the receptive female may be reinforced by tactile and taste stimuli and by the passive or receptive behavior of the female contrasting with aggressive reaction of other animals (R. G. Hartnoll, unpublished review).

In semiterrestrial and terrestrial crabs, chemical cues appear to be less important than visual, tactile, or acoustic signals (Bliss, 1968, 1979), which, like pheromones, may be expected to be species-specific. Acoustic and visual signals have been shown to be of major importance in territorial and courtship behavior of several species of temperate zone fiddler crabs, *Uca.* Salmon and Atsaides (1968) showed that *Uca* spp. employ species-specific characteristic patterns and timing of both cheliped waving and sound production for sex recognition.

b. Courtship. In copepods, mating behavior has been suggested to be precise, ritualized, and species-specific (Fleminger, 1967; Lee, 1972; Hopkins and Machin, 1977; Blades and Youngbluth, 1980). Sensory structures present on the integument may play an important role in mating encounters (Fleminger and Hulsemann, 1977; Mauchline, 1977). In some species, mating may be regulated by pheromones alone, while in others, a combination of cues may be used to advertise the receptiveness of the female, as for example, the presence or absence of spermatophores as well as chemical cues (Hopkins, 1977; Blades and Youngbluth, 1980). Males and females may exhibit different swimming behaviors in mate recognition (Katona, 1973; Kittredge *et al.*, 1974; Griffiths and Frost, 1976; Blades and Youngbluth, 1980).

Precopulatory pairing has been reported to occur in some branchiopods, amphipods, and brachyuran crabs. In the antarctic isopod, *Glyptonotus antarcticus,* the mature male carries the premolt female for as long as 190 days prior to mating (White, 1970). Among gammaridean amphipods, precopulatory pairing may last from several days to weeks during which males assume a riding position superior to the females (Kaestner, 1970). Male cumaceans continue holding females in riding position between copulations (Gnewuch and Croker, 1973).

Berry (1970) described in detail the courtship and mating behavior in *Panulirus homarus,* where the females become sexually attractive when the ovaries are in advanced stage of development. The presence of an attractive female in a tank results in the establishment of a reproductive hierarchy among the males, with one becoming dominant and showing aggression to all other males. All mating is done by the dominant male. In some species, no courtship display precedes copulation, e.g., the prawns *Macrobrachium rosenbergii* and *Macrobrachium australiense*. Males of these species recognize newly molted females as receptive by antennal touching (Wickins, 1976). In contrast, some other crustaceans exhibit elaborate courtship behavior a sort of prenuptial dance, e.g., the tanaidacean, *Heterotanais oerstaedi* (Kaestner, 1970), and the crab *Pachygrapsus crassipes* (Bovbjerg, 1960). Precopulatory behavior has also been observed in many hermit crabs involving stroking or tapping, while in others there may be no courtship behavior (Hazlett, 1968).

The importance of visual displays during courtship of semiterrestrial and terrestrial crabs has been well established (Crane, 1966; Bliss, 1968). The function of waving in particular was documented (Crane, 1966; Salmon and Atsaides, 1968; Schone, 1968; Wright, 1968). The interrelationship between waving displays and sound production (stridulation) has been shown for *Uca pugilator* and other fiddler crabs (Salmon, 1965). Bliss (1968, 1979) has suggested that evolution of semiterrestrial and terrestrial life in decapods is accompanied by increased reliance upon visual and acoustic signals to bring the sexes together for mating.

In brachyuran crabs, two types of courtship and mating occur (Hartnoll, 1973). In some species there is a lengthy courtship followed by the female molt and copulation, all controlled by chemical and tactile stimuli (e.g., in the Carcinidae, Portunidae). In other species, the males guard premolt females at the mouth of a sheltering burrow (e.g., *Menippe mercenaria*).

In other brachyuran crabs, courtship is brief. This is characteristic of semiterrestrial and terrestrial species where courtship is dependent upon auditory and visual stimuli and is reinforced by tactile stimuli in the final phases of courtship. In these crabs, females copulate during the intermolt stage of molt cycle when the exoskeleton is hard. Postcopulatory pairing is absent (e.g., Xanthidae, Majidae, Grapsidae, Ocypodidae) (Hartnoll, 1969). Auditory courtship signaling has been recorded for both Grapsidae and Ocypodidae, especially by males. This takes place within their burrows during the day or on the surface after dark (Salmon and Astaides, 1968).

Some species build special structures during courtship, such as the pyramids built by several species of ocypodes (Linsenmair, 1967) and the chimneys, pillars, and hoods constructed by some species of *Uca* (Crane, 1975), probably to signal the status of the owner (see Chapter 6, this volume). Male fiddler crabs, *Uca*, court from and defend burrows (Crane, 1975), while sexually receptive females approach and follow courting males down into their burrows (Salmon, 1965).

The time and the duration of courtship seems to vary for *Uca* spp. from temperate, subtropical, and tropical regions. Zucker (1978) indicates that there is an inverse relationship between the number of warm months in a given region and the amount of time devoted to courtship activities on a daily basis. Temperate species seem to hibernate in the winter, but in the warmer months males court regardless of tide periods (Crane, 1943; Salmon and Atsaides, 1968). In contrast, subtropical species court at low tide only during diurnal and early evening hours (Salmon, 1965). Tropical species seem to court in the daytime throughout the year (Crane, 1975). These differences in courtship behavior of fiddler crabs from different climatic regions are probably related to the patterns of annual reproductive cycles in the respective regions.

Although tropical species may exhibit courtship activities throughout the year, this activity may be restricted to certain periods. Courtship has been observed at low tide hours when it coincides with early morning hours (Crane, 1958), late morning hours (Feest, 1969), and late morning through mid-afternoon during full and new moon spring tides (Zucker, 1976) or full or new moon (von Hagen, 1970). While some species of *Uca* show receptivity around the full moon period (von Hagen, 1962; Zucker, 1978), others are receptive only once each month (von Hagen, 1970; Christy, 1978). These differences may again correlate with the patterns of reproductive cycles of the species. Studies on ovarian development of *Uca tangeri* (von Hagen, 1962) and *Uca musica terpsichores* (Zucker, 1978) showed that ovaries are most highly developed around the full moon. Von Hagen (1970) found that some species of tropical *Uca* exhibit two ovarian maxima, one the day before the new moon and a second on the day before the full moon.

Zucker (1978) found that among three sympatric species of Panamanian fiddler crabs, *Uca musica terpsichores* has a semimonthly courtship behavior, while *Uca latimanus* and *U. beebei* show a distinct monthly cycle of courtship activity. More males of all three species court during the week of full moon when compared to the week of new moon, and no courtship occurs during quarter moon. All three species wave at the same time, suggesting that the receptive females in the population of each species and their reproductive behaviors, as well as male waving patterns, are different as species-isolating mechanisms. The degree of courtship activity varies between species. Hood building by courting males also peaks around the full moon. In these tropical species, the courtship activities between sexes appear to be highly synchronized around the full moon rather than at some other phase of the lunar cycle or some other environmental event. Such a coordination and synchronization of courtship displays around new or full moon followed by copulation and oviposition is adaptive for timing the release of larvae in relation to the tidal cycle, thereby facilitating a transport of larvae (Christy, 1978; Zucker, 1978).

4. MATING, SPERM TRANSFER, AND OVULATION–OVIPOSITION

The temporal relationship between mating and sperm transfer, and ovulation and oviposition varies greatly for different species. Fertilization of ova, ovulation, and oviposition are, however, closely interrelated. The temporal differences among species regarding these events may be related to the molt cycle stage of females at mating, the method of sperm transfer and storage of sperm, and the manner by which fertilized eggs are brooded until hatching. These characteristics must obviously have evolved to ensure reproductive success of each species with respect to their life mode and the environment in which they live. The adaptive value of many of these differences to the survival of particular species remains to be established.

Precopulatory and copulatory behaviors and the method of sperm transfer have been extensively documented for branchiopods and ostracodes (Kaestner, 1970), copepods (Marshall and Orr, 1972; Corkett and McLaren, 1978; Blades and Youngbluth, 1980; Hopkins and Machin, 1977), barnacles (Barnes and Crisp, 1956), amphipods and isopods (Kaestner, 1970; Johnson, unpublished review), caridean and penaeid shrimps (Omori, 1974; Wickins, 1976), euphausiids (Mauchline and Fisher, 1969), homarid and palinurid lobsters (Berry, 1970; Atema *et al.*, 1979; Aiken and Waddy, 1980), brachyuran crabs (Bliss, 1968; Hartnoll, 1969; unpublished review), and stomatopods (Dingle and Caldwell, 1972).

Observations on mating of crustaceans under natural conditions are relatively rare, but there is some indication that certain species may be polygamous. In the copepod *Pseudocalanus minutus* (Corkett and McLaren, 1978), *Pareuchaeta norvegica* (Hopkins and Machin, 1977), and the diaptomids, attachment of several spermatophores to the females has been observed to indicate that several males may mate successfully with one female. Cases of a number of males attempting to copulate with one female have been observed in some species of copepods (Jacobs, 1961; Fleminger, 1967). In nephropid and palinurid lobsters, there is some evidence of a polygamous mating system, where the largest male does most of the mating and that females mate only once each time they are receptive (Berry, 1970). Mating of several males (either free-living or neotenic) with one female has also been observed in *Emerita talpoida* (Diaz, 1980) and *Emerita analoga* (Efford, 1967). In contrast, fidelity between mating pairs is high for the antarctic isopod, *Glyptonotus antarcticus* (White, 1970). For a discussion of the evolutionary significance of different mating systems and their adaptive values consult Orians (1970) and Emlen (1977).

The molt stage of females at the time of mating varies for different species. This relationship may be summarized as involving mating between either hard males and soft females, hard females, or females that have just hardened the exoskeleton. For many crustaceans, impregnation of the female is possible only during a short interval following the molt. In these species, molting of females and mating has to be synchronized.

Among isopods, molting occurs in two stages, with a delay between shedding of the posterior and anterior regions (Kaestner, 1970; W. S. Johnson, unpublished review). In the sand beach isopod *Excirolana chiltoni,* insemination takes place after the posterior molt of the female (Klapow, 1972). Molting is synchronized to a fortnightly cycle, with peak activity occurring prior to the new or full moon. The availability of many receptive females as a result of synchronization of molting may enhance successful mating by increasing sexual responsiveness of males at this time. Most Peracarida mate at or shortly after the female molt, which is also associated with the deposition of eggs into the marsupium (W. S. Johnson, unpublished review). In some isopods, *Porcellio laevis* and *Armadillidium vulgare,* copulation occurs before molting, and in *Armadillo officianalis,* mating occurs at any time.

In copepods, *Calanus finmarchicus* (Marshall and Orr, 1972) and *Pseudocalanus minutus* (Corkett and McLaren, 1978), mating occurs shortly after the female has molted into the adult stage.

The barnacle, *Balanus amphitrite,* produces several broods, one after another, during the breeding season. In such continually breeding species, copulation may take place at any time during the intermolt cycle, but it generally occurs soon after ecdysis (Patel and Crisp, 1961).

In carideans and penaeids with a complicated thelyca, mating occurs between hard-shelled males and females, but in those species with open thelyca (e.g., *Penaeus setiferus* and *Penaeus schmitt*) copulation can occur between hard-shelled males and soft females (Wickins, 1976). Among Macrura, *Homarus americanus* males molt several weeks before females and mating occurs between hard-shelled males and soft females (Aiken and Waddy, 1980). In the spiny lobster *Jasus lalandii,* mating occurs between hard males and soft females (von Bonde, 1936). In *Panulirus argus,* mating occurs subsequent to the initial hardening of the female exoskeleton (Berry, 1970). Among anomurans, copulation may take place during various times in the intermolt cycle of *Pagurus bonairensis* and *Paguristes tortugae* (Kaestner, 1970). When the female is soft, copulation is short, with a pair copulating only once. But if the female is hard, copulation may occur several times and the duration of copulation is longer. Hartnoll (1969) reviewed the relationship between molting and mating among the brachyuran crabs. The exoskeleton of males is always hard, but females of certain species mate while soft (Carcinidae, Portunidae) and others mate with hard exoskeletons (e.g., semiterrestrial and terrestrial crabs). In Pinnotheridae, the females mate at small size while in a special instar and then pass through a series of ecdyses before molting to the adult ovigerous instar, perhaps an adaptation to their commensal or parasitic mode of life (Pearce, 1966). In stomatopods, there does not appear to be an obligatory molt before oviposition (Reaka, 1979).

Mating season also varies for different species. Mating is followed by ovulation and oviposition in some species, while in others there may be long time intervals between these events, ranging anywhere from only a few minutes to several months. In those species where a delay occurs, it is difficult to know exactly when mating has occurred unless females are examined for evidence of the presence of sperm plugs or spermatophores. Generally, mating can occur at any time for species with an extended or year-round breeding, although it may be restricted to a certain season for those species with a defined breeding period.

Although these reproductive events vary for different species, it can be expected that these events may follow a particularly strict temporal pattern in species where synchronization is essential. In those environments where population density and food availability are low and sex ratios are biased toward females, as in the deep-sea, it would be advantageous for females to be able to mate opportunistically at any time during the molt cycle, and to store sperm for prolonged periods prior to ovulation. In these environments where it may be difficult to locate mates and where there is no need for synchronization of molting with environmental events, the temporal relationship between molting and mating may not be strictly synchronized. It would be of interest to learn of the mating systems and reproductive behaviors of deep-sea species and to see how they relate to reproductive strategies.

Sperm transfer seems to occur within a short time during copulation for most species. The mechanism of sperm transfer does vary for different species (Kaestner, 1970). In *Mesopodopsis,* sperm is shed by the male and is transported to the female brood pouch by male pleopods. Sperm is sprayed into the oostegites in *Hemimysis,* or deposited into the oostegites of isopods and amphipods or the thelycum of the female in euphausiids and penaeids. In barnacles, fertilization occurs in the brood chamber (Walker, 1980). In some copepods, spermatophores are attached to the female gonopore (Hopkins and Machin, 1977; Corkett and McLaren, 1978). Female limnetic copepods store sperm in a spermathecal sac and thus minimize the need for subsequent mating (Nilssen, 1980). In Thermosbaenacea, internal fertilization occurs with the development of eggs to the 12-cell stage (Kaestner, 1970). In decapods, fertilization is external for some species and internal for others. The Nephropidae store sperm internally, while the Palinuridae store externally. In some crabs, sperm plugs, which block the genital pores of the female, are formed by coagulation of secretions from genital ducts, and they disintegrate after a certain time. Some brachyuran crabs store sperm for prolonged periods. In some of these species, stored sperm may fertilize eggs for successive clutches during one intermolt cycle [e.g., *Menippe mercenaria* (Cheung, 1969); *Hemigrapsus penicillatus* (Pillay and Ono, 1978); *Microphrys bicornutus* (Hartnoll, 1963); and *Portunus sanguinolentus* (Ryan, 1967b)]. In other species, fertilization occurs much later when eggs are laid (e.g., *Orconectes, Cancer pagurus*).

Sperm transfer, ovulation, and oviosition appear to be coupled events for some species. In the barnacle *Semibalanus balanoides,* deposition of egg masses is induced by copulation (Clegg, 1957). For some species, the time interval between sperm transfer and ovulation may be only a few minutes (e.g., *Heterotanais oerstedi, Hemimysis,* gammaridean amphipods), hours [e.g., *Gammarus pulex., Sphaeroma, Glyptonotus antarcticus, Jaera* (Jones, 1974)], several days (*Talitrus saltator, Panulirus argus*), or months (Kaestner, 1970; White, 1970). In *Calanus finmarchicus,* there is a 6-month delay between mating in December–January to egg laying for the Greenland population, while in temperate regions, egg laying occurs in February–March following mating in December–January (Marshall and Orr, 1972). In some brachyuran crabs, the time between mating and egg laying is 2–9 months (e.g., *Callinectes sapidus*), 6 months (e.g., *Maja squinado*) or 12 months (e.g., *Cancer pagurus*). Oviposition in *Homarus americanus* may follow by as little as 1 month or as long as 2 years. Ovulation sometimes occurs when the females have not mated, but the unfertilized eggs are not attached securely to the abdomen (Aiken and Waddy, 1980).

In the absence of copulation the eggs are resorbed, or if they are laid, they fail to develop and disintegrate (Kaestner, 1970). In the terrestrial crab, *Gecarcinus lateralis,* ovulation does not occur except after copulation. For fertilization and normal development of eggs, copulation must take place within 6 hr to 9 days prior to ovulation (Klassen, 1975). This time relationship in successful mating is critical, since eggs within the ovaries of a female that did not succeed in copulation remain within the ovaries and degenerate (Bliss, 1968).

Fertilized eggs are rarely abandoned by female crustaceans. Exceptions include some species of copepods (*Calanus:* Marshall and Orr, 1972), euphausiids (Mauchline and Fisher, 1969), and penaeids except *Lucifer* (Omori, 1974) all of which shed their eggs freely into the environment. Some species lay their eggs on suitable substrates on the bottom (e.g., deep-sea, *Xanthocalanus*). A great majority of crustaceans carry fertilized eggs in a dorsal cavity or marsupium (e.g., Cladocera and Conchostraca), a special brood pouch (e.g., isopods and amphipods), or in ovisacs (certain species of copepods), or as egg masses fastened to the thoracopods (some euphausiids), or pleopods (e.g., stomatopods). The stimuli and the process of ovulation, and the mechanism of egg or ovisac attachment in copepods is discussed by Blades and Youngbluth (1980), and the cementing of egg masses to the pleopods in decapods is considered by Cheung (1966), Templeman (1937), Aiken and Waddy (1980), and R. G. Hartnoll (unpublished review).

D. Breeding Patterns

1. BREEDING PERIOD

Breeding implies that females are ovigerous (or shedding eggs), and the breeding period is the time when females of a population are ovigerous. Information available on the breeding period of crustaceans is extensive. A few examples are cited here to illustrate the immense inter- and intraspecific variation in breeding patterns among the various groups inhabiting different environments.

a. *Pelagic Environment.* Breeding periods of coastal and neritic pelagic copepods vary in relation to latitude as well as regionally in relation to prevailing hydrographical conditions. In the higher latitudes, herbivorous copepods undergo an ascending spring migration and breed coinciding with the time of intense primary production in the surface waters during spring or summer; then they descend in the autumn for overwintering. In the antarctic region, *Rhincalanus gigas, Calanoides acutus,* and *Calanus propinquus* show latitudinally related variation in the time period of migration, with populations lagging behind in the south compared to north (Voronina, 1970). These three species also show temporal and spatial differences in the breeding period, dominance of copepodite stages, and the time of summer biomass maximum, which may minimize interspecific competition. In the north Atlantic, *Calanus finmarchicus* ascends from the deeper waters to the surface waters in December and January, where molting and copulation occurs (Marshall and Orr, 1972). In temperate waters off England, populations of *C. finmarchicus* begin breeding in March–April, which coincides with the spring diatom increase, and there may be two or three egg-laying periods in March–April, April–May, and June–July, before autumnal migration into the deeper waters. In comparison, the population of *C. finmarchicus* in East Greenland has an annual cycle and the life cycle takes one whole year. Over the distributional range, life cycle varies in temporal details, with the breeding taking place earlier in the south compared to north. Regional differences in breeding period are also found between populations as in the inner and outer parts of Oslo Fijord, and in the western and eastern Gulf of Maine. Similar regional and geographical differences in the timing of breeding period and number of generations are also found for the populations of other copepods, e.g., *Microsetella norvegica* (Fish, 1961) and *Pseudocalanus minutus* (Corkett and McLaren, 1978).

The timing of the breeding period of herbivorous copepods coincides with the phytoplankton blooms in spring or summer in the higher latitudes (Mullin, 1967). However, carnivorous species may breed at different times throughout the year. The predatory copepod *Pareuchaeta norvegica* has a half year life cycle with a summer and winter generation. Breeding takes place intensively in the mid-winter and in the summer off the western Norway coast (Båmstedt, 1979). These examples indicate that breeding periods of planktonic copepods may vary regionally and geographically, relative to the adult and larval food requirements with species not directly linked with primary production breeding outside the bloom periods.

Many tropical copepods breed throughout the year, although peaks in breeding intensity do occur at certain times of the year. Copepods near Barbados (13 11′N) may breed sporadically or continuously, but also with peaks of intensive breeding (Moore and Sanders, 1976). In the same area, the harpacticoid copepod *Miracia efferata,* which spends most of its life on the bottom and swarms to the surface during mating and egg production time, has a recognizable periodicity. In the Gulf of Mannar off South India, a number of copepods breed throughout the year with distinct peaks in some months, while others breed in only certain months (Ummerkutty, 1967).

Epipelagic micronekton, including euphausiids, penaeids, and carideans, also show an increase in the length of breeding season with the decrease in latitude (Table I). In temperate waters off Japan, shallow-water penaeids breed in the summer, and carids in the spring and autumn (Omori, 1974). Some penaeids have a longer winter generation and a shorter summer generation, but the duration of breeding of both generations overlaps. Some carids in temperate regions breed in early summer and winter (e.g., *Pasiphaea sivado*), while others breed throughout the year (e.g., *Pasiphea multidentata*).

Compared to the restricted breeding season of populations in the epipelagic zones of the higher latitudes, euphausiids, carids, and penaeids in the lower mesopelagic and bathypelagic regions may breed for an extended period or all year round (Mauchline and Fisher, 1969; Omori, 1974; Marshall, 1979). Generally, these species release larvae at an advanced stage of development that quickly permits the adult mode of feeding (Nemoto, 1965). Some euphausiids that carry eggs will breed for a more extended period compared to those species which shed their eggs into the water (Rudd, 1936). The absence of seasonal breeding in the lower mesopelagic and bathypelagic regions appears to be related to the feeding habits of the larvae of species. The larvae not directly dependent upon plankton blooms in the euphotic zone feed on a wide range of organisms as adults. According to Marshall (1979) deep-sea copepods that remain mesopelagic during their entire life have no naupliar migration as do herbivorous species in epipelagic regions of the higher latitudes. Rather, these copepods produce large yolky eggs that sustain development through naupliar stages, with feeding beginning in copepodite stage I. Such species may not have a restircted breeding season.

b. Deep-Sea Benthopelagic and Benthic Environments. Breeding patterns of benthopelagic and deep-sea benthic crustaceans are not well known. Rokop's (1977) study on the deep-sea benthic isopods, *Eurycope californiensis* and *Ilyarachna profunda* (450–1300 m), and the amphipod, *Harpiniopsis excavata* (400–5100 m), sampled throughout the year from the same location at a depth of 1240 m from the San Diego Trough off California, shows that these species breed all year round.

c. Subtidal and Intertidal Marine Environment. Breeding patterns of antarctic benthic crustaceans are not well known. The amphipod, *Bovallia gigantea,* lays eggs in February and March, and they hatch to release young 7 months later in September–October (Thurston, 1970). This pattern of autumn mating, and spring hatching and release of young appears common to antarctic amphipods. The hippolytid prawn *Chorismus antarcticus* and the carangonid shrimp *Notocrangon antarcticus* from South Georgia, pro-

TABLE I

Breeding Period of Euphausiids, Carideans, and Penaeids from the Epipelagic and Upper Mesopelagic Regions[a]

Species	Geographic distribution or location	Latitude (°)	Depth distribution (m)	Breeding notes
Sergestes arcticus	Norway		0–1300	December to January and June–July
Pasiphaea multidentata	Off Norway coast			Throughout the year with peaks between May–June and October
	Gulf of Maine			March to May and August to September
Thysanoessa raschii	Firth of Clyde, U.K.	40 N to 75 N	0–200	February to middle of May and again August to middle of September
Meganyctiphanes norvegica	Firth of Clyde, U.K.	30 N to 70 N	0–400	April to July, peak in April and June–July, produces two broods
Euphausia pacifica	North Pacific off Oregon coast, U.S.A.	35 N to 50 N	0–400	Throughout the year
Nyctiphanes couchii	Eastern Atlantic	30 N to 50 N		Beginning of spring to end of summer, produces few eggs throughout the year

Sergestes similis	Off Oregon coast		0–1300	Throughout the year
	Monterey Bay, U.S.A.			December to January, and June–July
Sergia lucens	Sagura Bay, Japan	35 N	0–500	May to mid-November, peak in June–August
Acetes japonicus	Yellow sea, Korea	33 N to 37 N		July to August
	Ariake Sea, Japan	33 N		May to October
Euphausia krohnii	Atlantic and Mediterranean	65 N to 0 S	0–400	Prolonged breeding season
Nematoscelis atlantica	Atlantic and Pacific	40 N to 40 S	0–400	Very long breeding season
Nematoscelis megalops	S. Pacific, S. Atlantic, and Indian ocean	20 S to 55 S	0–300	Prolonged breeding season
Acetes erythraeus	Off S. India; Ambaro Bay, Madagascar	8 N to 12 N, 13 S		January to April November to March
Acetes australis	Australia	33 S		December to February
Euphausia lucens	Southern temperate zone, North of Antarctic convergence zone	30 S to 50 S		Beginning of spring to end of summer, produce few eggs throughout the year
Nyctiphanes australis	Off New Zealand and S.E. coast of Australia	35 S to 50 S		Extended breeding season for 8 to 11 months, several broods per year
Euphausia superba	Antarctic, south of convergence zone	55 S to 75 S	0–900	January to March

[a] From Mauchline and Fisher, 1969; Omori, 1974.

duce large yolky eggs and breed for prolonged periods (Maxwell, 1977), but individuals of each species produce a brood only once a year (Clarke, 1979). In comparison, the antarctic isopod, *Glyptonotus antarcticus,* breeds throughout the year at South Orkney Island and McMurdo Sound (Dearborn, 1967; White, 1970).

In temperate waters, individuals of *Homarus americanus* and *Homarus gammarus* take about 2 years to produce a brood (Aiken and Waddy, 1980), but populations of both species release larvae once annually. Oviposition by *H. americanus* is initiated between May and July and ends between July and October. Oviposition occurs in different months in different geographical areas. Eggs are carried for about 11 months until hatching in the summer of the next year from May to July, with peak intensity in late June and early July (Hughes and Matthiessen, 1962). Oviposition of *H. gammarus* occurs between early July and September (Branford, 1978) and for the Norway lobster *Nephrops norvegicus* in August (Thomas, 1964). Among temperate zone crabs, *Cancer irroratus* breeds from October to August, with a peak between February and May in Maine, while a population further south in Chesapeake Bay breeds from September to June with a peak between March–April (Haefner, 1976). In Puget Sound, *Cancer oregonensis* breeds from late November through February, with occasional ovigerous individuals occurring in April and June. *Cancer productus* from the same locality breeds from September to April (Knudsen, 1964).

The commensal crab *Pinnotheres ostreum* breeds from July to September off the New Jersey coast (Christensen and McDermott, 1958). In Puget Sound, *Pinnixa faba* and *Pinnixa littoralis,* which are sympatrically distributed with the host *Tresus capax,* breed twice a year in late spring and summer, and again in winter (Pearce, 1966). *Pinnixa littoralis* becomes ovigerous earlier than *P. faba*. Another commensal crab, *Fabia subquadrata,* breeds from early December through the middle of February.

The spiny lobster *Panulirus argus,* which occurs on the Bahama Banks, breeds from spring through summer. The breeding activity terminates with the onset of winter (Kanciruk and Herrnkind, 1976). A high percentage of ovigerous individuals are found at depths of from 10 to 25 m, and of the various populations in different geographical regions, breeding periods vary. *Panulirus homarus,* off the South African coast, breeds in summer from November to April, with a peak between November and January (Heydorn, 1969a,b).

Sublittoral Caridea in temperate waters off the North Umberland coast of England show a considerable interspecific variation in the breeding patterns in terms of timing, number of broods per female, and length of incubation period (Allen, 1966a). In some species, individuals aggregate before the breeding period. For example, *Spirontocaris* sp. *Pontophilus spinosus,* and

Caridion gordoni aggregate in late autumn, lay eggs synchronously in November or December, and incubate the eggs to release larvae in April. Some other species which show regular onshore–offshore seasonal movements do not aggregate before breeding (e.g., *Crangon* sp., *Hippolyte varians*). These species lay more than one brood of eggs in each breeding season. The breeding period of these species is in the summer, with asynchronous peaks of egg laying. Some other Caridea, *Pandalus montagui* and *Pandalus borealis,* do not aggregate before the breeding period, but they lay only one brood each year synchronously in November and December. Off the North Umberland coast, winter-breeding species lay eggs at different times, but the larvae of all the species are released at about the same time at the end of March, because of the differences in the length of their incubation period. Larvae of winter breeding species are released to coincide with the spring phytoplankton peak, while those of summer breeders are released before the end of autumn phytoplankton peak. After September, when the autumn phytoplankton peak ends, none of the carideans breed for about 8–10 weeks (Allen, 1966a).

A number of sublittoral crabs from tropical waters off southwest India breed for an extended period or all year round (e.g., *Portunus sanguinolentus, Charybdis cruciata,* and *Charybdis hoplites pusilla*). *Portunus pelagicus* breeds for about 9 months from August to April, with a maximum between December and January (Pillay and Nair, 1968), and continuously throughout the year on the east coast (Rahaman, 1967). The Indo-West Pacific crab, *Portunus sanguinolentus,* breeds all year round in Hawaiian waters, with a peak from October to February (Ryan, 1967b).

The arctic caprellid, *Caprella septentrionalis,* breeds from late April to the middle of June. By the middle of July, the over-wintering adults disappear. The spring hatchlings grow during the summer and autumn, experience a decline of growth rate during the winter, and breed the following spring (Heptner, 1963). This type of life cycle also occurs for some other boreoarctic amphipods, e.g., *Gammarus setosus, Gammarus wilkitzki,* and *Gammarellus angulosus* (V. J. Steele and Steele, 1970; D. H. Steele and Steele, 1975). Alternatively, some gammarid amphipods from the temperate regions exhibit a reproductive cycle with two breeding periods, i.e., spring and late summer, and a complete cessation or decline of breeding in the winter (Rees, 1975; Gable and Croker, 1977). *Pseudoprotella phasma* breeds in spring and summer, with individuals producing more than one brood. Young hatch in 1 year over winter to breed the next year. On the west coast of Wales, England, *Bathyporeia pilosa* breeds throughout the year with peaks in April and May and in October, causing an overlap of two generations (Fish and Preece, 1970). Northern populations of the species do not breed in December, but southern populations breed throughout the year. *Bathyporeia pelagica* from the same area in Wales has also two generations with peaks of breeding activity in spring and autumn. Ovigerous females are rarely found in the winter months from November to February. Populations of this species, which migrate into estuarine areas, continue breeding into December, while that on open coast terminates breeding in October (Fish and Fish, 1978).

Interspecific variation is also illustrated in the breeding patterns of some amphipods from the North Carolina coast. The sand beach amphipod, *Neohaustorius schmitzi,* breeds from February to October, with peaks in spring and summer. The populations are composed of a winter and summer generation (Dexter, 1971). Individuals of this species produce only one brood in their lifetime, with the adults dying 1 week after the release of the young. The amphipod *Ampelisca abdita* breeds twice a year north of Cape Hatteras, and breeding is continuous in the south (Keith, 1971). Females of *A. abdita* populations produce more than one brood during their lifetime, which is less than 1 year. In comparison, the temperate–tropical cosmopolitan amphipod *Caprella penantis* breeds all year round on the North Carolina coast, with a decline but no cessation of breeding activity in the winter (Bynum, 1978). Similarly, *Caprella californica* on the California coast breeds throughout the year.

Breeding periods of a number of isopods summarized by Carefoot (1973) show that sublittoral and intertidal species in the temperate regions live for about 1 year and may breed within restricted periods ranging from only a few months to several months, or continuously all year round. Species with restricted breeding periods usually breed in the spring and summer. *Idotea emarginata* (Naylor, 1955) and *Ligia oceanica* breed all year round in England, while *Ligia pallasii* breeds from May to July in British Columbia, Canada (Carefoot, 1973). *Sphaeroma* sp. breeds all year round on the southwest coast of India (Pillai, 1967).

Benthic infaunal cumaceans from the arctic and subarctic oceans at depths ranging from littoral to 200 m also show considerable interspecific variation in their breeding patterns (Corey, 1981). Littoral and shallow-water species breed during spring, spring–summer, spring and fall, or continuously. Generally, deep-water species (>20 m) breed during the winter, or if two breeding seasons occur, one of these is in the winter. The exceptions to the above are the temperate species *Diastylis lucifera* and arctic species *Diastylis goodsiri,* which breed in the summer. In none of the species is breeding season related to depth and temperature, except for *Diastylis rathkei,* which breeds from October to February at 13–20 m and from May to April at 30–180 m depth (Corey, 1981).

The meiofaunal mystacocarid, *Derocheilocaris algolensis,* breeds throughout the year, but with distinct peaks in breeding activity on the sand beaches off the South African Coast. This species develops from egg to a reproductively active adult in about 60 days, and the remaining 6 months of its life span are spent in reproductive activities (McLachlan, 1977). Another species, *Derocheilocaris remanei* from the same coast, also breeds continuously throughout the year (Delamare Deboutteville, 1960), while *Derocheilocaris typica* from the Northeastern American coast breeds only in spring and summer (Hall and Hessler, 1971).

Intertidal barnacles show considerable interspecific variation in the breeding patterns in the same geographical area and intraspecifically over the distributional range. Boreoarctic barnacles, *Semibalanus balanoides* and *Balanus balanus,* have a well-defined annual cycle, produce a synchronous annual brood, and release larvae to coincide with the spring diatom increase (Barnes *et al.,* 1963; Crisp and Patel, 1969). In comparison, *Balanus crenatus,* which has a circumboreal distribution, produces several broods annually. Warm-temperate species produce broods during the summer. The length of breeding season of the temperate species is apparently limited only by the duration of suitable temperatures and the production of broods is limited only by food availability, the length of time for recovery of the gonads to maturity, and the incubation period of the embryos in the mantle chamber (Patel and Crisp, 1960). *Elminius modestus* on the coast of England produces 12 to 20 broods per year (Crisp and Davies, 1955), and *Balanus pacificus,* a subtidal species in southern California, produces 23 to 33 broods per year (Hurley, 1973). Both species breed continuously all year round. In the tropics, *Balanus amphitrite communis* breeds from September to June on Bombay's coast. Breeding of this population is interrupted by the low salinities and reduced food availability during the monsoon period (Karande, 1967), while populations on both east and west coasts of South India breed throughout the year (Pillay and Nair, 1972).

Breeding patterns of the barnacles, *Chthamalus fissus, Balanus glandula,* and *Tetraclita squamosa* vary in Morro Bay, California (Hines, 1978, 1979). *Chthamalus fissus* breeds from March to October, *B. glandula* from December or January to May, and *T. squamosa* from June to September. The populations of these species differ in the time and duration of the breeding period in different parts of the distributional range. For example, *B. glandula* breeds for an extended period from November to May further south of Morrow Bay in California (Hines, 1978) and northern populations of *Pollicipes polymerus* breed in the summer in Washington, over a more extended period in the northern and central regions of California, and erratically year round in Morro Bay. Further south, populations breed actively in the winter and at a low frequency in the summer. Populations of closely related species occurring in about the same latitudes in the northern and southern hemispheres may exhibit a mirror image of their breeding periods. *Pollicipes poly-*

merus in Monteray Bay (36° 40′) breeds from May to December, while *Pollicipes spinosus* in Ottago peninsula (44 52′) in New Zealand breeds from December to June (Hilgard, 1960).

Several anomuran and brachyuran crabs inhabiting the intertidal zones of coastal and estuarine waters in temperate and tropical regions breed for an extended period or even continuously throughout the year (Table II). There is much inter- and intraspecific variations in breeding patterns, both regionally and geographically. Generally, populations in temperate waters breed for a relatively shorter time compared to those in subtropical and tropical regions, but there are some exceptions. For example, continuous breeding occurs for *Pugettia producta* and *Petrolisthes cinctipes* in Montery Bay, California, and for *Petrolisthes elongatus* in Haukari Gulf (37° S), New Zealand, while restricted breeding occurs for *Calcinus tibicen* and *Clibanarius tricolor* in Barbados (13° N), Jamaica. In the tropics, a great majority of species breed for a very prolonged period or continuously, but breeding peaks occur within that period (Goodbody, 1965; Ahmed and Mustaquim, 1974; Ameyaw-Akumfi, 1975).

Generally, the breeding period is longer for populations of a species in the South compared to that in North (e.g., *Hemigrapsus nudus* and *Pugettia producta* in Puget Sound and Montery Bay; *Petrolisthes elongatus* on the New Zealand coast). However, populations of *Helice crassa* in Avon Heathcote estuary (43° 33′ S) breed for a longer period compared to populations at higher (Dundein, 45° 50′ S) and lower (Wellington, 41° S) latitudes (Jones, 1980). Sometimes variation in breeding period may also occur for populations in the same geographical area. For example, *Emerita talpoida* breeds from January to September on some beaches of North Carolina, while breeding occurs from June to September on other beaches (Diaz, 1980).

Populations inhabiting the high and low intertidal zones may also exhibit differences in breeding patterns. Generally, those of the low intertidal zone breed for a longer period compared to those in the high intertidal zone. Hence, congeneric species occupying different habitats in an area may breed for different lengths of time over the year, but their breeding periods may overlap (e.g., *Hemigrapsus oregonenesis* and *H. nudus* in Puget Sound). Overlap of breeding periods between April and mid-September occurs for *Sesarma cinereum,* which lives in supralittoral zone of higher salinities and *Sesarma reticulatum,* which lives in the eulittoral zone of the lower salinity portions of estuaries in marshes of North Carolina coast (Seiple, 1979). Within the overlapping breeding season, *S. cinereum* shows a lunar cycle in the number of ovigerous animals present at a given time, whereas *S. reticulatum* breeds with a lunar period from 1 to 2 lunar months in duration.

In some tropical Crustacea, when breeding of congeneric species over-

TABLE II

Breeding Period of Intertidal Anomuran and Brachyuran Crabs

Species	Habitat	Location and/or latitude (°)	Breeding period	Reference
Pisidia longicornis		Norway	May to October	Samuelson, 1970
		England	April to October	Smaldon, 1972
Porcellana platycheles		England	April to September	Smaldon, 1972
Scyra acutifrons	Intertidal to 40 fathoms, pilings and rocky areas	Puget Sound, U.S.A. 49 N	January to March, June–July, November– December	Knudsen, 1964
Petrolisthes eriomerus	Intertidal rocks	Puget Sound, U.S.A. 49 N	February to April; May to August	Knudsen, 1964
Pachycheles rudis	Low intertidal rocks, oyster and mussel beds	Puget Sound, U.S.A. 49 N	December to early summer	Knudsen, 1964
Hapalogaster mertensii	Under rocks	Puget Sound, U.S.A. 49 N	November to January, April	Knudsen, 1964
Hemigrapsus nudus	Rock and sand cover on sandy substrate	Puget Sound, U.S.A. 49 N	January to April	Knudsen, 1964
		Monterey Bay, U.S.A. 36–30 N	October to May	Boolootian *et al.*, 1959
Hemigrapsus oregonensis	Muddy portion of coast line	Puget Sound, U.S.A. 49 N	Mid-February to April, August	Knudsen, 1964
Lophopanopeus bellus bellus	Buried under sand, mud, and rocks	Puget Sound, U.S.A. 49 N	January to March; September	Knudsen, 1964
Pugettia gracilis	Intertidal to 40 fathoms	Puget Sound, U.S.A. 49 N	Alternate months throughout the year	Knudsen, 1964
Pugetta producta	Varying habitats, open kelp beds, unprotected rocky coast to bays and estuaries	Puget Sound, U.S.A. 49 N	January to April, June to August, November to December	Knudsen, 1964
		Monterey Bay, U.S.A. 36–30 N	Continuous	Boolootian *et al.*, 1959

(*continued*)

TABLE II *(Continued)*

Species	Habitat	Location and/or latitude (°)	Breeding period	Reference
Pachygrapsus crassipes	Narrow crevices of rocks, high intertidal to 12 fathoms	Monterey Bay, U.S.A. 36–30 N	March to August	Boolootian *et al.*, 1959
		Palos Verdes, U.S.A.	March to July	Boolootian, 1965
Emerita analoga	Sandy beaches	Monterey Bay, U.S.A.	April to October	Boolootian *et al.*, 1959
Petrolisthes cinctipes	Under flat rocks, middle to intertidal zone	Monterey, Bay, U.S.A.	Continuous	Boolootian *et al.*, 1959
Sesarma pictum	Upper intertidal zone	Tatar-Umi estuary, Japan	May to September	Pillay and Ono, 1978
Hemigrapsus penicillatus	Low intertidal	Tatar-Umi estuary, Japan	March to November	Pillay and Ono, 1978
Sesarma cinereum	Tidal marshes	Beaufort, U.S.A., 34–37 N	April to August	Seiple, 1979
Sesarma reticulatum	Tidal marshes	Beaufort, U.S.A.	Mid-April to September	Seiple, 1979
Emerita talpoida	Sandy beaches	Beaufort, U.S.A.	January to September	Diaz, 1980
Petrolisthes boscii	High water mark under rocks	Karachi, Pakistan 24–30 N	February to October	Ahmed and Mustaquim, 1974
Petrolisthes rufescens	High water mark under rocks	Karachi, Pakistan	January to November	Ahmed and Mustaquim, 1974
Pachycheles tomentosus	Low water mark, crevices of rocks	Karachi, Pakistan	Continuous	Ahmed and Mustaquim, 1974
Pachycheles natalensis	Low water mark, crevices of rocks	Karachi, Pakistan	Continuous	Ahmed and Mustaquim, 1974
Calcinus laevimanus	Intertidal	Keneohe Bay, Hawaii 21–30 N	February to early October	Reese, 1968
Calcinus latens	Intertidal	Kaneohe Bay, Hawaii 21–30 N	December-early January to September	Reese, 1968

Clibanarius zebra	Supralittoral	Kaneohe Bay, Hawaii 21–30 N	Late January to mid-August	Reese, 1968
Emerita portoricensis	Sandy beaches	Jamaica, W. Indies 18 N	Continuous	Goodbody, 1965
Mysidium columbiae		Jamaica, W. Indies 18 N	Continuous	Goodbody, 1965
Calcinus tibicen		Barbados, 13 N	April to June	Lewis, 1960
Clibanarius tricolor		Barbados, 13 N	April to June	Lewis, 1960
Emerita holthuisi	Sandy beaches	S.W. Coast of India, 10 N	Continuous	Subramoniam, 1979
Clibanarius chapini		Prampram, Ghana 6 N	Continuous	Ameyaw-Akumfi, 1975
Clibanarius senegalensis		Prampram, Ghana 6 N	Continuous	Ameyaw-Akumfi, 1975
Petrolisthes elongatus	Mid to shallow intertidal	Hauraki Gulf, New Zeland, 37 S	Continuous	Jones, 1977
		Wellington, 41 S	July to March	Greenwood, 1965
		Kaikoura Peninsula 42 S	October to March	
Helice crassa	Burrows, enclosed beaches, mangrove swamps	Avon and Heathcote estuary, New Zeland 43–33 S	August–September to March	Jones, 1980
		Wellington, 41 S	August to January	
		Dundein, 45–50 S	September to February	Nye, 1977
Heterozius rotundifrons	Intertidal, burrows in stone and pebble	Kaikoura Gulf, 42–25 S New Zeland	March to January	Jones, 1978

laps, the peak occurrences of ovigerous individuals for a species may be out of phase. Such temporal separation of breeding activities of closely related species may reduce interspecific competition between larvae. For example, *Calcinus laevimanus* breeds from May to August, while *C. latens* breeds from February to March and again in August on the coast of Hawaii (Reese, 1968). On the coast of Karachi, Pakistan, *Petrolisthes boscii* has a peak of ovigerous animals in April and a low in June–July, while the peak for *Petrolisthes rufescens* is in September–October and a low in April–May (Ahmed and Mustaquim, 1974).

Many tropical gonodactylid stomatopods breed for either a very prolonged period or all year round; oviposition typically occurs according to lunar and tidal rhythms (Reaka, 1976).

d. Semiterrestrial and Terrestrial Environments. Breeding patterns of semiterrestrial and terrestrial crustaceans follow the same general trends observed for marine forms. There is much interspecific variation, but populations in the warmer regions breed for an extended period compared to those in colder regions of the north. Several terrestrial isopods breed for a restricted period in the spring and summer in temperate regions (Carefoot, 1973). Fiddler crabs, *Uca pugilator,* breed from May through October in south Florida (Herrnkind, 1968), while the land crab *Cardisoma guanhumi* breeds from June to December, with peaks in September and October (Gifford, 1962). Tropical fiddler crabs, *Uca lactea annulipes* and *Uca vocans pacificensis,* breed for an extended period with several distinct peaks on the southwest coast of India (Pillay and Nair, 1968, 1971). There is no seasonality evident in breeding pattern of the populations of *Uca maracoani, Uca musica terpsichores,* and *Uca beebei* inhabiting the intertidal and flats of Panama Canal zone, but each species exhibits synchronization of the peak breeding activity to the full moon period (Zucker, 1978).

e. Freshwater Environment. Freshwater crustaceans show interspecific and geographical variation in the breeding patterns as seen in those inhabiting other environments. In temperate, dimictic, oligotrophic lakes, breeding of cyclopoid copepods coincides with the spring and autumn phytoplankton peaks (Nilssen, 1980). Generally, populations living in unpredictable environments reproduce frequently, while those in predictable environments reproduce fewer times. Under stable conditions in the tropics, breeding is continuous for cyclopoid copepods, which results in a constant age structure for the populations throughout the year (Burgis, 1971). In temperate lakes, calanoid copepods are univoltine or multivoltine or sometimes are acyclic (Allan and Goulden, 1980). Multivoltine populations that inhabit water bodies subjected to anoxia, very low temperatures, or drying are monocyclic. Extreme multivoltinism occurs in subtropical and tropical regions.

Periodicity in breeding of some freshwater ostracodes is found in colder climatic regions, while others breed year around. *Cyprinotus carolinensis* inhabiting temporary bodies of water in St. Louis, Missouri, breeds from January to February, and *Candona orangeburgensis* breeds from March to May (Ferguson, 1967). *Cypridopsis vidua,* in a small pond in Urbana, Illinois, breeds all year round, but the survival rate of the individuals is greater during spring and summer than in winter (Kesling, 1951). Generally, benthic ostracodes breed all year round compared to planktonic species. Not much is known about the breeding patterns of ostracodes in the tropics.

The freshwater prawn *Macrobrachium rosenbergii,* in Hoogly River, India, breeds from December to July, with peaks in March and May following the northeast monsoon, while a population in south India breeds from August to December, with a peak in October–November following the southwest monsoon (Raman, 1967). Dwarf crayfish *Cambarellus shufeldti* breeds throughout the year in Louisiana, but peak activity occurs in late winter and again in early summer (Lowe, 1961).

2. DETERMINANTS OF BREEDING PATTERNS

Proximate (exogenous and endogenous) factors synchronizing the reproductive cycles to time the breeding period and ultimate causes for timing the breeding period (Baker, 1938; Dunbar, 1968; Giese and Pearse, 1974) are of interest to population and community ecologists and evolutionary ecologists. Many of these aspects are not well understood.

Orton (1920) suggested that breeding activities of marine organisms are controlled by the temperature of the sea ("Orton's Rule"), and that marine animals in the tropics breed continuously or over an extended period of the year compared to species in colder temperate and boreal regions. Crisp (1954), working with the barnacles, extended this idea to include availability of planktonic food and suggested that tropical species inhabiting cooler waters confine their breeding season to warmer months, while the boreoarctic species breed only during cooler months ("Crisp's Rule"). Much of the recent literature indicates that breeding patterns may not simply be determined by one or only a few environmental factors, but by a complex coordination of the endogenous events within the individuals of populations and within the environment. With respect to life history, individuals of short-lived species may produce a single brood or several broods during a lifetime typically less than 1 year and then die (semelparous). Individuals of long-lived species may produce only one brood or several broods during each breeding season over a lifetime (iteroparous). With respect to the environment, the life span and the number of broods may vary for both semelparous and iteroparous species and for populations of a species in different parts of the distributional range.

The inter- and intraspecific variations in breeding may be explained by considering the breeding pattern as a genotypic response to the environment for successful reproduction (Sastry, 1975). The environmental limits within which a species lives are genetically determined, and within these limits reproduction occurs over a narrower or broader range.

The period of the oogenic cycle, the length of time required for incubation of eggs, and the duration of any quiescent period are intrinsic to a species in a given environment. The timing of these events may take place within the same or different environmental regimes suitable for the reproductive processes of a species. When the length of time for oogenesis and incubation of eggs is short relative to the duration of favorable environmental conditions, including food availability, the individuals of a population may go through successive oogenic-incubation cycles. Such species may produce more than one brood and hence breed for the extended period that the environment is within a favorable range. On the other hand, if the period of oogenesis-incubation cycles is as long as the period during which favorable conditions prevail in the environment for the respective events, the number of oogenic cycles is limited, and the frequency of brood production is correspondingly reduced. From these basic relationships a variety of breeding patterns may be visualized.

The complex interactions between endogenously controlled events and the environment may result in different breeding patterns for species inhabiting different environments. Breeding patterns of species may thus vary relative to the life-history traits—life span, genetically determined tolerances, requirements for survival and reproduction, zoogeographical affinities, intrinsic rate of oogenesis, length of egg incubation time, type of development, adult and larval or juvenile food requirements, and the environmental conditions in different climatic zones. The interactions determining the reproductive success of populations in a given environment may also determine biogeographical limits of a species and reproductive failure of expatriate populations.

The genotype–environment interactions in regulation of the timing and duration of breeding period are most evident for sympatric species and for geographically separated populations. Variations among geographically separated populations may be phenotypic or genetic (incipient speciation) or both (Sastry, 1975). Thus, the initiation and duration of breeding period of geographically separated populations may primarily be determined by the occurrence of favorable environmental conditions. Hence, intraspecific variation in the breeding period may occur as an adaptive response for reproductive success in different environments.

Breeding patterns of populations may be examined in terms of factors controlling the initiation and duration of the breeding period, and synchrony among members of a population. Proximate factors controlling the breeding period may allow release of larvae or juveniles to coincide with the times when environmental conditions are favorable for development and growth of the progeny. The ultimate causes determining the time of breeding may maximize the probabilities for survival of the progeny to ensure species continuity. Both are interrelated, but their causation may be different (Dunbar, 1968).

a. *Proximate Factors.* Crustacean breeding periods may be synchronized with one or a combination of environmental factors. Synchronization of breeding is most evident for populations in seasonally varying environments, but it can also occur for populations in other environments where there are no pronounced seasonal changes. Breeding periods may be synchronized with the day-length, the time of food abundance for larvae or juveniles, temperature, lunar and tidal rhythms (e.g., intertidal animals), changes in salinity (e.g., estuarine animals), and other factors. However, synchronization of breeding may be achieved through a complex interaction of the internal events entrained long before the actual time of breeding, and coordinated with the environment. The interactions between environmental, behavioral, physiological, neuroendocrine aspects (proximate mechanisms), and their influence on oogenesis, ovulation, and oviposition in coordination and synchronization are evident. The temporal sequence of the events of the reproductive cycle of individuals with respect to the environment may determine the timing and synchronization of a breeding pattern. Although breeding activities may be correlated with one or a combination of factors, the relative importance of any particular factors may vary for different species and in different environments. Little study has been done to clarify the relative importance of different factors in the synchronization of breeding activities. A better understanding of the factors synchronizing breeding activities may provide insights into the causes of fluctuations in annual abundance of populations so common in the marine environment.

Synchronization of breeding to coincide with the food abundance is most evident for species with planktotrophic larvae, especially those occurring at higher latitudes (Marshall and Orr, 1972; Voronina, 1970; Allen, 1966a; Thorson, 1950; Barnes, 1957; Giese, 1959). For example, the antarctic euphausiids, *Euphausia superba,* shed their eggs near the surface waters from January to March. The eggs sink to depths (500–1500 m) to hatch as nauplii, which then develop while migrating to the surface waters to begin feeding during summer phytoplankton blooms (Fig. 4). Thus, the breeding

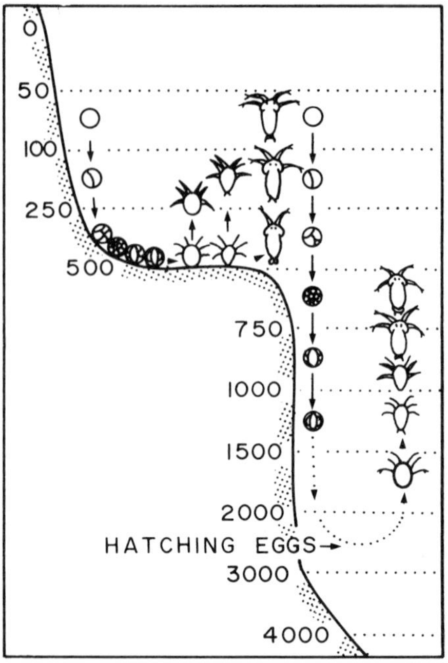

Fig. 4. The relationship between the time of breeding and food availability for larvae of the antarctic euphasusiid, *Euphausia superba,* suggesting synchronization of reproductive events with the environment. Eggs shed into the water sink to depths and nauplius and metanauplius develop utilizing yolk reserves in their bodies while ascending to surface waters to begin feeding in the calyptopsis stage on the phytoplankton blooms in the summer. (Originally from Marr, 1962, source: Mauchline and Fisher, 1969.)

period of this species is timed so that larval appear in the surface waters when food is expected to be abundant at a future time interval. The breeding season of many species having larvae trophically linked with primary production is short in the higher latitudes, and larval release is coincident with phytoplankton blooms. However, the breeding period of some species may not be timed so that larval release coincides with phytoplankton blooms. The environmental conditions suitable for larval development and the larval food requirements may vary for species (Sastry and Vargo, 1977; Phillips and Sastry, 1980) so that the relative importance of the factors of the breeding period may be different for different species in different amounts.

Recent studies with some subtropical and tropical groups, such as *Uca* (Feest, 1969; von Hagen, 1970; Zuker, 1976), gonodactylid stomatopods (Reaka, 1979), the land crab *Cardisoma guanhumi* (Gifford, 1962), and the mangrove tree crab *Aratus pisonii* (Warner, 1967), have shown that breeding occurs either for a prolonged period or all year round. However, there are breeding cycles within the breeding period and these are synchronized with the tidal and lunar cycles. Synchronization with the tidal and lunar cycles is obviously adaptive, especially for shallow intertidal, semiterrestrial, or terrestrial species with planktotrophic larvae to facilitate seaward dispersal of the larvae.

In crustaceans inhabiting the lower mesopelagic and deep-sea benthic environments, there are no obvious physical environmental cues for the synchronization of breeding activities, but food availability may be one cue which may vary relative to the events taking place in the upper epipelagic zone. In these food-poor environments, oogenesis-incubation cycles may basically progress at an intrinsic rate characteristic to each species, but it is also possible that oogenesis may be influenced by the availability of food and ovulation by successful mating of the individuals. Observation of shallow-water species indicates that the oogenic cycle may be entrained or delayed depending on food availability (Barnes and Barnes, 1967) and energy storage status of the individuals, and oviposition will be affected by the temporal relationship between mating and ovulation (see Section III,C,4). Individuals of these populations may thus breed at any time, but the time when they may breed could be determined by the duration of oogenesis-incubation. Individuals of these populations may breed asynchronously or sporadically during parts or all of the year.

b. Ultimate Factors. Proximate control mechanisms of breeding patterns must have evolved and been maintained through natural selection to ensure continuity of species in its environment (Baker, 1938; Dunbar, 1968). For an understanding of the ultimate determinants of breeding patterns, an evolutionary outlook of reproduction is needed. Not a great deal is known about the ultimate factors controlling the breeding patterns; however, a variety of parameters has been suggested as factors to which reproductive processes may be adapted for decapod crustaceans with planktotrophic larvae. Boolootian *et al.* (1959) and Reese (1968) have indicated that breeding cycles are adapted to the availability of food for larvae, whereas there are no obvious correlations between these cycles and temperature, or adult food requirements among those species which they have examined. Variation in food supply to the adults has been suggested important in timing of breeding of tropical crabs (Ahmed and Mustaquim, 1974). Ameyaw-Akumfi (1975) suggested that competition plays an important role in breeding of tropical crabs.

The adaptive value of the breeding patterns is obviously to time the release of offspring as larvae or juveniles into an environment suitable for their development and growth, and hence to increase the probabilities for their survival, which ensures species continuity. The ultimate causes determining the breeding patterns may be better understood by considering the selective pressures on the life history as a whole, and by considering the niche of species in their respective communities. Factors responsible for the selection of breeding strategies may be different for species with different life history patterns and reproductive traits, inhabiting different environments. Selective pressures may include primarily biotic factors, such as food availability, competition, and predation in constant environments, while both biotic and abiotic factors may be involved in variable environments. Studies examining life history and reproductive patterns of species and relating them to their niche may provide insights into the ultimate causes in selection of species-specific breeding patterns in different environments.

3. POPULATION SYNCHRONY

Breeding activities among members of a population in a given environment may be synchronous, or they may show synchrony followed by asynchrony. In other populations, breeding actimitus may be asynchronous, continuous, or sporadic within a year. In synchronously breeding populations, all individuals of the population are ovigerous at the same time and breeding is usually restricted to a short period (e.g., *Semibalanus balanoides*). At any given time of the year, the whole population will be composed of individuals with ovaries in about the same stage of development, or eggs in the same stage of development. In addition, the populations will be composed of individuals representing discrete year classes. Generally, a highly synchronous breeding pattern seems to occur only for species with planktotrophic larva linked with the peak primary production in higher latitudes.

For some species, breeding is restricted to a certain time of the year and the first brood is synchronous for the whole population, but later the individuals become asynchronous in the production of the subsequent broods (e.g., *Calanus finmarchicus, Balanus glandula, Verruca stroemia*). This may occur for populations of species with a short oogenic-incubation period that inhabit the warmer part of their distributional range, where environmental conditions are favorable for a longer period than one oogenic-incubation period, but not for the entire year.

Asynchronous breeding within a restricted period of the year occurs in several crustaceans from temperate, subtropical, and tropical regions (Section III,D,1). Individuals in these populations are usually out of phase, but a population as a whole begins and stops breeding at well defined times of the year (see Knudsen, 1964, for examples). Asynchrony may result from production of several clutches by the individuals or by production of a single brood by each individual of the population, but out of phase. Within the breeding season, the proportion of ovigerous individuals in the population may increase to a maximum and then decline. In some of these species, the oogenic cycle may be longer than the breeding season, but time for incubation may be short, so that individuals are out of phase. A year class of these populations may be composed of several size groups.

Generally, continuous breeding throughout the year occurs in a number of crustaceans in the tropics and in the deep-sea, but this pattern is also seen in some species in temperate and polar regions (see Section III,D,1). Generally, recruitment to these populations is continuous and the age structure of the population is constant throughout the year. Populations may be composed of an equal proportion of individuals undergoing oogenesis, oviposition, and developing embryos and gonads in a quiescent stage. However, peaks in breeding occur for some species, so that the relative proportion of individuals in different phases of the reproductive cycle may change in different times of the year (Ameyaw-Akumfi, 1975). Individuals of continuously breeding populations may produce several clutches within a short incubation period. The period of oogenesis-incubation may be short, or oogenesis may be longer compared to incubation periods, or each incubation may be followed by a quiescent period before another oogenic cycle is activated. In contrast, some species may breed only sporadically during parts or all of the year (e.g., some copepods in tropical waters: Moore and Sanders, 1976).

The only broad generalization that seems to emerge is that crustaceans with planktotrophic larvae linked with peak primary production breed synchronously in the higher latitudes and asynchronously or continuously with the decrease of latitude. In comparison, species with planktotrophic larva not directly dependent upon primary production and those species with direct development may breed asynchronously or continuously regardless of the climatic region.

IV. LIFE HISTORY PATTERNS AND REPRODUCTIVE TRAITS

The immense variation in crustacean life history patterns and reproductive traits has obviously contributed to their evolutionary success both in number of species and in habitats that they have colonized in marine, terrestrial, and freshwater environments. The selective pressures shaping the life history patterns of crustaceans are of interest to ecology and evolution, particularly considering their importance to the structure and dynamics of communities.

A. Theory

The general theoretical problem to predict what life history traits evolve in organisms in a particular environment has received much attention in recent years (Stearns, 1976, 1977). Crustaceans as a group provide challenges and also unique opportunities to both theoreticians and empiricists to develop a generalized theory of life history evolution. The adaptive value of life history patterns and reproductive traits is that it provides a design for survival of progeny to develop to the adult stage so that they may reproduce for representation by future generations. Within the past two decades, there has been a growing recognition of organisms as an adapted complex, not only in terms of molecular, cellular, physiological, and organ system responses to the environment (proximate) (Prosser, 1973), but also in terms of the adaptation of a species through the evolution of its life history as a whole (ultimate) in the total environmental context (Cole, 1954; Williams, 1966; Wilbur *et al.*, 1974; and others). The adaptive significance of life histories are now being examined from a theoretical viewpoint using a holistic approach to explain the selective pressures for their evolution. Contributors to this field have formulated several theoretical ideas (MacArthur and Wilson, 1967; Lewontin, 1965; Gadgil and Bossert, 1970; Murphy, 1968; Schaffer, 1974). Many of the theoretical expectations remain to be validated by empirical studies. MacArthur and Wilson (1967) introduced the theory of *r*- and *K*-selection to ecology based on concepts originally discussed by Dobzhansky (1950). This theory has stimulated many studies on the life history patterns, including some on crustaceans (Allan, 1976; Allan and Goulden, 1980). MacArthur and Wilson (1967) originally theorized that in the absence of competition, species that reproduce rapidly are favored (*r*-selection), and when resources are limiting, the advantage goes to those species that are able to compete best (*K*-selection). The terms are derived from the logistic equation of population growth, where *r* refers to the intrinsic rate of population growth, (*r*-max), and *K* refers to the carrying capacity. Since its original formulation, the meaning of this theory has been broadened (Parry, 1981), and a number of correlates of *r*- and *K*-selection have been advanced. Different interpretations of the theory (Wilbur *et al.*, 1974; Allan, 1976) and empirical studies to test the theory can be found (Atkinson, 1979; Taylor and Condra, 1980). For a comprehensive discussion of the rationale and theoretical expectations, consult reviews by Pianka (1970, 1976), Demetrius (1975), Stearns (1976, 1977), and Giesel (1976). Pianka (1970) summarized *r*- and *K*-selection, and also visualized *r*–*K* continuum. Stearns (1976), in summarizing the concept, notes that *r*- and *K*-selection predicts an association of life history traits into two groups: *r*-selection—early age at maturity, large number of young, semelparity, absence of parental care, and a large reproductive effort; *K*-selection—delayed maturity, small number of young, iteroparity, parental care, and a small reproductive effort. Central to this concept is reproductive effort (Hirshfield and Tinkle, 1975; Pianka, 1976), which has been defined as the proportion of assimilated energy allocated to reproduction over the lifetime of an individual (Section III,B,2). A comparative measure of reproductive effort is generally unavailable for most species, including crustaceans from different environments.

Life history and reproductive strategies of species appear to be selective compromises with respect to a combination of biotic and abiotic factors in a particular environment. The traits may have evolved for each species to cope with the environmental problems of food availability, overall unpredictability/variability or predictability/stability of the physical and chemical factors in the environment, competition, and predation. The life history and reproductive strategies of species appear dynamic rather than fixed with respect to both life history and the environment. The prediction of which traits will evolve in organisms living in different environments, and what selective pressures will be acting to mold life history patterns continues to be an unresolved problem. A critical evaluation of the life history patterns in detail with respect to the diverse environments crustaceans inhabit or in the context of theoretical expectations is not attempted here, but a few examples are considered to illustrate some of the general trends and problems.

B. Observations

In the marine pelagic environment, physical and chemical factors, food availability, and biotic interactions such as competition and predation may vary regionally and geographically in the epipelagic zone, and with depth from the epipelagic to bathypelagic region. Mauchline (1972) compared aspects of life history and reproductive traits of copepods, mysids, and euphausiids from the epipelagic to bathypelagic regions. The information presented here is derived largely from Mauchline, a review on pelagic shrimp by Omori (1974), and a book on deep-sea biology by Marshall (1979). The mesopelagic and bathypelagic regions are characterized by essentially no seasonal variation in physical and chemical factors, they experience several atmospheres of hydrostatic pressure, and they possibly experience seasonal fluctuations in food abundance relative to the events that may be occurring in the upper epipelagic zone. The relative amount of food available and its quality decreases in relation to increasing depth. Many crustaceans in the lower mesopelagic zone are omnivores tending toward carnivory, while those in bathypelagic regions are carnivorous. The density of organisms also decreases with the increase of depth. In these food-poor environments, competitive interactions may be keen and predation pressure may be high. Copepods, mysids, euphausiids, and carideans from these regions are characterized by large body size, long life span, sex ratios biased toward females, delayed maturity, possibly continuous breeding, fewer large eggs per female, brood care, and release of young as larvae at a larger and more advanced stage of development compared to the species from epipelagic zones.

The longer life span, delayed maturity, and fewer eggs per female may result in a slower replacement rate and lower population density. The life history and reproductive traits of crustaceans from these environments correlate well with the characteristics predicted for *K*-selected species by the theory.

Not much information is available on the life histories and reproductive traits of benthopelagic and deep-sea benthic crustaceans. These species may also exhibit characteristics of *K*-selected species. In the two dimensional deep-sea benthic environment, the selection pressures may be somewhat different from those in lower mesopelagic and bathypelagic regions (Mauchline, 1972). There is some evidence to indicate that body size of some Crustacea at hadal depths is smaller compared to those in the lower mesopelagic and bathypelagic regions. However, bathyal and abyssal species of asellote isopods produce a small brood of large eggs compared to a large brood of small eggs by the boreal and littoral species (Wolff, 1962). Rokop's (1977) study on the deep-sea benthic isopods also indicates that these species produce a few large eggs that develop directly to hatch juveniles.

Shallow water benthic Crustacea in the polar regions attain a relatively larger size compared to those from temperate and tropical regions (Dunbar, 1968). In polar species there is an increasing tendency for the production of large yolky eggs and nonpelagic development compared to the temperate and tropical species, which produce small eggs that hatch at a relatively early stage (Thorson, 1950; Dunbar, 1968; Knox, 1970). Some polar species in the shallower waters are characterized by slow growth, delayed maturity, large adult size, and small number of large size eggs. The eggs are hatched to release advanced larvae or juveniles (Dell, 1972; Clarke, 1977). The antarctic isopod, *Glyptonotus antarcticus,* oviposits about 512 eggs per female; these eggs take about 577–626 days to hatch. After release of young, the female normally becomes moribund and dies within a few days without feeding. In comparison, the amphipod *Bivallia gigantea* produces relatively small eggs which also take a long time to develop to hatching (Thurston, 1970). The hippolytid prawn, *Notocrangon antarcticus,* also exhibits features of reduced fecundity and large yolky eggs, which develop into demersal shizopod larvae that have an abbreviated plankton phase (Makarov, 1973; Maxwell, 1977). Clarke (1979) has examined the reproductive traits of two species pairs of carideans, each species pair containing one polar and one temperate species. In each pair, the polar species *Chorismus antarcticus* and *Notocrangon antarcticus* displayed traits associated with *K* strategists, compared to the temperate species, *Pandalus montagui* and *Crangon crangon*. They differ in brood size, egg weight, and maturity of newly hatched larvae. Annual reproductive effort (measured as grams wet weight of fresh eggs per gram fresh weight of females) and total life time reproductive effort were significantly less for the polar species compared to those of temperate species. Not all antarctic species appear to be *K* strategists. Variation in the reproductive traits is evident among species that are trophically linked with plankton blooms and those with a direct development. In polar regions, both *r* and *K* strategists may coexist in the same habitat, but a large number of antarctic benthic crustaceans seem to tend more toward *K* rather than *r* strategy (Clarke, 1979).

Variation in life history patterns and reproductive traits is also found in freshwater cladocerans and marine and freshwater copepods, both regionally and geographically (Marshall and Orr, 1972; Allan, 1976; Allan and Goulden, 1980; Corkett and McLaren, 1978). Allan and Goulden (1980) indicate that copepods have a long generation time and brief reproductive period, during which several broods are produced; the adults die at the end of reproductive period (semelparity). In comparison, cladocerans mature rapidly, have a short generation time, indeterminate growth, and iteroparity. These authors suggest that cladocerans have *r*-selected characteristics, while copepods do not. Among these small-sized and short-lived groups, there are other differences in energy investment for egg production, mode of reproduction, and growth patterns. These authors point out that the relationship between intrinsic rate of population growth (*r*-max) and body size is complicated by differences between species in response to changing conditions, and they caution against the idea that *r* strategists have a high *r*-max and small body size compared to *K* strategists. Instead, a classification of life histories in terms of opportunism is suggested (Hutchinson, 1967). Studies on some limnetic and marine copepods indicate that reproductive traits, age at first reproduction, average brood size, and number of eggs per brood per lifetime are affected by temperature and food conditions (Weglenska, 1971; Corkett and McLaren, 1978; Smyly, 1973), season (Røen, 1957), eutrophic or oligotrophic environments, and latitude (Nilssen, 1980).

The duration of the life span and the time required to reach maturity increases with an increase in latitude inter- and intraspecifically for marine copepods, amphipods, and euphausiids (Dunbar, 1968; Corkett and McLaren, 1978; Marshall and Orr, 1972; Lasenby and Langford, 1972). Within a species, for example, the populations of *Calanus finmarchicus* and *Pseudocalanus minutus,* individuals may live for 1 year or more in northern areas compared to those in warmer waters. Copepods may take up to 2 years or even more to produce a brood in the northern locations, while they produce one brood or more per year in temperate regions. The life span may also vary among individuals of different generations within a population in the same area. For example, the overwintering generation of *Calanus fin-*

marchicus may live for about 7 months compared to 2–3 months for the spring generation. Generally, tropical and subtropical populations produce several generations.

Harpacticoid copepods are one of the more important components of the meiobenthic community. The reproductive potential, expressed by the intrinsic rate of growth (*r*-max) for *Tachidius discipes* and *Paronychocamptus nanus,* differs considerably depending on the prevailing temperature and food conditions (Heip and Smol, 1976), Hicks (1980), examinig the reproductive cycles of harpacticoid copepods from epibenthic, interstitial, and phytal habitats, found that the duration of the breeding period is dependent on the availability of food resources and the differential capacity of various species to exploit these resources. Energy limitations appear to be related to the habitat and have an influence on the size and number of eggs. Interstitial species with discontinuous breeding tend to produce a low number of small eggs, while phytal species with continuous breeding produce a large number of small eggs. Generally, harpacticoids have been considered as *r*-selected species, but *K*-selected attributes are found among different species and also within the life cycle of a single species. According to this author, harpacticoid copepods do not appear to be easily classified as being clearly *r* or *K* selected; rather, specific components of their life histories may be ranked as representing *r* or *K* characteristics along the *r*–*K* continuum.

Benthic infaunal cumaceans examined from both sides of the North Atlantic and Canadian arctic and subarctic areas from littoral to 200 exhibit variable life history and reproductive traits (Corey, 1981). In the arctic, *Diastylis goodsiri* has a long life span, delayed maturity, attains larger adult size, and produces only a few eggs. In comparison, life history patterns and reproductive traits are extremely variable for species from temperature regions. All these cumaceans inhabit soft bottom substrates, but they have evolved different combinations of life span, number of eggs, and size of eggs with no apparent relationship to the time of breeding and depth of occurrence.

Life history patterns and reproductive traits of amphipods from the temperate regions vary both inter- and intraspecifically. The gammaridean amphipods inhabiting the low intertidal region produce more offsprings per female compared to those inhabiting high intertidal regions (Borowsky, 1980). Intraspecifically, both life span and reproductive traits vary in relation to seasonal changes in the environment and habitat. For example, the winter generation of the sand beach amphipod, *Neohaustorius schmitzi,* lives for 8 months compared to only a 4-month life span of the summer generation (Dexter, 1971). Summer animals grow faster, reproduce over a shorter period, and produce fewer eggs compared to the winter animals. The maturity of *Caprella penantis* is delayed for the winter animals compared to the summer animals (Bynum, 1978). Females of *Bathyporeia pelagica* inhabiting the estuarine areas produce more eggs than females of the same size, which inhabit the open coast (Fish and Preece, 1970).

Life cycle and breeding characteristics also vary for maine, freshwater, and terrestrial isopods (Carefoot, 1973). Isopods living in freshwater ponds and salt lakes have a 1-year life-cycle compared to the longer life span of terrestrial isopods of up to 4 years. Geographic differences also exist in size at maturity and the number of eggs within a given group of isopods. The terrestrial isopod, *Armadillidium vulgare,* breeds after 1 year in California (Paris and Pitelka, 1962), with the 1-year-old animal producing a single brood, while the larger 2- and 3-year-old individuals produce two broods per season (Lawlor, 1976). In contrast, a population of this species in England does not initiate breeding until after 2 years (Heeley, 1941).

Intertidal barnacles, *Chthamalus fissus, Balanus glandula,* and *Tetraclita squamosa,* which occupy different but overlapping intertidal ranges in Morrow Bay, California, vary in the components of life history and reproductive traits (Hines, 1979). Reproductive strategy of the three species seems to vary relative to the degree of environmental stress, competition, and predation, to influence the dynamics of populations and their ecology. *Chthamalus fissus* begins reproduction at an early age and has a short generation time. This species opportunistically exploits the dynamic conditions in its environment and rapidly colonizes ephemeral open space in the lower intertidal zone. In contrast, *B. glandula* has a longer life span and delayed onset of reproduction. This species produces a series of broods and spatially dominates the lower intertidal zone for parts of the year, but populations in the high intertidal zone subjected to less competition and predation, survives better. The third species, *Tetraclita squamosa,* has a long generation time, delayed onset of reproduction, large body size, and a low yearly allocation of energy to egg production. Rapid growth to a large size allows this species to effectively compete for space and to avoid predators.

Variation in the life history patterns of coral dwelling tropical gonodactylid stomatopods appears to be primarily influenced by body size among different species (Reaka, 1979). Differences in body size among different species has an important consequence influencing the size of eggs, size of larvae at hatching, and planktonic larval development time. Life history patterns of these stomatopods do not conform to the correlates predicted for *r* and *K* selection or bet hedging (Reaka, 1979).

There is considerable inter- and intraspecific variation among anomuran and brachyuran crabs in the number and size of eggs produced. Typically, many crabs produce a large number of eggs, incubate them to hatching, and release planktotrophic larvae. Fewer large size eggs are produced by species with an abbreviated larval development. Specific components of these species may be again ranked as *r* or *K* characteristics along the *r–K* continuum. Pillay and Ono (1978) have reported that *Sesarma pictum* from the high intertidal zone has delayed maturity and produces fewer eggs over a short breeding season compared to *Hemigrapsus penicillatus* from the low intertidal zone. *Sesarma cinereum* from high salinity and supralittoral zone produces small eggs, four to six broods, and incubates eggs for about 28 days to hatch larvae that go through four stages to complete development in about 25–30 days (Seiple, 1979). In comparison, the congeneric species *Sesarma reticulatum* from eulittoral zones produces only two to three broods, and larger eggs, and it incubates eggs for about 45 days hatching larvae that go through three stages to complete development in about 15–30 days.

Semiterrestrial and terrestrial decapods that seek marine and fresh water for the breeding purposes vary in their reproductive strategies. Those species that return to the sea to release larvae produce a large number of small eggs as marine forms (Bliss, 1968, 1979). In contrast, those species that seek freshwater produce only a small number of large eggs, which develop directly to hatch young as miniature adults (Koba, 1936). The characteristics of these freshwater species seem to correlate with the predictions for *K*-selected species by the theory.

C. Current Status

Life history and reproductive traits of crustaceans do not always seem to conform to the correlations predicted by the *r*- and *K*-selection theory. The principle exceptions to this are only the crustaceans living in the lower mesopelagic, bathypelagic, and deep-sea benthic environments, and certain freshwater and terrestrial decapods that inhabit freshwater during part of their life. The characteristics of these species correlate with those predicted for *K*-selected species. In those inhabiting all other environments, specific life history components and reproductive traits of the crustaceans may be ranked as *r*- or *K*- selected along the *r–K* continuum. Furthermore, these characteristics do not seem fixed for a species, but they vary for geographical populations and for different generations in the same area. Any evaluation of the life history and reproductive strategies of crustaceans based on only correlations of the observed traits with the predictions of the *r*- and *K*-selection theory alone may be misleading. Other factors, including environmental conditions, habitat stability, need for dispersal, and predation may also be involved in shaping the life history patterns (Wilbur *et al.*, 1974; Stearns, 1977; Taylor and Condra, 1980; Parry, 1981).

Although, life history patterns and reproductive traits of crustaceans in the same or different environment are extremely variable, their development, however, essentially follows two basic patterns. Development with a larva or direct development to hatch juveniles occurs in the life cycle whether a species is semelparous or iteroparous. In between these extremes, all sorts of gradations occur for species with no apparent phylogenetic, habitat, or environmental relationships. However, all species with a planktotrophic larva in the life cycle produce large numbers of small eggs, while those with direct development produce only a few large eggs. This relationship between the type of development of a species and the number and size of eggs per female appears consistent to all crustaceans inhabiting the different climatic zones.

It is still speculative whether reproductive effort of species with different developmental patterns is the same or not, and what adaptive value a particular type of development may have for a species for reproductive success in its environment. Generally, planktotrophic larva in the life cycle of intertidal and benthic species is considered to be advantageous for dispersal, gene exchange, to maintain genetic variability essential for adaptability, colonization of diverse habitats over the geographical range, and to overcome unpredictable habitat disturbances in the adult environment (Crisp, 1974). However, some infaunal crustaceans have a benthic larva (e.g., harpacticoid copepods, cephalocarids), and epipelagic crustaceans have a planktonic larva in their life cycles. Direct development is considered to be advantageous since young are released at a larger size and at an advanced stage of development that can employ an adult mode of feeding and can avoid predators. Crustaceans with both developmental patterns occur in almost all the climatic zones except in deep-sea, suggesting that within the same region the selective pressures acting on different species in the evolution of their life history patterns and reproductive traits may be different.

The life history patterns and reproductive traits of species have doubtless evolved within constraints of phylogenetic relationships as well as environmental circumstances of a species and habitat, with selection favoring optimal exploitation and utilization of resources through all phases of the life cycle in relation to food availability, competition, and predation, thus maximizing the probabilities of reproductive success. Hence, the selection of life history and reproductive traits may relate to niche of a species and resource availability, rather than to specific biotic or abiotic factors in their environment.

V. CONCLUSIONS

At one extreme of biological organization, reproduction can be studied on the molecular level; at the other extreme, it can be studied from the standpoint of influences on population dynamics or community organization and function. The biological aspects of crustacean reproduction are being studied at such contrasting levels for one species or another, but there is as yet no synthesis of the findings to provide a thorough understanding of the factors initiating and regulating the reproductive cycle within individuals or within populations. Accordingly, many of our views on the adaptive value of the reproductive patterns to enhance survival of species remain speculative.

Little is known about the mechanisms by which diverse reproductive processes are initiated or controlled within an individual, including the interaction of an organism with its environment that results in observed temporal patterns of reproduction. The biochemical, physiological, and neuroendocrine mechanisms are well understood for certain decapods, while the effects of environmental factors have been examined in great detail for barnacles. However, it still remains to be established how far conclusions drawn for one group apply to another.

Reproduction in all crustaceans, whether living in constant or variable environments, is dynamically coupled with the organism's physical and chemical environmental conditions, food availability, and the presence of competitors and predators. The relative importance of proximate factors controlling the reproductive activities may vary for different species in the same environment, and also for different species inhabiting different environments.

Availability of food for adults to maintain, grow, and reproduce, and for larvae and/or juveniles to develop, grow, and survive may be one of the more important factors in the synchronization and coordination of reproductive activities in a particular environment, and also in the evolution of life history and reproductive strategies. Many crustaceans change their trophic position and hence their food requirements in different stages of the life cycle. For a given life history stage, the quality and quantity of available food, competition, and predation may also vary temporally and spatially, and may hence act to exert a selective pressure in the evolution of reproductive traits.

Such a holistic view of organisms as an adapted complex in the context of its entire life history may facilitate our elucidating the proximate and ultimate factors controlling the reproductive patterns. Accordingly, integrated studies on reproductive ecology, population dynamics, and community ecology should provide insights into the evolution of life history patterns and reproductive strategies of crustaceans which inhabit different environments.

ACKNOWLEDGMENTS

The author would like to thank Dr. W. S. Johnson (Peracarida), Dr. R. G. Hartnoll (Brachyura), Prof. G. Hartman (Ostracoda), and Dr. A. E. F. Heydorn (Macrura) for generously allowing me to read copies of their manuscripts of unpublished reviews. I would like to thank Dr. Don C. Miller for his helpful comments in the manuscript.

REFERENCES

Adiyodi, K. G., and Adiyodi, R. G. (1970). Endocrine control of reproduction in Decapod Crustaceans. *Biol. Rev. Cambridge Philos. Soc.* **45,** 121–165.

Adiyodi, R. G. (1968a). Protein metabolism in relation to reproduction and molting in the crab *Paratelphusa hydrodromous* (Herbst). Part I. Electrophoretic studies on the mode and utilization of soluble proteins during vitellogenesis. *Indian J. Exp. Biol.* **6,** 144–147.

Adiyodi, R. G. (1968b). Protein metabolism in relation to reproduction and molting in the crab *Paratelphusa hydrodromous* (Herbst). Part II. Fate of conjugated proteins during vitellogenesis. *Indian J. Exp. Biol.* **6,** 200–203.

Adiyodi, R. G. (1969). On the storage and mobilization of organic resources in the hepatopancreas of a crab (*Paratelphusa hydrodromous*). *Experientia* **25,** 43–44.

Ahmed, M., and Mustaquim, J. (1974). Population structure of four species of porcellanid crabs (Decapoda: Anomura) occurring on the coast of Karachi. *Mar. Biol. (Berlin)* **26,** 173–182.

Aiken, D. E. (1969a). Photoperiod endocrinology and the Crustacean molt cycle. *Science* **164,** 149–55.

Aiken, D. E. (1969b). Ovarian maturation and egg laying in the crayfish *Orconectes virilis:* Influence of temperature and photoperiod. *Can. J. Zool.* **48,** 931–935.

Aiken, D. E., and Waddy, S. L. (1976). Controlling growth and reproduction in the American lobster. *Proc. Annu. Meet.—World Maric. Soc.* **7,** 415–430.

Aiken, D. E., and Waddy, S. L. (1980). Reproductive Biology. *In* "The Biology and Management of Lobsters" (J. S. Cobb and B. F. Phillips, eds.), Vol. 2, Chapter 4, pp. 215–276. Academic Press, New York.

Allan, J. D. (1976). Life history patterns in zooplankton. *Am. Nat.* **110,** 165–180.

Allan, J. D., and Goulden, C. E. (1980). Some aspects of reproductive variation among fresh water zooplankton. *Spec. Symp.—Am. Soc. Limnol. Oceanogr.* **3,** 388–410.

Allen, J. A. (1966a). The dynamics and interrelationships of mixed population of caridea found off the north-east coast of England. *In* "Some Contemporary Studies in Marine Science" (H. B. Barnes, ed.), pp. 45–66. Allen & Unwin, London.

Allen, J. A. (1966b). The rhythms and population dynamics of decapod Crustacea. *Oceanogr. Mar. Biol. Annu. Rev.* **4,** 247–265.

Allen, W. V. (1972). Lipid transport in the Dungeness crab, *Cancer magister* Dana. *Comp. Biochem. Physiol. B* **43B,** 193–207.

Ameyaw-Akumfi, C. (1975). The breeding biology of two sympatric species of tropical intertidal hermit crabs, *Clibanarius chapini* and *C. senegalensis. Mar. Biol. (Berlin)* **29,** 15–28.

Atema, J., and Engström, D. G. (1971). Sex pheromone in the lobster *Homarus americanus. Nature (London)* **232,** 261–263.

Atema, J., Jacobson, S., Karnofsky, E., Oleszko-Szuts, S., and Stein, L. S. (1979). Pair formation in the lobster, *Homarus americanus:* Behavioral, developmental, pheromones and mating. *Mar. Behav. Physiol.* **2,** 15–20.

Atkinson, W. D. (1979). A comparison of the reproductive strategies of domestic species of *Drosophila. J. Anim. Ecol.* **48,** 53–64.

Baker, J. R. (1938). The evolution of breeding seasons. *In* "Evolution: Essays on Aspects of Evolutionary Biology" (G. E. DeBeer, ed.), pp. 161–177. Oxford Univ. Press, London and New York.

Båmstedt, U. (1979). Reproductive bioenergetics within the summer and winter generations of *Euchaeta norvegica* (Copepoda). *Mar. Biol. (Berlin)* **54,** 135–142.
Banta, A. M. (1939). "Studies on the Physiology, Genetics, and Evolution of Some Cladocers," Pap. No. 39. Dept. of Genetics, Carnegie Institution of Washington, Washington, D.C.
Barnes, H. (1957). Processes of restoration and synchronization in marine ecology. The spring diatom increase and the spawning of the common barnacle, *Balanus balanoides. Annee Biol.* **33,** 67–85.
Barnes, H. (1963). Light, temperature and the breeding of *Balanus balanoides. J. Mar. Biol. Assoc. U.K.* **43,** 717–727.
Barnes, H. (1972). The seasonal changes in body weight and biochemical composition of the warm-temperate cirripede *Chthamalus stellatus* (Poli). *J. Exp. Mar. Biol. Ecol.* **8,** 89–100.
Barnes, H. (1975). Reproductive rhythms and some marine invertebrates: An introduction. *Publ. Stn. Zool. Napoli* **39,** Suppl., 8–25.
Barnes, H., and Barnes, M. (1962). The distribution and general ecology of *Balanus balanoides* together with some observations on *Balanus improvisus* in the waters around the coast of Denmark, southern Sweden and north-east Germany. *Lunds. Univ. Arsskr., Avd. 2* [N.S.] **58**(8), 1–41.
Barnes, H., and Barnes, M. (1967). The effect of starvation and feeding on the time of production of egg masses in the boreo-arctic Cirripede *Balanus balanoides* (L.). *J. Exp. Mar. Biol. Ecol.* **1,** 1–6.
Barnes, H., and Barnes, M. (1968). Egg numbers, metabolic efficiency of egg production and fecundity: Local and regional variations in a number of common cirripedes. *J. Exp. Mar. Biol. Ecol.* **2,** 135–153.
Barnes, H., and Barnes, M. (1969). Seasonal changes in the acutely determined oxygen consumption and the effect of temperature for the common cirripedes, *Balanus balanoides* (L.), *B. balanus* (L.) and *Chthamalus stellatus* (Poli). *J. Exp. Mar. Biol. Ecol.* **4,** 36–50.
Barnes, H., and Crisp, D. J. (1956). Evidence of self fertilization in certain species of barnacles. *J. Mar. Biol. Assoc. U.K.* **35,** 631–639.
Barnes, H., Barnes, M., and Finlayson, D. M. (1963). The seasonal changes in body weight, biochemical composition, and oxygen uptake of two common boreo-arctic cirripedes *Balanus balanoides* and *B. balanus. J. Mar. Biol. Assoc. U.K.* **43,** 185–211.
Berry, P. F. (1969). The biology of *Nephrops andamanicus* Wood-Mason (Decapoda, Reptantia). *Oceanogr. Res. Inst. (Durban), Invest. Rep.* **22,** 1–55.
Berry, P. F. (1970). Mating behavior, oviposition and fertilization in the spiny lobster *Panulirus homarus* (Linnaeus). *Oceanogr. Res. Inst. (Durban), Invest. Rep.* **31,** 1–27.
Blades, P. I., and Youngbluth, M. J. (1980). Morphological, physiological, and behavioral aspects of mating in calanoid copepods. *Spec. Symp. Am. Soc. Limnol. Oceanogr.* **3,** 39–51.
Bliss, D. E. (1966). Physiological processes and neurosecretion as related to ecdysis and reproduction. Introduction: Relation between reproduction and growth in decapod Crustaceans. *Am. Zool.* **6,** 231–233.
Bliss, D. E. (1968). Transition from water to land in decapod Crustaceans. *Am. Zool.* **8,** 355–392.
Bliss, D. E. (1979). From sea to tree: Saga of a land crab. *Am. Zool.* **19,** 385–410.
Boolootian, R. A. (1965). Aspects of reproductive biology in the striped shore crab *Pachygrapsus crassipes. Bull. South. Calif. Acad. Sci.* **64,** 43–49.
Boolootian, R. A., Giese, A. C., Farmanfarmian, A., and Tucker, J. (1959). Reproductive cycles of five west coast crabs. *Physiol. Zool.* **32,** 213–220.
Borowsky, B. (1980). Reproductive patterns of three intertidal salt-marsh gammaridean amphipods. *Mar. Biol. (Berlin)* **55,** 327–334.

Bourdon, R. (1962). Observations préliminaries sur la ponte des xanthidae. *Bull. Soc. Lorraine Sci.* **2,** 2–27.

Bovbjerg, R. V. (1960). Courtship behavior of the lined shore crab, *Pachygrapsus crassipes* Randall. *Pac. Sci.* **14,** 421–422.

Bowman, T. E., and Kornicker, L. S. (1967). Two new crustaceans: The parasitic Copepod *Sphaeronellopsis monothrix* (Choniostomatidae) and its myodocopid ostracod host *Parasterope pollex* (Cylindroleberidae) from the southern New England coast. *Proc. U.S. Natl. Mus.* **123**(3613), 1–28.

Branford, J. F. (1978). Incubation period for the lobster *Homarus gammarus* at various temperatures. *Mar. Biol. (Berlin)* **47,** 363–368.

Broekhuysen, G. J. (1941). The life history of *Cyclograpsus punctatus,* H.Edw.: Breeding and growth. *Trans. R. Soc. S. Afr.* **28,** 331–368.

Buchanan, J. (1963). The biology of *Calocaris macandreae* (Crustacea: Thalassinidea). *J. Mar. Biol. Assoc. U.K.* **43,** 729–747.

Burgis, M. J. (1971). The ecology and production of copepods, particularly *Thermocyclops byalinus,* in the tropical Lake George, Uganda. *Freshwater Biol.* **1,** 169–192.

Bynum, K. H. (1978). Reproductive biology of *Caprella penantis* Leach, 1814 (Amphipoda: Caprellidae) in North Carolina, U.S.A. *Estuarine Coastal Mar. Sci.* **7,** 473–485.

Calow, P. (1979). The cost of reproduction—a physiological approach. *Biol. Rev. Cambridge Philos. Sci.* **54,** 23–40.

Carefoot, T. H. (1973). Studies on the growth, reproduction, and life cycle of the sublittoral isopod *Ligia pallasii. Mar. Biol. (Berlin)* **18,** 302–311.

Carlisle, D. B. (1953). Studies on *Lysamata seticaudata* Risso (Crustacea: Decapoda). V. The ovarian inhibiting hormone and the hormonal inhibition of sex-reversal. *Publ. Stn. Zool. Napoli* **24,** 355–372.

Carlisle, D. B. (1957). On the hormonal inhibition of moulting in decapod Crustacea. 2. The terminal anecdysis in crabs. *J. Mar. Biol. Assoc. U.K.* **36,** 291–307.

Carlisle, D. B. (1959a). On the sexual biology of *Pandalus borealis.* II. The termination of the male phase. *J. Mar. Biol. Assoc. U.K.* **38,** 381–395.

Carlisle, D. B. (1959b). On the sexual biology of *Pandalus borealis.* III. The initiation of the female phase. *J. Mar. Biol. Assoc. U.K.* **38,** 481–492.

Charniaux-Cotton, H. (1960). Sex determination. *In* "The Physiology of Crustacea" (T. H. Waterman, ed.), Vol. 1, pp. 411–447. Academic Press, New York.

Charniaux-Cotton, H. (1962). Androgenic gland of Crustaceans. *Gen. Comp. Endocrinol., Suppl.* **1,** 241–247.

Charniaux-Cotton, H. (1965). Hormonal control of sex differentiation in invertebrates. *In* "Organogenesis" (R. L. DeHann and H. Ursprung, eds.), pp. 701–740. Holt, New York.

Charniaux-Cotton, H. (1975). Hermaphroditism and gynandromorphism in malacostracan Crustacea. *In* "Inter-sexuality in the Animal Kingdom" (R. Reinboth, ed.), pp. 91–105. Springer-Verlag, Berlin and New York.

Charniaux-Cotton, H., and Kleinholz, L. H. (1964). Hormones in invertebrates other than insects. *In* "The Hormones" (G. Pincus, K. V. Thimann, and E. B. Astwood, eds.), Vol. 4, pp. 135–198. Academic Press, New York.

Charnov, E. L. (1979a). Natural selection and sex change in pandalid shrimp: Test of life history theory. *Am. Nat.* **113,** 715–734.

Charnov, E. L. (1979b). Simultaneous hermaphroditism and sexual selection. *Proc. Natl. Acad. Sci. U.S.A.* **76,** 2480–2484.

Charnov, E. L. (1981). Sex reversal in *Pandalus borealis:* Effect of a shrimp fishery? *Mar. Biol. Lett.* **2,** 53–57.

Cheung, T. S. (1966). The development of egg membranes and egg attachment in the shore crab, *Carcinus maenas,* and some related decapods. *J. Mar. Biol. Assoc. U.K.* **46,** 373–400.

Cheung, T. S. (1969). The environment and hormonal control of growth and reproduction in the adult female stone crab, *Menippe mercenaria* (Say). *Biol. Bull. (Woods Hole, Mass.)* **136,** 327–346.

Chittleborough, R. G. (1976). Breeding of *Panulirus longipes cygnus* George under natural and controlled conditions. *Aust. J. Mar. Freshwater Res.* **27,** 499–516.

Christensen, A. M., and McDermott, J. J. (1958). Life history and biology of the oyster crab, *Pinnotheres ostreum* Say. *Biol. Bull. (Woods Hole, Mass.)* **114,** 146–179.

Christy, J. H. (1978). Adaptive significance of reproductive cycles in the fiddler crab *Uca pugilator:* A hypothesis. *Science* **199,** 453–455.

Clarke, A. (1977). Seasonal variations in the total lipid content of *Chorismus antarcticus* (Pfeffer) (Crustacea: Decapoda) at South Georgia. *J. Exp. Mar. Biol. Ecol.* **27,** 93–106.

Clarke, A. (1979). On living in cold water: K-strategies in Antarctic benthos. *Mar. Biol. (Berlin)* **55,** 111–119.

Clegg, D. J. (1957). Some observations on pairing in *Balanus balanoides. Rep. Challenger Soc.* **3,** 18–19.

Clutter, R. I. (1969). The microdistribution and social behavior of some pelagic mysid shrimps. *J. Exp. Mar. Biol. Ecol.* **3,** 125–155.

Cole, L. C. (1954). The population consequences of life history phenomena. *Q. Rev. Biol.* **29,** 103–137.

Conover, R. J. (1967). Reproductive cycle, early development, and fecundity in laboratory populations of the copepod *Calanus hyperboreus. Crustaceana* **13,** 61–72.

Conover, R. J. (1978). Transformation of organic matter. *In* "Marine Ecology" (O. Kinne, ed.), Vol. 4, pp. 221–449. Wiley-Interscience, New York.

Corey, S. (1981). Comparative fecundity and reproductive strategies in seventeen species of the Cumacea (Crustacea: Peracarida). *Mar. Biol. (Berlin)* **62,** 65–72.

Corkett, C. J., and McLaren, I. (1978). The biology of *Pseudocalanus. Adv. Mar. Biol.* **15,** 1–231.

Crane, J. (1943). Display, breeding and relationships of fiddler crabs (Brachyura, genus *Uca*) in the Northeastern United States. *Zoologica (N.Y.)* **28,** 217–223.

Crane, J. (1958). Aspects or social behavior in fiddler crabs with special reference to *Uca maracoani* (Latreille). *Zoologica (N.Y.)* **43,** 113–130.

Crane, J. (1966). Combat, display and ritualization in fiddler crabs (Ocypodidae, genus *Uca*). *Philos. Trans. R. Soc. London* **251,** 459–472.

Crane, J. (1975). "Fiddler Crabs of the World." Princeton Univ. Press, Princeton, New Jersey.

Crisp, D. J. (1954). The breeding of *Balanus porcatus* (da Costa). *J. Mar. Biol. Assoc. U.K.* **33,** 473–494.

Crisp, D. J. (1960). Factors influencing growth rates in *Balanus balanoides. J. Anim. Ecol.* **29,** 95–116.

Crisp, D. J. (1974). Factors influencing the settlement of marine invertebrate larvae. *In* "Chemoreception in Marine Organisms" (P. T. Grant and A. M. Mackie, eds.), pp. 177–265. Academic Press, New York.

Crisp, D. J., and Davies, P. A. (1955). Observations *in vivo* on the breeding of *Eliminus modestus* grown on glass slides. *J. Mar. Biol. Assoc. U.K.* **34,** 357–380.

Crisp, D. J., and Patel, B. (1969). Environmental control of the breeding of three boreo-arctic cirripedes. *Mar. Biol. (Berlin)* **2,** 283–295.

Daborn, G. (1975). Life history and energy relations of the giant fairy shrimp *Branchinecta gigas* Lynch, 1937 (Crustacea: Anostraca). *Ecology* **56,** 1025–1039.

Dahl, E., Emanuelsson, H. E., and von Macklenberg, G. (1970). Pheromone transport and reception in an amphipod. *Science* **170,** 739–740.
Dean, J. M., and Vernberg, F. J. (1965). Variations in the blood glucose level of Crustacea. *Comp. Biochem. Physiol.* **14,** 29–34.
Dearborn, J. (1967). Food and reproduction of *Glyptonotus antarcticus* (Crustacea: Isopoda) at McMurdo Sound, Antarctica. *Trans. R. Soc. N.Z.* **8,** 163–168.
Delamare Deboutteville, C. (1960). "Biologie des aeux souterraines littorales et continentales." Hermann, Paris.
Dell, R. K. (1972). Antarctic benthos. *Adv. Mar. Biol.* **10,** 1–216.
Demetrius, L. (1975). Reproductive strategies and natural selection. *Am. Nat.* **109,** 243–249.
Démeusy, N. (1965). New results concerning the relations between somatic growth and reproductive function in the Brachyuran decapod *Carcinus maenas* L. Case of females followed during winter. *C. R. Hebd. Seances Acad. Sci.* **260,** 323–326.
Dexter, D. M. (1971). Life history of the sandy-beach amphipod *Neohaustorius schmitzi* (Crustacea: Haustoriidae). *Mar. Biol. (Berlin)* **8,** 232–237.
Diaz, H. (1980). The mole crab *Emerita talpoida* (Say): A case of changing life history pattern. *Ecol. Monogr.* **50,** 437–456.
Dingle, H., and Caldwell, R. L. (1972). Reproductive and maternal behavior of the mantis shrimp *Gonodactylus bredini* Manning (Crustacea: Stomatopoda). *Biol. Bull. (Woods Hole, Mass.)* **142,** 417–426.
Dobzhansky, T. (1950). Evolution in the tropics. *Am. Sci.* **38,** 209–221.
Dunbar, M. J. (1968). "Ecological Development in Polar Regions: A Study in Evolution." Prentice-Hall, Englewood Cliffs, New Jersey.
Dunham, P. J. (1978). Sex pheromones in Crustacea. *Biol. Rev. Cambridge Philos. Soc.* **53,** 555–583.
Eales, A. J. (1974). Sex pheromone in the shore crab *Carcinus maenas* and the site of its release from females. *Mar. Behav. Physiol.* **2,** 345–355.
Edmondson, W. T. (1968). A graphical model for evaluating the use of the egg ratio for measuring birth and death rates. *Oecologia* **1,** 1–37.
Efford, I. E. (1967). Neoteny in sand crabs of the genus *Emerita* (Anomura, Hippidae). *Crustaceana* **13,** 81–93.
Elner, R. W., and Hughes, R. N. (1978). Energy maximization in the diet of the shore crab, *Carcinus maenas. J. Anim. Ecol.* **47,** 103–116.
Emlen, J. M. (1977). "Ecology: An Evolutionary Approach." Addison-Wesley, Reading, Massachusetts.
Enright, J. T. (1975). Orientation in time: Endogenous clocks. *In* "Marine Ecology" (O. Kinne, ed.), Vol. 2, Part 2, pp. 917–944. Wiley-Interscience, New York.
Feest, J. (1969). Morphological studies on the ontogeny and sexual biology of *Uca annulipes* and *Uca triangularis* as compared to *Ilyoplax gangetica. Forma Functio* **1,** 159–225.
Ferguson, E. (1967). Morphology and taxonomy of freshwater Ostracoda. *Proc. Symp. Crustacea, 1965* Ser. II, Part 2, pp. 497–505.
Fhyn, U. E. H., and Costlow, J. D. (1977). Histology and histochemistry of the ovary and oogenesis in *Balanus amphitrite* L. and *B. eburneus* Gould (Cirripedia; Crustacea). *Biol. Bull. (Woods Hole, Mass.)* **151,** 351–359.
Fish, C. J. (1961). Observations on the biology of *Microsetella norvegica. Deep-Sea Res., Suppl.* **3,** 242–249.
Fish, J. D., and Fish, S. (1978). Observations on an annual migration of *Bathyporeia pelagica* (Amphipoda, Haustoriidae). *Crustaceana* **35,** 215–221.
Fish, J. D., and Preece, G. S. (1970). The annual reproductive patterns of *Bathyporeia pilosa*

and *Bathyporeia pelagica* (Crustacea: Amphipoda). *J. Mar. Biol. Assoc. U.K.* **50,** 475–488.

Fleminger, A. (1967). Taxonomy, distribution, and polymorphism in the *Labidocera jollae* group with remarks on evolution within the group (Copepoda: Calanoida). *Proc. U.S. Natl. Mus.* **120,** 1–61.

Fleminger, A., and Hulsemann, K. (1977). Geographical range and taxonomic divergence in North Atlantic *Calanus (C. helgolandicus, C. finmarchicus,* and *C. glacialis). Mar. Biol. (Berlin)* **40,** 233–238.

Florkin, M. (1960). Blood chemistry. *In* "The Physiology of Crustacea" (T. H. Waterman, ed.), Vol. 1, pp. 141–159. Academic Press, New York.

Gable, M. F., and Croker, R. A. (1977). The salt marsh amphipod, *Gamarus palustris* Bousfield, 1969 at the northern limit of its distribution. I. Ecology and life cycle. *Eastuarine Coastal Mar. Sci.* **5,** 123–135.

Gadgil, M., and Bossert, W. H. (1970). Life history consequences of natural selection. *Am. Nat.* **104,** 1–24.

George, J. C., and Patel, B. S. (1956). The seasonal variation in the fat content of the liver and gonads in a marine and freshwater decapod. *J. Anim. Morphol. Physiol.* **3,** 49–55.

Gerritsen, J. (1980). Sex and parthenogenesis in sparse populations. *Am. Nat.* **115,** 718–742.

Ghiselin, M. T. (1974). "The Economy of Nature and the Evolution of Sex." Univ. of California Press, Berkeley.

Giese, A. C. (1959). Comparative physiology: Annual reproductive cycles of marine invertebrates. *Annu. Rev. Physiol.* **21,** 547–576.

Giese, A. C. (1966). Lipids in the economy of marine invertebrates. *Physiol. Rev.* **46,** 244–298.

Giese, A. C., and Pearse, J. S. (1974). Introduction: General principles. *In* "Reproduction of Marine Invertebrates' (A. C. Giese and J. S. Pearse, eds.), Vol. 1, pp. 1–49. Academic Press, New York.

Giesel, J. (1976). Reproductive strategies as adaptations to life in temporally heterogeneous environments. *Annu. Rev. Ecol. Syst.* **7,** 57–79.

Gifford, C. (1962). Some observations on the general biology of the land crab, *Cardisoma guanhumi* (Latreille), in South Florida. *Biol. Bull. (Woods Hole, Mass.)* **123,** 207–223.

Gnewuch, W. T., and Croker, R. A. (1973). Macroinfauna of northern New England sand. I. The biology of *Manocuma stellifera* Zimmer, 1943 (Crustacea: Cumacea). *Can. J. Zool.* **51,** 1011–1023.

Gomez, R., and Nayar, K. K. (1965). Certain endocrine influences in the reproduction of the crab *Paratelphusa hydrodromous. Zool. Jahrb., Abt. Allg. Zool. Physiol. Tiere* **71,** 694–701.

Goodbody, I. (1965). Continuous breeding in populations of two tropical Crustaceans, *Mysidium columbiae* (Zimmer) and *Emerita portoricensis* (Schmitt). *Ecology* **46,** 195–197.

Goodwin, T. W. (1951). Carotenoid metabolism during development of lobster eggs. *Nature (London)* **7,** 559.

Gray, G. W., and Powell, G. C. (1966). Sex ratios and distribution of spawning King crabs in Alitak Bay, Kodiak Island, Alaska (Decapoda, Anomura, Lithoidae). *Crustaceana* **10,** 303–309.

Greenwood, J. G. (1965). The larval development of *Petrolisthes elongatus* (H. Milne Edwards) and *Petolisthes novaezelandiae* Filhol (Anomura, Porcellanidae) with notes on breeding. *Crustaceana* **8,** 285–307.

Griffiths, A. M., and Frost, B. W. (1976). Chemical communication in the marine planktonic Copepods *Calanus pacificus* and *Pseudocalanus* sp. *Crustaceana* **30,** 1–8.

Haefner, P. A. (1976). Distribution, reproduction, and molting of the rock crab, *Cancer irroratus* Say 1917, in the mid-Atlantic Bight. *J. Nat. Hist.* **10,** 377–397.

Haefner, P. A. (1977). Aspects of the biology of the Jonah crab, *Cancer borealis* Stimpson, 1859 in the mid-Atlantic Bight. *J. Nat. Hist.* **11,** 303–320.

Hall, J. R., and Hessler, R. R. (1971). Aspects in the population dynamics of *Derocheilocaris typica* (Mystacocarida, Crustacea). *Vie Millieu, Ser. A* **22,** 305–326.

Hartnoll, R. G. (1963). The biology of Manx spider crabs. *Proc. Zool. Soc. London* **141,** 423–496.

Hartnoll, R. G. (1969). Mating in the Brachyura. *Crustaceana* **16,** 161–181.

Hartnoll, R. G. (1973). Factors affecting the distribution and behaviour of the crab *Dotilla fenestrata* on east African shores. *Estuarine Costal Mar. Sci.* **1,** 137–152.

Hayward, T. L. (1980). Spatial and temporal feeding patterns from the north Pacific central gyre. *Mar. Biol. (Berlin)* **58,** 295–309.

Hazlett, B. A. (1968). The sexual behavior of some European hermit crabs. *Publ. Stn. Zool. Napoli* **36,** 238–252.

Heath, J. R., and Barnes, H. (1970). Some changes in the biochemical composition with season and during the moulting cycle of the common shore crab, *Carcinus maenas* (L.). *J. Exp. Mar. Biol. Ecol.* **5,** 199–233.

Hebert, P. D. N. (1980). The genetics of Cladocera. *Spec. Symp.—Am. Soc. Limnol. Oceanogr.* **3,** 329–336.

Heeley, W. (1941). Observations on the life histories of some terrestrial Isopods. *Proc. Zool. Soc. London, Ser. B* **111,** 79–149.

Heinle, D. R., Harris, R. P., Ustach, J. F., and Flemer, D. A. (1977). Detritus as food for estuarine Copepods. *Mar. Biol. (Berlin)* **40,** 341–353.

Heinrich, A. K. (1962). The life histories of plankton animals and seasonal cycles of plankton communities in the oceans. *J. Cons. Perm. Int. Explor. Mer.* **27,** 15–24.

Heip, C., and Smol, N. (1976). Influence of temperature on the reproductive potential of two brackish-water harpacticoids (Crustacea: Copepoda). *Mar. Biol. (Berlin)* **35,** 327–334.

Hepper, B. T., and Gough, C. J. (1978). Fecundity and rate of embryonic development of the lobster *Homarus gammarus* (L.), off the coast of North Wales. *J. Cons., Cons. Int. Explor. Mer* **38,** 54–57.

Heptner, M. V. (1963). Reproductive biology and life cycle of *Caprella septentrionalis* Kroyer (Amphipoda, Caprellidae) in the White Sea. *Zool. Zh.* **42,** 1169–1630 (in Russian, with English summary).

Herring, P. J. (1973). Depth distribution of the carotenoid pigments and lipids of some oceanic animals. 2. Decapod Crustaceans. *J. Mar. Biol. Assoc. U.K.* **53,** 539–562.

Herrnkind, W. F. (1968). The breeding of *Uca pugilator* (Bosc) and the mass rearing of the larvae with comments on the behavior of the larval and early crab stages (Brachyura, Ocypodidae). *Crustaceana, Suppl.* **2,** 214–224.

Herrnkind, W. F. (1980). Spiny lobsters: Patterns of movement. *In* "The Biology and Management of Lobsters" (J. S. Cobb and B. F. Phillips, eds.), Vol. 2, pp. 349–407. Academic Press, New York.

Heydorn, A. E. F. (1965). The rock lobster of the south west coast *Jasus lalandii* (H. Milne-Edwards). 1. Notes on the reproductive biology and the determination of minimum size limits for commercial catches. *Invest. Rep. Div. Sea Fish S. Afr.* **53,** 1–32.

Heydorn, A. E. F. (1969a). Notes on the biology of *Panulirus homarus* and on length/weight relationships of *Jasus lalandii*. *Invest. Rep. Div. Sea Fish. S. Afr.* **69,** 1–27.

Heydorn, A. E. F. (1969b). The rock lobster of the south African west coast *Jasus lalandii* (H. Milne-Edwards). 2. Population studies, behaviour, reproduction, molting, growth and migration. *Invest. Rep. Div. Sea Fish. S. Afr.* **71,** 1–52.

Hicks, G. R. F. (1980). Patterns and strategy in the reproductive cycles of benthic harpacticoid copepods. *In* "Cyclic Phenomena in Marine Plants and Animals" (E. Naylor and R. G. Hartnoll, eds.), pp. 139–147. Pergamon, Oxford.

Hilgard, G. H. (1960). A study of reproduction in the intertidal barnacle, *Mitella polymerus,* in Monterey Bay, California. *Biol. Bull. (Woods Hole, Mass.)* **119,** 169–188.

Hill, B. J. (1979). Aspects of the feeding strategy of the predatory crab *Scylla serrata. Mar. Biol. (Berlin)* **55,** 209–214.

Hines, A. H. (1978). Reproduction in three species of intertidal barnacles from central California. *Biol. Bull. (Woods Hole, Mass.)* **154,** 262–281.

Hines, A. H. (1979). The comparative reproductive ecology of three species of intertidal barnacles. *In* "Reproductive Ecology of Marine Invertebrates" (S. E. Stancyk, ed.), pp. 213–234. Univ. of South Carolina Press, Columbia.

Hirshfield, M. F., and Tinkle, D. W. (1975). Natural selection and evolution of reproductive effort. *Proc. Natl. Acad. Sci. U.S.A.* **72,** 2227–2231.

Hoestlandt, H. (1948). Recherches sur le biologie de l'*Eriocheir sinensis* en France (Crustacé, Brachyoure-. *Ann. Inst. Océanogr. (Monaco)* **24,** 1–116.

Holland, D. L. (1978). Lipid reserves and energy metabolism in the larvae of benthic marine invertebrates. *Biochem. Biophys. Perspect. Mar. Biol.* **4,** 85–123.

Hopkins, C. C. E. (1977). The relationship between maternal body size and clutch size, development time and egg morality in *Euchaeta norvegica* (Copepoda: Calanoida) from Loch Etive, Scotland. *J. Mar. Biol. Assoc. U.K.* **57,** 723–733.

Hopkins, C. C. E., and Machin, D. (1977). Patterns of spermatophore distribution and placement in *Euchaeta norvegica* (Copepoda: Calanoida). *J. Mar. Biol. Assoc. U.K.* **57,** 113–131.

Hughes, J. T., and Matthiessen, G. C. (1962). Observations on the biology of the American lobster *Homarus americanus. Limnol. Oceanogr.* **7,** 414–421.

Hughes, R. N., and Elner, R. W. (1979). Tactics of a predator, *Carcinus maenas,* and morphological response of the prey, *Nucella lapillus. J. Anim. Ecol.* **48,** 65–78.

Hurley, A. C. (1973). Fecundity of the acorn barnacle *Balanus pacificus* Pilsbury: A fugitive species. *Limnol. Oceanogr.* **18,** 386–393.

Hutchinson, G. E. (1967). "A Treatise on Limnology," Vol. 2, Wiley, New York.

Ikeda, T. (1974). Nutritional ecology of marine zooplankton. *Mem. Fac. Fish., Hokkaido Univ.* **22,** 1–97.

Jacobs, J. (1961). Laboratory cultivation of the marine copepod (*Pseudodiaptomus coronatus* Williams). *Limnol. Oceanogr.* **6,** 443–446.

Jensen, J. P. (1958). The relation between body size and the number of eggs in marine Malacostrakes. *Medd. Dan. Fisk.-Havunders* [N.S.] **2,** No. 19, 1–25.

Jones, M. B. (1974). Breeding and seasonal population changes of *Jaera nordmanni nordica* Lemercier (Isopoda: Asellota). *J. Mar. Biol. Assoc. U.K.* **54,** 727–736.

Jones, M. B. (1977). Breeding and seasonal population changes of *Petrolisthes elongatus* (Crustacea, Decapoda, Anomura) at Kaikoura, New Zealand. *J. R. Soc. N. Z.* **7,** 259–272.

Jones, H. B. (1978). Aspects of the biology of the big-handed crab, *Heterozius rotundifrons* (Decapoda: Brachyura), from Kaikoura, New Zealand. *N. Z. J. Zool.* **5,** 783–794.

Jones, M. B. (1980). Reproductive ecology of the estuarine burrowing mud crab *Helice crassa* (Grapsidae). *Estuarine Costal Mar. Sci.* **11,** 433–443.

Junera, H., Zerbib, C., Martin, M., and Meusy, J. J. (1977). Evidence for control of vitellogenin synthesis by an ovarian hormone in *Orchestia gammarella* (Pallas) (Crustacea, Amphipoda). *Gen. Comp. Endocrinol.* **31,** 456–462.

Kaestner, A. (1970). "Invertebrate Zoology," Vol. III. Wiley-Interscience, New York.

Kanciruk, P., and Herrnkind, W. F. (1976). Autumnal reproduction in *Panulirus argus* at Bimini, Bahamas. *Bull. Mar. Sci.* **26,** 417–432.

Karande, A. A. (1967). On cirriped crustaceans (barnacles), an important fouling group in Bombay waters. *Proc. Symp. Crustacea, 1965* Ser. II, Part 4, pp. 1245–1253.

Katona, S. K. (1973). Evidence of sex pheromones in planktonic copepods. *Limnol. Oceanogr.* **18,** 574–583.

Keith, D. E. (1971). Substrate selection in caprellid amphipods of southern California, with emphasis on *Caprella californica* Stimpson and *Caprella equilibra* Say (Amphipoda). *Pac. Sci.* **25,** 387–394.

Kesling, R. V. (1951). The morphology of Ostracod molt stages. *Ill. Biol. Monogr.* **21,** 1–323.

Kinne, O. (1964). The effects of temperature and salinity on marine and brackish water animals. II. Salinity and temperature combinations. *Oceanogr. Mar. Biol.* **2,** 281–339.

Kittredge, J. S., Takahashi, F. T., Lindsey, J., and Lasker, R. (1974). Chemical signals in the sea: Marine allelochemics and evolution. *Fish. Bull.* **72,** 1–11.

Klaassen, F. (1975). Ökologische und ethologische untersuchungen zur fortpflanzungsbiologie von *Gecarcinus lateralis* (Decapoda, Brachyura). *Forma Functio* **8,** 101–174.

Klapow, L. A. (1972). Fortnightly molting and reproductive cycles in the sand-beach Isopod, *Excirolana chiltoni. Biol. Bull. (Woods Hole, Mass.)* **143,** 568–591.

Kleinholz, L. H. (1976). Crustacean neurosecretory hormones and physiological specificity. *Am. Zool.* **16,** 151–166.

Knox, G. A. (1970). Antarctic marine ecosystem. *In* "Antarctic Ecology" (M. W. Holgate, ed.), Vol. 1, pp. 69–96. Academic Press, New York.

Knudsen, J. W. (1964). Observations on the reproductive cycles and ecology of the common Brachyura and crab-like Anomura of the Puget Sound, Washington, *Pac. Sci.* **18,** 3–33.

Koba, K. (1936). Preliminary notes on the development of *Geotelphusa dehaani* (White). *Proc. Imp. Acad. (Tokyo)* **12,** 105–107.

Kon, T., and Honma, Y. (1970a). Studies on the maturity of the gonad in some marine invertebrates. IV. Seasonal changes in the testes of the tanner crab. *Bull. Jpn. Soc. Sci. Fish.* **36,** 1028–1033.

Kon, T., and Honma, Y. (1970b). Studies on the maturity of the gonad in some marine invertebrates. III. Seasonal Changes in the ovary of the tanner crab. *Bull. Jpn. Soc. Sci. Fish.* **36,** 1021–1027.

Kuhl, H. (1966). Der abfluss der Elbe im Jahr 1965 und seine Wirkung auf Saltzgehalt, Plankton auf Bewuchsbildung bei Cuxhaven. *Veröff. Inst. Meeresforsch. Bremerhaven* **10,** S. 61–70.

Lachaise, F., and Hoffmann, J. A. (1977). Ecdysone and ovarian development in a decapod, *Carcinus maenas. C. R. Hebd. Seances Acad. Sci., Ser. D* **285,** 701–704.

Lang, K. (1953). *Apseudes hermaphroditicus* n. sp. a hermaphroditic tanaid from the Antarctic. *Ark. Zool.* **2**(4), 341–350.

Lasenby, D. C., and Langford, R. R. (1972). Growth, Life history and respiration of *Mysis relicta* in an Arctic and temperate lake. *J. Fish. Res. Board Can.* **29,** 1701–1708.

Laverack, M. S. (1964). The antennular sense organs of *Panulirus argus. Comp. Biochem. Physiol.* **13,** 301–321.

Laverack, M. S. (1968). On the receptors of marine invertebrates. *Oceanogr. Mar. Biol. Annu. Rev.* **6,** 249–324.

Lawlor, L. R. (1976). Molting, growth and reproductive strategies in the terrestrial isopod, *Armadillidium vlugare. Ecology* **57,** 1179–1194.

Lee, C. M. (1972). Structure and function of the spermatophore and its coupling device in the Centropagidae (Copepoda: Calanoida). *Bull. Mar. Ecol.* **8,** 1–20.

Lewis, J. B. (1960). The fauna of rocky shores of Barbados, West Indies. *Can. J. Zool.* **38,** 391–435.

Lewontin, R. C. (1965). Selection for colonizing ability. *In* "The Genetics of Colonizing Species" (H. G. Baker and G. L. Stebbins, eds.), pp. 77–94. Academic Press, New York.

Linford, E. (1965). Biochemical studies on marine zooplankton. II. Variations in the lipid content of some Mysidacea. *J. Cons., Cons. Int. Explor. Mer* **30,** 16–27.
Linsenmair, K. E. (1967). Konstruktion und signalfunktion der sandpyramide der Reiterkrabbe *Ocypode saratan* Forsk (Decapoda, Brachyura, Ocypodidae). *Z. Tierpsychol.* **24,** 403–456.
Little, G. (1968). Induced winter breeding and larval development in the shrimp, *Palaemonetes pugio* Holthuis (Caridae, Palaemonidae). *Crustaceana, Suppl.* **2,** 19–26.
Longhurst, A. R. (1976). Vertical migration. *In* "The Ecology of the Seas" (D. H. Cushing and J. J. Walsh, eds.), pp. 116–137. Saunders, Philadelphia, Pennsylvania.
Lowe, M. E. (1961). The female reproductive cycle of the crayfish *Cambarellus shufeldti:* The influence of environmental factors. *Tulane Stud. Zool.* **8,** 157–176.
Lynch, M. (1980a). The evolution of cladoceran life histories. *Q. Rev. Biol.* **55,** 23–42.
Lynch, M. (1980b). Predation, enrichment, and the evolution of cladoceran life histories: A theoretical approach. *Symp.—Am. Soc. Limnol. Oceanogr.* **3,** 367–376.
MacArthur, R. H., and Wilson, E. O. (1967). "The Theory of Island Biogeography." Princeton Univ. Press, Princeton, New Jersey.
Mackie, A. M. (1975). Chemoreception. *Biochem. Biophys. Perspect. Mar. Biol.* **2,** 69–105.
Makarov, R. R. (1973). Larval development of *Notocrangon antarcticus* (Decapoda, Crangonidae). *Zool. Zh.* **52,** 1149–1155.
Marr, J. W. S. (1962). The natural history and geography of Antarctic krill (*Euphausia superba* Dana). *Discovery Rep.* **32,** 33–464.
Marshall, N. B. (1979). "Developments in Deep Sea Biology." Blanford Press Poole, Dorset.
Marshall, S. M. (1973). Respiration and feeding in Copepods. *Adv. Mar. Biol.* **11,** 57–120.
Marshall, S. M., and Orr, A. P. (1952). On the biology of *Calanus finmarchicus.* VII. Factors affecting egg production. *J. Mar. Biol. Assoc. U.K.* **30,** 527–547.
Marshall, S. M., and Orr, A. P. (1972). "The Biology of a Marine Copepod." Springer-Verlag, Berlin and New York.
Mauchline, J. (1968). The development of the eggs in the ovaries of euphasuiids and estimation of fecundity. *Crustaceana* **14,** 155–163.
Mauchline, J. (1972). The biology of bathypelagic organisms, especially Crustacea. *Deep-Sea Res.* **19,** 753–780.
Mauchline, J. (1977). The integumental sensilla and glands of pelagic Crustacea. *J. Mar. Biol. Assoc. U.K.* **57,** 973–994.
Mauchline, J., and Fisher, L. R. (1969). The biology of euphausiids. *Adv. Mar. Biol.* **7,** 1–454.
Maxwell, J. G. H. (1977). The breeding biology of *Chorismus antarcticus* (Pfeffer) and *Notocrangon antarcticus* (Pfeffer) (Crustacea, Decapoda) and its bearing on the problems of the impoverished Antarctic Decapod fauna. *In* "Adaptations Within Antarctic Ecosystems" (G. A. Llano, ed.), pp. 335–342. Smithsonian Institution, Washington, D.C.
Maynard-Smith, J. (1978). "The Evolution of Sex." Cambridge Univ. Press, London and New York.
Mayr, R. (1963). "Animal Species and Evolution." Harvard Univ. Press, Cambridge, Masscahusetts.
McLachlan, A. (1977). The larval development and population dynamics of *Derocheilocaris algoensis* (Crustacea, Mystacocarida). *Zool. Afr.* **12,** 1–14.
McLaren, I. A. (1965). Some relationships between temperature and egg size, body size, development rate, and fecundity of the copepod *Pseudocalanus. Limnol. Oceanogr.* **10,** 528–538.
McQueen, D. J., and Steel, C. G. H. (1980). The role of photoperiod and temperature in the initiation of reproduction in the terrestrial isopod *Oniscus asellus* (Linnaeus). Can. J. Zool. **8,** 58, 235–240.

Miyawaki, M., and Tanoue, G. S. (1962). Electron microscopy of the hepatopancreas in the crayfish *Procambarus clarkii*. *Kumamato J. Sci.* **6,** 1–12.

Moore, E., and Sander, F. (1976). Quantitative and qualitative aspects of the zooplankton and breeding patterns of copepods at two Caribbean coral reef stations. *Estuarine Coastal Mar. Sci.* **4,** 589–607.

Morgan, G. R. (1972). Fecundity in the western rock lobster *Panulirus longipes cygnus* (George) (Crustacea: Decapoda: Palinuridae). *Aust. J. Mar. Freshwater Res.* **23,** 133–141.

Mullin, M. M. (1967). On the feeding behavior of planktonic marine copepods and separation of their ecological niches. *Proc. Symp. Crustacea. 1965,* Ser. II, Part 3, pp. 955–964.

Murphy, G. I. (1968). Patterns in life history and the environment. *Am. Nat.* **102,** 390–403.

Naylor, E. (1955). The life cycle of the isopod *Idotea emarginata* (Fabricius). *J. Anim. Ecol.* **24,** 270–281.

Nemoto, T. (1965). Feeding pattern of euphausiids, comparative morphology, allomorphosis and ecology. *Inf. Bull. Planktol. Jpn.* **12,** 24–36.

Nilssen, J. P. (1980). When and how to reproduce: A dilemma for limnetic cyclopoid copepods. *Spec. Symp.—Am. Soc. Limnol. Oceanogr.* **3,** 418–426.

Nye, P. (1977). Reproduction, growth and distribution of the grapsid crab *Helice crassa* (Dana, 1851) in the southern part of New Zealand. *Crustaceana* **33,** 75–89.

Omori, M. (1974). The biology of pelagic shrimps in the ocean. *Adv. Mar. Biol.* **12,** 233–324.

Orians, G. H. (1970). On the evolution of mating systems in birds and mammals. *Am. Nat.* **103,** 589–603.

Orton, J. H. (1920). Sea temperature, breeding and distribution of marine animals. *J. Mar. Biol. Assoc. U.K.* **12,** 339–366.

Otsu, T. (1963). Bihormonal control of sexual cycle in the freshwater crab, *Potamon dehaani*. *Embryologica* **8,** 1–20.

Pandian, T. J. (1975). Mechanisms of heterotrophy. *In* "Marine Ecology" (O. Kinne, ed.), Vol. 3, pp. 61–249. Wiley-Interscience, New York.

Panouse, J. B. (1943). Influence de l'ablation, de pedoncule oculaire sur la croissance de l'ovarie chez la crevette *Leander serratus*. *C. V. Hebd. Seances Acad. Sci.* **217,** 553–555.

Panouse, J. B. (1947). The sinus gland and maturation of genital products in shrimp. *Bull. Biol. Fr. Belg., Suppl.* **33,** 160–163.

Paris, O. H., and Pitelka, F. A. (1962). Population chracteristics of the terrestrial isopod *Armadillidium vulgare* in California grasslands. *Ecology* **43,** 229–248.

Parry, G. D. (1981). The meanings of r- and K-selection. *Oecologia* **48,** 260–264.

Passano, L. M. (1960). Molting and its control. *In* "The Physiology of Crustacea" (T. H. Waterman, ed.), Vol. 1, pp. 473–536. Academic Press, New York.

Patel, B., and Crisp, D. J. (1960). The influence of temperature on the breeding and moulting activities of some warm-water species of operculate barnacles. *J. Mar. Biol. Assoc. U.K.* **39,** 667–680.

Patel, B., and Crisp, D. J. (1961). Relation between the breeding and moulting cycles in cirripedes. *Crustaceana* **2,** 89–107.

Pearce, J. B. (1966). On *Pinnixa faba* and *Pinnixa littoralis* (Decapoda: Pinnotheridae) symbiotic with the clam, *Tresus capax* (Pelecypoda: Mactridae). *In* "Some Contemporary Studies in Marine Science" (H. B. Barnes, ed.), pp. 565–589. Allen & Unwin, London.

Perez, C. (1929). Charactères sexuels chez un crabe oxyrhynque (*Macropodia rostrata* L.). *C. R. Hebd. Seances Acad. Sci.* **188,** 91–93.

Phillips, B. F., and Sastry, A. N. (1980). Larval ecology *In* "The Biology and Management of Lobsters" (J. S. Cobb and B. F. Phillips, eds.), Vol. 2, pp. 11–57. Academic Press, New York.

Pianka, E. R. (1970). On "r" and "K" selection. *Am. Nat.* **106,** 592–597.

Pianka, E. R. (1976). Natural selection and optimal reproductive tactics. *Am. Zool.* **16,** 775–784.

Pillai, N. K. (1967). The role of Crustacea in the destruction of submerged timber. *Proc. Symp. Crustacea, 1965* Ser. II, Part V, pp. 1274–1283.

Pillay, K. K., and Nair, N. B. (1968). Observations on the reproductive cycles of some crabs from south-west coast of India. *J. Mar. Biol. Assoc. India* **10,** 384–385.

Pillay, K. K., and Nair, N. B. (1971). The annual reproductive cycles of *Uca annulipes, Portunus pelagicus* and *Metapenaeus affinis* (Decopoda: Crustacea) from the south-west coast of India. *Mar. Biol. (Berlin)* **11,** 152–166.

Pillay, K. K., and Nair, N. B. (1972). Reproductive biology of the sessile barnacle, *Balanus amphitrite communis* (Darwin), of the south-west coast of India. *Indian J. Mar. Sci.* **1,** 8–16.

Pillay, K. K., and Nair, N. B. (1973). Observations on the biochemical changes in gonads and other organs of *Uca annulipes, Portunus pelagicus* and *Metapenaeus affinis* (Decapoda: Crustacea) during the reproductive cycle. *Mar. Biol. (Berlin)* **18,** 167–198.

Pillay, K. K., and Ono, Y. (1978). The breeding cycles of two species of grapsid crabs (Crustacea: Decapoda) from the north coast of Kyushu, Japan. *Mar. Biol. (Berlin)* **45,** 237–248.

Prosser, C. L., ed. (1973). "Comparative Animal Physiology," Vol. 1. Saunders, Philadelphia, Pennsylvania.

Prus, T. (1971). The assimilation efficiency of *Asellus aquaticus* L. (Crustacea: Isopoda). *Freshwater Biol.* **1,** 287–305.

Rahaman, A. A. (1967). Reproductive and nutritional cycles of the crab *Portunus pelagicus* (Linnaius) (Decapoda: Brachyura) of the Madras coast. *Proc.—Indian Acad. Sci., Sect. B* **65,** 76–82.

Raman, K. (1967). Observations on the fishery and biology of the giant freshwater prawn *Macrobrachium rosenbergii* De Man. *Proc. Symp. Crustacea, 1965* Ser. II, Part 2, pp. 649–669.

Rangneker, P. V., and Deshmukh, R. D. (1968). Effect of eyestalk removal on the ovarian growth of the marine crab, *Scylla serrata* (Forskall). *J. Anim. Morphol. Physiol.* **15,** 116–126.

Raven, C. P. (1961). "Oogenesis: The Storage of Development Information," Pergamon, Oxford.

Raymont, J. E. G., Austin, J., and Linford, E. (1966). Biochemical studies on marine zooplankton. III. Seasonal variation in the biochemical composition of *Neomysis integer. In* "Some Contemporary Studies in Marine Science" (H. B. Barnes, ed.), pp. 597–605. Allen & Unwin, London.

Reaka, M. L. (1976). Lunar and tidal periodicity of molting and reproduction in stomatopod Crustaceans: A selfish herd hypothesis. *Biol. Bull. (Woods Hole, Mass.)* **150,** 468–490.

Reaka, M. L. (1979). The evolutionary ecology of life history patterns in stomatopod Crustacea. *Belle W. Baruch Libr. Mar. Sci.* **9,** 235–260.

Rees, C. P. (1975). Life cycle of the amphipod *Gammarus palustris* Bousfield. *Estuarine Coastal Mar. Sci.* **3,** 413–419.

Reese, E. R. (1968). Annual breeding seasons of three sympatric species of tropical intertidal hermit crabs, with a discussion of factors controlling breeding. *J. Exp. Mar. Biol. Ecol.* **2,** 308–318.

Richman, S. (1958). The transformation of energy by *Daphnia pulex. Ecol. Monogr.* **28,** 273–291.

Røen, V. (1957). Contributions to the biology of some Danish free-living freshwater copepods. *Biol. Skr.—K. Dan. Vidensk. Selsk.* **9,** 1–101.

Rokop, F. J. (1977). Patterns of reproduction in the deep-sea benthic crustaceans: A reevaluation. *Deep-Sea Res.* **24,** 683–692.

Rudd, J. T. (1936). Euphausiacea. *Mediterr.* **2,** 1–86.

Ryan, E. P. (1960). Pheromone: Evidence in a Decapod Crustacean. *Science* **151,** 340–341.

Ryan, E. P. (1967a). Structure and function of the reproductive system of the crab *Portunus sanguinolentus* (Herbst) (Brachyura: Portunidae). I. The male system. *Proc. Symp. Crustacea, 1965* Ser. II, Part 2, pp. 506–521.

Ryan, E. P. (1967b). Structure and function of the reproductive system of the crab *Portunus sanguinolentus* (Herbst) (Brachyura: Portunidae). II. The female system. *Proc. Symp. Crustacea, 1965* Ser. II, Part 2, pp. 522–524.

Salmon, M. (1965). Waving display and sound production in the courtship behavior of *Uca pugilator,* with comparisons to *U. minax* and *U. pugnax. Zoologica (N.Y.)* **50,** 123–150.

Salmon, M., and Atsaides, S. (1968). Visual and acoustic signalling during courtship of fiddler crabs (Genus *Uca*). *Am. Zool.* **8,** 623–639.

Samuelson, T. J. (1970). The biology of six species of Anomura (Crustacea, Decapoda) from Raunfjordan, western Norway. *Sarsia* **45,** 25–52.

Sastry, A. N. (1975). Physiology and ecology of reproduction in marine invertebrates. *Belle W. Baruch Libr. Mar. Sci.* **3,** 277–299.

Sastry, A. N., and Vargo, S. L. (1977). Variation in the physiological responses of crustacean larvae to temperature. *In* "Physiological Responses of Marine Biota to Pollutants" (F. J. Vernberg, A. Calabrese, F. J. Thurberg, and W. B. Vernberg, eds.), pp. 401–423. Academic Press, New York.

Schaffer, W. M. (1974). Selection for optimal life histories: the effects of age structure. *Ecology* **55,** 291–303.

Schoener, T. W. (1971). Theory of feeding strategies. *Annu. Rev. Ecol. Syst.* **2,** 369–404.

Schöne, H. (1968). Agonistic and sexual displays in aquatic and semi-terrestrial brachyuran crabs. *Am. Zool.* **8,** 641–654.

Seiple, W. (1979). Distribution, habitat preferences and breeding periods in the crustaceans *Sesarma cinereum* and *S. reticulatum* (Brachyura: Decapoda: Grapsidae). *Mar. Biol. (Berlin)* **52,** 77–86.

Sergestråle, S. G. (1970). Light control of the reproductive cycle of *Pontoporeia affinis* Lindström (Crustacea: Amphipoda). *J. Exp. Mar. Biol. Ecol.* **5,** 272–275.

Sheader, M. (1977). Production and population dynamics of *Ampelisca tenuicornis* (Amphipoda) with notes on the biology of its parasite *Sphaeronella longipes* (Copepoda). *J. Mar. Biol. Assoc. U.K.* **57,** 955–968.

Sheader, M., and Chia, F. S. (1970). Development, fecundity and brooding behavior of the amphipod, *Marinogammarus obtusatus. J. Mar. Biol. Assoc. U.K.* **50,** 1079–1099.

Shelton, R. G. J., and Laverack, M. S. (1968). Observations on a redescribed crustacean cuticular sense organ. *Comp. Biochem. Physiol.* **25,** 1049–1059.

Smaldon, G. (1972). Population structure and breeding biology of *Pisidia longicornis* and *Porcellana platycheles. Mar. Biol. (Berlin)* **17,** 171–179.

Smyly, W. J. P. (1973). Bionomics of *Cyclops strenuus abyssorum* Sars (Copepoda: Cyclopoida). *Oecologia* **11,** 163–186.

Sochasky, J. B. (1973). Failure to accelerate molting following eyestalk ablation in decapod crustaceans: A review of the literature. *Fish. Res. Board Can.* **431,** 1–127.

Stearns, S. C. (1976). Life history tactics: A review of the ideas. *Q. Rev. Biol.* **51,** 3–47.

Stearns, S. C. (1977). The evolution of life history traits: A critique of the theory and review of the data. *Annu. Rev. Ecol. Syst.* **8,** 145–171.

Steel, C. G. H. (1980). Mechanisms of coordination between moulting and reproduction in terrestrial isopod Crustacea. *Biol. Bull. (Woods Hole, Mass.)* **159,** 206–218.

Steele, D. H., and Steele, V. J. (1975). The biology of *Gammarus* (Crustacea, Amphipoda) in the north-western Atlantic. XI. Comparison and discussion. *Can. J. Zool.* **53,** 1116–1126.

Steele, V. J., and Steele, D. H. (1970). The biology of *Gammarus* (Crustacea, Amphipoda) in the north-western Atlantic, 2. *Gammarus setosus* Dementieva. *Can. J. Zool.* **48,** 659–671.

Steele, V. J., Steele, D. H., and Macpherson, B. R. (1977). The effect of photoperiod on the reproductive cycle of *Gammarus setosus* Dementieva, 1931. *Crustaceana, Suppl.* **4,** 58–63.

Stephens, G. J. (1952). Mechanisms regulating the reproductive cycle in the crayfish, *Cambarus. I. The female cycle. Physiol. Zool.* **25,** 70–84.

Stickler, J. R., and Bal, A. K. (1973). Setae of the first antennae of the copepod *Cyclops scutifer* (Sars): Their structure and importance. *Proc. Natl. Acad. Sci. U.S.A.* **70,** 2656–2659.

Stross, R. G., and Hill J. C. (1965). Diapause induction in *Daphnia* requires two stimuli. *Science* **150,** 1462–1464.

Subramoniam, T. (1979). Some aspects of reproductive ecology of a mole crab *Emerita asiatica* Milne-Edwards. *J. Exp. Mar. Biol. Ecol.* **36,** 259–268.

Suko, T. (1958). Studies on the development of the crayfish. V. The histological changes of the developmental ovaries influenced by the condition of darkness. *Sci. Rep. Saitama Univ., Ser. B* **3,** 67–78.

Taylor, B. E., and Saltkin, M. (1981). Estimating birth and death rates of zooplankton. *Limnol. Oceanogr.* **26,** 143–158.

Taylor, C. E., and Condra, C. (1980). r- and K-selection in *Drosophila pseudoobscura. Evolution* **34,** 1183–1193.

Telford, M. (1968). The identification and measurement of sugars in the blood of three species of Atlantic crabs. *Biol. Bull. (Woods Hole, Mass.)* **135,** 574–584.

Templeman, W. (1936). Local differences in the life history of the lobster (*Homarus americanus*) on the coast of the Maritime Provinces of Canada. *J. Biol. Board Can.* **2,** 41–88.

Templeman, W. (1937). Egg-laying and hatching postures and habits of the American lobster (*Homarus americanus*). *J. Biol. Board Can.* **3,** 339–342.

Teytaud, A. R. (1971). The laboratory studies of sex recognition in the blue crab *Callinectes sapidus* (Rathburn). *Univ. Miami Sea Grant Program, Sea Grant Tech. Bull.* **15,** 1–63.

Thomas, H. J. (1964). The spawning and fecundity of the Norway lobsters (*Nephrops norvegicus*) around the Scottish coast. *J. Cons. Int. Explor. Mer.* **29,** 221–229.

Thompson, M. H., and Farragut, R. N. (1966). Amino acid composition of the Chesapeake Bay blue crab *Callinectes sapidus. Comp. Biochem. Physiol.* **17,** 225–233.

Thorson, G. (1950). Reproductive and larval ecology of marine bottom invertebrates. *Biol. Rev. Cambridge Philos. Soc.* **25,** 1–45.

Thurston, M. H. (1970). Growth in *Bovallia gigantea* Pfeffer (Crustacea: Amphipoda). *In* "Antarctic Ecology" (M. W. Holgate, ed.), Vol. 1, pp. 269–278. Academic Press, New York.

Tighe-Ford, D. J. (1967). Possible mechanisms for the endocrine control of breeding in a cirripede. *Nature (London)* **216,** 920–921.

Ummerkutty, A. N. P. (1967). Observations on the breeding and seasonal abundance of ten species of planktonic Copepods of the Gulf of Mannar. *Proc. Symp. Crustacea, 1965* Ser. 2, Part II, pp. 685–697.

Valentin, C., and Anger, K. (1977). *In situ* studies on the life cycle of *Diastylis rathkei* (Cumacea: Crustacea). *Mar. Biol. (Berlin)* **39,** 71–76.

Vedel, J. P., and Clarac, F. (1976). Hydrodynamic sensitivity by cuticular organs in the rock lobsters *Palinurus vulgaris:* Morphological and physiological aspects. *Mar. Behav. Physiol.* **3,** 235–251.

Vernberg, F. J., and Vernberg, W. B., (1970). "The Animal and the Environment." Holt, New York.

Vinogradov, M. E. (1962). Feeding of the deep-sea zooplankton. *Proc. Cons. Int. Explor. Mer.* **153,** 114–120.

von Bonde, C. (1936). The reproduction, embryology and metamorphosis of the cape crawfish (*Jasus lalandii*) (Milne-Edwards) Ortman. *Invest. Rep. Fish. Mar. Biol. Surv. Div. S. Afr.* **6,** 1–25.

von Hagen, H. O. (1962). Freilandstudien zur sexual und fortplanzungs biologie von *Uca tangeri* in Anadalusien. *Z. Morphol. Oekol. Tiere* **51,** 611–725.

von Hagen, H. O. (1970). Adaptations to the special intertidal level of habitat in Ocypodid crabs (Decapoda, Brachyura). *Forma Functio* **2,** 361–413.

Voronina, N. M. (1970). Seasonal cycles of some antarctic Copepod species. *In* "Antarctic Ecology" (M. W. Holgate, ed.), Vol. 1, pp. 162–172. Academic Press, New York.

Walker, G. (1980). A study of the oviducal glands and ovisacs of *Balanus balanoides* (L), together with comparative observations on the ovisacs of *Balanus hameri* (Ascanius) and the reproductive biology of the two species. *Philos. Trans. R. Soc. London, Ser. B* **291,** 147–162.

Wallace, R. A., Walker, S. L., and Hauschka, P. V. (1967). Crustacean lipovitellin: Isolation and characterization of the major high-density lipoprotein from the eggs of decapods. *Biochemistry* **6,** 1582–1590.

Warner, G. F. (1967). The life history of the mangrove tree crab, *Aratus pisoni*. *J. Zool.* **153,** 321–335.

Waterman, T. H., ed. (1961). "The Physiology of Crustacea," Vol. 2. Academic Press, New York.

Watras, C. J., and Haney, J. F. (1980). Oscillations in the reproductive condition of *Diaptomus leptopus* (Copepoda: Calanoida) and their relation to egg-clutch production. *Oeocologia* **45,** 94–103.

Wear, R. G. (1974). Incubation in British decapod Crustacea, and the effect of temperature on the rate and success of embryonic development. *J. Mar. Biol. Assoc. U.K.* **54,** 745–762.

Weglenska, T. (1971). The influence of various concentrations of natural food on the development, fecundity and production of planktonic filtrators. *Ekol. Pol.* **19,** 427–473.

White, M. G. (1970). Aspects of the breeding biology of *Glyptonotus antarcticus* (Eights) (Crustacea, Isopoda) at Signy Island, South Orkney Islands. *In* "Antarctic Ecology" (M. W. Holgate, ed.), Vol. 1, pp. 279–285. Academic Press, New York.

Wickins, J. F. (1976). Prawn biology and culture. *Oceanogr. Mar. Biol. Annu. Rev.* **14,** 435–507.

Wilbur, H. M., Tinkle, D. W., and Collins, J. P. (1974). Environmental certainty, trophic level, and resource availability in life history evolution. *Am. Nat.* **108,** 805–817.

Williams, G. C. (1966). "Adaptation and Natural Selection." Princeton Univ. Press, Princeton, New Jersey.

Williams, G. C. (1975). "Sex and Evolution." Princeton Univ. Press, Princeton, New Jersey.

Wolcott, T. T. (1978). Ecological role of ghost crabs *Ocypode quadrata* on an ocean beach: Scavengers or predators? *J. Exp. Mar. Biol. Ecol.* **31,** 67–82.

Wolff, T. (1962). The systematics and biology of bathyal and abyssal Isopoda *Asellota*. *Galathea Rep.* **6,** 1–320.

Wright, H. O. (1968). Visual displays in Brachyuran crabs: field and laboratory studies. *Am. Zool.* **8,** 655–665.

Wu, R. S. S., and Levings, C. D. (1978). An energy budget for individual barnacles (*Balanus glandula* Darwin). *Can. J. Zool.* **55,** 643–647.

Yaldwyn, P. J. C. (1966). Protandric hermaphroditism in the decapod prawns of the families Hippolytidae and Campylonotidae. *Nature (London)* **209,** 1366.

Yamaguti, S. (1963). "Parasitic Copepoda and Branchyura of Fishes." Wiley-Interscience, New York.

Yamamoto, Y. (1960). Effect of eyestalk hormones on the mid-gut gland in the crayfish, *Procambarus clarkii. Zool. Mag.* **69,** 195–201.

Young, J. P. W. (1979). Enzyme polymorphism and cyclic parthenogenesis in *Daphnia magna.* I. Selection and clonal diversity. *Genetics* **92,** 953–970.

Zucker, N. (1976). Behavioral rhythms in the fiddler crab *Uca terpsichores. In* "Biological Rhythms in the Marine Environment" (P. J. DeCoursey, ed.), pp. 145–159. Univ. of South Carolina Press, Columbia.

Zucker, N. (1978). Monthly reproductive cycles in three sympatric hood-building tropical fiddler crabs (genus *Uca*). *Biol. Bull. (Woods Hole, Mass.)* **155,** 410–424.

6

Terrestrial Adaptations

LAWRENCE W. POWERS
and
DOROTHY E. BLISS

THE BIOLOGY OF CRUSTACEA, VOL. 8

ISBN 0-12-106408-5

I. INTRODUCTION

This chapter reviews aspects of crustacean biology that pertain to the successful exploitation of terrestrial habitats. Other chapters in this treatise present information on specific morphological features and physiological systems, including material on some terrestrial species. We provide an overview of various groups of terrestrially adapted Crustacea, making comparisons with aquatic crustaceans and other terrestrial arthropods; and we discuss the environmental variables that influence existence on land. In an earlier review, Edney (1960) considered physiological aspects of crustacean terrestriality, emphasizing data on isopods. A symposium organized by Bliss and Mantel (1968) summarized much contemporary research on this subject, including material on ecology and behavior; this collection of papers provided an important reference source for the present review. Recent monographs on terrestrial arthropods (Edney, 1977) and woodlice (Cloudsley-Thompson, 1977) emphasize the complex relationships among water loss, temperature, and humidity, and among various ecological and behavioral factors. Partly because of the examples employed in previous summaries and partly because of the focus of our own research, we concentrate on ecological and behavioral adaptations of decapods, especially brachyuran and anomuran crabs. On a more speculative note, we assess the relative "success" of various crustacean groups on land.

II. TERRESTRIALNESS

Although there is no consensus among workers as to a defined terrestrial crustacean, several stages of adaptation can be delineated. Species that are completely independent of standing water for breeding, development, and dispersal would qualify as "completely terrestrial," whereas dependence on the sea or a body of freshwater for any of these activities would represent some lesser grade of terrestrialness (Table I). Some species in the literature have been designated "semiterrestrial," but this term has been applied to such ecologically different genera as *Gecarcinus,* land crabs that occupy semiarid sand dunes and grasslands of the extralittoral zone, and *Pachygrapsus,* crevice-seeking crabs of the rocky intertidal zone. Some subtidal crabs such as the portunid swimming crabs, *Callinectes,* occasionally venture from the water; they can survive many hours of aerial exposure at favorable conditions of humidity and temperature. These temporary excursions are either accidental or opportunistic, rather than representative of sustained terrestrial activity, which is the primary concern of this chapter.

We designate five grades of terrestrial adaptation, based on requirements

TABLE I

Descriptive Scale of Terrestrial Adaptations in Crustaceans[a]

Grade	Zone	General description	Typical representatives[b]
1	Midlittoral	Habitat frequently inundated; activities greatly influenced by tidal periodicity; organisms may be active underwater; use both aerial and aquatic respiration; tolerate dehydration poorly	(I) *Pentidotea stenops* (A) *Neohaustorius schmitzi* (D) *Neopanope sayi*
2	High Littoral	Habitat usually damp, occasionally inundated; activities often regulated by tidal cycle; some organisms migrate with changes in water level; terrestrial activity and aerial respiration prevail; many species are diurnal, tolerate mild dehydration	(I) *Cirolana harfordi* *Ligia occidentalis* (A) *Orchestoidea corniculata* (D) *Uca pugilator* *Ocypode pallidula*
3	Supralittoral	Damp substrates and microclimate of high humidity; organisms require occasional access to standing water; many species are nocturnal, tolerate moderate dehydration; decapods usually have aquatic dispersed larval stages	(I) *Alloniscus perconvexus* *Ligia occidentalis* (A) *Talitrus saltator* (D) *Uca pugilator* *Ocypode ceratophthalmus*
4	Extralittoral	Habitat typically mesic but microclimates have high humidities; organisms often independent of standing water, but some have aquatic dispersed larval stages; many species are nocturnal; some tolerate prolonged dehydration	(I) *Porcellio scaber* (A) *Orchestia patersoni* (D) *Gecarcinus lateralis*
5	Terrestrial	Habitat typically mesic, some are xeric; orga-	(I) *Venezillo arizonicus* (A) *Orchestia hurleyi*

(continued)

TABLE I (*Continued*)

Grade	Zone	General description	Typical representatives[b]
		nisms independent of permanent standing water; activity often nocturnal; decapods have direct larval development; cryptozoic habits are common	(D) *Holthuisana transversa*

[a] Definitional notes: littoral (intertidal) = between low and mean high water marks; midlittoral = between low and high neap tide water marks; high littoral = between mean midwater and extreme high water marks; supralittoral (supratidal) = between extreme high water and highest level influenced by windborn spray; extralittoral = area beyond supralittoral subject to effects of aquatic (marine) climate and groundwater; terrestrial = no direct relation to large body of saltwater or freshwater; diurnal = predominantly active during daylight hours; mesic = conditions of moderate moisture or precipitation; xeric = arid, generally lacks moisture; standing water = body of saltwater or freshwater that persists for at least a season or few months and occurs with predictable regularity.

[b] (I) Isopods; (A) amphipods; (D) decapods.

for a source of standing water and on resistance to desiccation. Although we provide this classification as a descriptive convenience for correlates of terrestrialness, we also believe that standardization of these or some other set of terms is desirable. However, a few qualifications are in order. Just as Edney (1968) has cautioned against conclusions regarding evolutionary relationships from existing adaptations of terrestrial isopods, no linear hierarchy or phyletic progression should be inferred from the examples of Table I. Crustaceans have undoubtedly invaded the land numerous times, and any one phyletic group has probably utilized several different approaches (Wright, as discussed by Bliss and Mantel, 1968). Thus, contemporary adaptations represent various modifications to habitats previously occupied by each group as it emerged from the sea (e.g., intermediate brackish and freshwater environments versus direct adaptation to land via rocky or sandy intertidal zones); common adaptations may be derived from diverse phylogenetic origins. Furthermore, not all characteristics noted for a particular terrestrial grade are equally applicable to all groups of crustaceans (e.g., larval development may differ markedly among decapods from different habitats, whereas isopods and amphipods retain a relatively uniform pattern of epimorphic direct development). Some species occupy a broad range of habitats, as defined by specific ecological factors (e.g., availability of food and water, competition for resources that define a local niche) and location within the geographical range of the species. Assessments of terrestrialness for organisms such as the rocky shore isopod *Ligia occidentalis* and the ocypodid crabs *Ocypode quadrata* and *Uca pugilator* depend on characteristics of each local population; factors include the influence of climatic conditions, amount and type of vegetative cover, local topography, and presence of potential predators. These factors are considered in greater detail in Section IV of this chapter. Finally, although we emphasize a progression of increasing terrestrialness from marine shore to inland environment, an analogous continuum exists for emergence from estuarine and freshwater habitats, albeit with different consequences for physiological processes, such as osmotic regulation.

III. MAJOR ADAPTATIONS BY GROUPS OF CRUSTACEA

Three groups of Crustacea (Amphipoda, Isopoda, and Decapoda) are usually credited with varying degrees of success on land (Bliss and Mantel, 1968). Species of these groups have adapted to the extremes of such diverse habitats as the abyssal depths of the oceans, montane freshwater springs, xeric terrestrial plains, and desert areas subject to prolonged droughts. Some features of these organisms have been preadaptive for the successful transition to land (e.g., dorsoventral compression in isopods and retention of the marsupium in amphipods), and other features have evolved independently and converged as common solutions to the selective pressures of terrestrial existence (e.g., direct larval development has arisen in unrelated genera of several brachyuran families). In addition to these three eumalacostracan groups, limited success in marginally terrestrial habitats has been achieved by a few species of cladocerans, ostracodes, and copepods (Table II).

A. Cladocera

Among the bryophytic coverings of cloud forests and rain forests of Puerto Rico, Venezuela, and New Zealand are chydorid cladocerans that are unique in their inability to swim. These newly described species (genus *Bryospilus*) crawl slowly over the wet surfaces of liverworts, mosses, and other forest vegetation and litter, utilizing their trunk limbs; they have been collected from sites as high as 3–5 m above the forest floor, presumably a result of their ability to climb trees (Frey, 1980). Frey discussed the problems of accounting for dispersal of these moss cladocerans. Resting eggs are buried in dense masses of vegetation, so that opportunities for passive dispersal by animals would appear to be limited. The adults are very sensitive to environmental stresses, limiting active dispersal from one damp site to another. Despite these restrictions on genetic exchange and sufficient time to diverge morphologically, there is remarkable similarity between populations of one species, *B. repens,* at sites in Puerto Rico, Venezuela, and New

TABLE II

Systematic List of Terestrial Crustacea[a]

Taxa	Grade	References
Class Branchiopoda		
Order Cladocera		
Family Chydoridae	2+	Frey, 1980
Class Ostracoda		
Subclass Podocopa		
Family Cyprididae	2+	Harding, 1953
Class Maxillopoda		
Subclass Copepoda		
Order Harpacticoida	2+	Harding, 1953
Class Malacostraca		
Subclass Eumalacostraca		
Superorder Peracarida		
Order Amphipoda		
Suborder Gammaridea		
Family Talitridae	5	Hurley, 1968; Duncan, 1969
Family Hyalidae	2+	Bousfield, 1968
Family Hyalellidae	2+	Bousfield, 1968
Order Isopoda		
Superfamily Oniscoidea		
Family Ligiidae	3	Miller, 1938
Family Trichoniscidae	3	Edney, 1968
Family Oniscidae	3	Edney, 1968; Warburg, 1968
Family Armadillididae	4	Edney, 1968; Warburg, 1968
Family Armadillidae	5	Warburg, 1968
Family Porcellionidae	5	Warburg, 1968; Cloudsley-Thompson, 1977
Superorder Eucarida		
Order Decapoda		
Suborder Pleocyemata		
Infraorder Caridea		
Family Hippolytidae	2+	Abele, 1970
Infraorder Astacidea		
Family Parastacidae	4	Bishop, 1967; Suter and Richardson, 1977
Family Cambaridae	2+	Crocker and Barr, 1968
Infraorder Thalassinidea		
Family Thalassinidae	2+	Pearse, 1911
Infraorder Anomura		
Family Coenobitidae	4	Johnson, 1965; de Wilde, 1973
Infraorder Brachyura		
Family Xanthidae	2+	Knudsen, 1960
Family Mictyridae	2+	Cameron, 1966
Family Hymenosomatidae	5	Bishop, 1967
Family Ocypodidae	3	Crane, 1975; Vannini, 1976b
Family Gecarcinidae	4	Bliss, 1968; Bliss *et al.*, 1978

TABLE II (*Continued*)

Taxa	Grade	References
Family Grapsidae	5	Johnson, 1965; Abele and Means, 1977
Families of freshwater crabs[b]	5	Bishop, 1967; Greenaway and Taylor, 1976; Williams, 1980

[a] Only taxa containing species existing at terrestrial grades above 2 (Table I) are listed. Marginally terrestrial species that inhabit damp microhabitats, such as those under leaf mold, are indicated by 2+; the highest grade known for each taxon is indicated. No evolutioanry significance should be inferred from the order in which these families are listed.

[b] Families: Gecarcinucidae, Parathelphusidae, Sundathelphusidae, Deckeniidae, Potamidae, Isolapotamidae, Potamonautidae, Sinopotamidae, Potamocarcinidae, Pseudothelphusidae, and Trichodactylidae; these freshwater families, once listed under the single family Potamonidae, are considered as a composite group in this table. Little ecological information is available for some of these families; some genera of the Parathelphusidae have attained grade 5.

Zealand. Similar questions of evolution and adaptation can be asked of wet forest ostracodes and copepods.

B. Ostracoda

At least one genus, *Mesocypris,* of the family cyprididae shows tendencies toward terrestrial life. Harding (1953) collected one species from mossy areas along mountain streams and a second species from forest humus. Although most ostracodes avoid desiccation passively by shell closure, the species found in forest humus was able to migrate within a Berlese funnel as the soil dried. The forcal rami and tips of the second antennae are equipped with strong claws, and the appendages appear to be well-adapted for terrestrial locomotion. Many ostracodes are parthenogenetic, a feature of possible advantage to a semiterrestrial species confined to patchy microhabitats of leaf mold and damp moss. The ability of ostracodes to withstand extreme desiccation is well known. Unfortunately, nothing is known about the life history of *Mesocypris.*

C. Copepoda

A few species of freshwater harpacticoid copepods have also adapted to living in wet moss. Because of their small size, most of these organisms maintain close contact with the watery film present in the moss; but a few species, such as *Maroenobiotus vejdovski,* have been collected from material that is occasionally moistened by rain (cited in Harding, 1953). Of all copepods, harpacticoids display the best preadaptations for benthic locomotion. Yet, invasion of terrestrial habitats by this group is apparently limited to the marginal leaf mold and mossy patches of vegetation that border pools, streams, and other freshwater sources.

D. Amphipoda

Three suborders compose the Amphipoda, but only the Gammaridea contain terrestrial species (Table II). Although many gammaridean families are represented in the intertidal zone, fully terrestrial (grade 5) amphipods are restricted to the Talitridae. A few species of *Hyale* (family Hyalidae) are semiterrestrial, and a few species of *Chiltonia* (family Hyalellidae) may exist in damp margins of freshwater habitats (discussed by Bousfield, 1968). The Talitridae are nearly cosmopolitan in distribution, but the most terrestrial species (grades 4 and 5) are largely confined to oceanic areas of the Southern Hemisphere. Amphipods are moderately dependent on an external source of chloride ions; thus, they may be restricted to maritime regions (Bousfield, 1968). Two morphological groups of talitrids can be distinguished. One group lives adjacent to seashores, has pronounced sexual dimorphism, retains primitive features associated with aquatic forms, and has nonspecialized pleopods of variable reduction or development. A second group occupies inland habitats, exhibits little or no sexual dimorphism, and has reduced, often highly modified, pleopods (Bousfield, 1968). Species with exceptions to these patterns exist among both groups, however, and the evolutionary factors responsible for these morphological patterns are not understood.

Although no talitrids are known from below the low tide mark, several species can be found in the intertidal and supralittoral zones. Four genera, comprising some 50 species, are regarded as terrestrial (our grades 4 and 5) by Hurley (1959, 1968). The terrestrial forms may have evolved from supralittoral species that were able to enter the extralittoral leafmold environment directly, with little modification of existing adaptations (Table III). Indeed, many morphological features necessary for the success of amphipods on land already exist in littoral species of talitrids (e.g., powerful posterior pereopods and shortening of the urosome for jumping), serving as preadaptations for a cryptozoic existence on land (Hurley, 1968).

In general, amphipods have adapted to land more by behavioral mechanisms than by specific physiological or morphological means (Edney, 1954). For example, in aquatic species such as *Gammarus pulex* and *G. chevreuxi,* courtship usually occurs several days prior to ecdysis of the female. Ecdysis

TABLE III

Characteristics of Terrestrial Amphipods

Preadaptations	Modifications in terrestrial species[a]
STRUCTURAL CHARACTERISTICS	
Lateral compression of body: utilization of crevices Integument: thin, facilitates evaporative cooling and gas exchange; light weight Pleopods: for swimming, respiratory current; feeding current in some Pereopods: strong for jumping, crawling, and burrowing	Exoskeleton: relatively fragile and smooth; no development of waxy or protective layer; high water loss tolerated for evaporative cooling[1] Pleopods: reduced or lost in many species; retained by some terrestrial species, probably for circulation of burrow water; occasional locomotory or respiratory use in supralittoral forms[1] Gills: surface area usually increased; pleopod reduction correlated with expansion of gills on pereopods[1]
PHYSIOLOGICAL CHARACTERISTICS	
Development: direct (epimorphic) to an immature adult; retention and protection of developing embryos in brood chamber (marsupium); (Amphipoda not physiologically well-adapted for life on land)[6]	Development: decreased number of eggs/brood and development time; egg size increases; number of broods/year may decrease, depending on climate[1] Reproduction: egg release from ovary at delayed interval after molt; long-lived sperm stored by female until eggs are released[2] Excretion: nitrogen metabolism partially suppressed; decreased nitrogen excretion[4] Copper metabolism: increased extraction efficiency from leaf litter; enhanced storage and binding in hepatopancreas[3]
ECOLOGICAL CHARACTERISTICS	
Feeding: scavenging or consumption of detritus by most species; intertidal forms often living in algal mats Microhabitats: utilization of crevices, burrows; reliance on camouflage and immobility	Feeding: commonly on algal mats and drift material in supralittoral zone; feeding by terrestrial species in leaf-mold habitat[1] Microhabitats: typically low light, high humidity; all terrestrial forms are cryptozoic

(continued)

TABLE III (*Continued*)

Preadaptations	Modifications in terrestrial species
	Freshwater submergence: active avoidance of inundation by terrestrial species; risk of predation when climbing vegetation[1]
BEHAVIORAL CHARACTERISTICS	
Courtship: before female ecdysis, use of enlarged second gnathopod of male to carry female until postecdysal copulation; copulatory sessions may be lengthy[2]	Courtship: after female ecdysis, carrying behavior is minimal or absent; antennal holding of females by males in some species during brief copulation; "rocking" behavior during mating in supralittoral species[2]
	Brood care: combing of brood plates and turning of eggs with second gnathopod by female; second gnathopod with well-developed pellucid lobes (probable sensory structure)[5]

[a] References: (1) Hurley, 1968; (2) Williamson, 1951; (3) Wieser, 1967; (4) Dresel and Moyle, 1950; (5) McIntyre, 1954; (6) Edney, 1954.

is followed immediately by copulation and the eggs are laid still later. The premolt association in these aquatic species is realized by a form of carrying behavior: the male possesses greatly enlarged second gnathopods, and the female is stroked, restrained, and carried about until she completes ecdysis. In contrast, supralittoral species, such as *Talitrus saltator,* lack a premolt association, and both courtship and copulation occur very soon after the female sheds her exoskeleton. Egg deposition (descent from the ovary to the brood pouch) in the supralittoral species is delayed, occurring at some fixed time relative to ecdysial events rather than relative to copulation (Williamson, 1951). Whether this modification in timing is adaptive for terrestrial species remains questionable. Further research on physiological processes involved in ovarian development of terrestrial amphipods, on the interactions of their molting, courtship, and copulation, and on the timing of their reproductive cycle is needed. Also needed is research on the fate of eggs if the female fails to mate within the fixed postecdysial period.

Among British species of *Orchestia,* several trends correlate well with increasing terrestrialness of habitat: (1) growth rate decreases by means of an increase in intermolt interval, without a change in growth increment per molt; (2) sexual maturity is reached at an earlier stage of growth; (3) smaller body size at sexual maturity is associated with reduction in number of offspring that are produced; (4) sex ratio is biased toward females, compensating for reduction in number of offspring per batch; (5) females reach a greater size compared with males (Wildish, 1979). However, features associated either with marine/supralittoral species or with terrestrial species are not always correlated with terrestrialness of habitat. *Orchestia hurleyi* can occupy dryer grasslands than can *O. patersoni;* but *O. patersoni* produces fewer eggs with a shorter development time and a resultant decrease in mortality rate. Vestigial pleopods, larger eggs, shorter antennae, and a high ecological efficiency in *O. patersoni* are coupled with retention of the enlarged second gnathopods and the habit of female-carrying by the male prior to ecdysis and copulation (Duncan, 1969). This combination of presumed terrestrial and marine/littoral traits probably reflects the independent paths of evolution followed by various terrestrial species.

Although terrestrial genera of amphipods include supralittoral species, freshwater genera are more closely related to marine forms than to supralittoral and terrestrial species (Hurley, 1959). However, a freshwater littoral route for migration onto land cannot be ruled out for all species of amphipods. The close relationship between some freshwater/estuarine species, such as *Orchestia cavimana,* and some marine/estuarine species, such as *O. mediterranea,* has been noted by Wildish (1979). Further studies are needed on osmoregulation in taxonomically similar groups of amphipods that represent a full spectrum of habitats.

E. Isopoda

Isopods are generally regarded as the most successful group of crustaceans with respect to terrestrial exploitation (Edney, 1968). Like other land crustaceans, terrestrial isopods are restricted to moist microhabitats, and most species are nocturnal or cryptozoic, or both. They possess the best adaptive features of amphipods (direct development in a maternal brood pouch) and decapods (versatile locomotory appendages suited to a variety of substrates). We have restricted our comments, based on the availability of ecological and physiological data, to the six families that are listed in Table II. Recent systematic revisions of isopods (Vandel, 1973; Schmalfuss and Ferrara, 1978) have established additional families of terrestrial oniscoids, and future research should reveal further evidence of terrestrial diversity in this group.

Based partly on ecological studies by Vandel (1946, 1954), Edney (1960) ranked oniscoid families in order of ability to survive in dry air. Survival times under different conditions of humidity and temperature varied with respect to the method employed and resulted in different rankings of these families in earlier studies (Miller, 1938). Differences among genera depend on the criteria and experimental techniques used. In addition, families that are represented by only a single species are unlikely to reflect the true characteristics of the family. A broader summary of behavioral and ecological factors related to terrestrial survival result in the following ascending order of terrestrialness: Ligiidae, Tylidae, Oniscidae, Armadillidiidae, Armadillidae, and Porcellionidae (Warburg, 1968) (Table IV).

Isopods probably invaded land several times during their evolution as a group (Vandel, 1964), and most evidence suggests that they adapted to land by way of the marine littoral and supralittoral zones (Edney, 1960). The Oniscoidea may be polyphyletic (Vandel, 1965), accounting for some of the differences between terrestrial taxa. Since isopods generally lack specific characteristics that would adapt them to the rigors of aerial exposure, Paris (1963) has argued that they invaded land in order to exploit the trophic niche of dead vegetation; this postulate is similar to the adaptive hypothesis presented for amphipods (Hurley, 1968). Similarly, the zoogeographical success of *Armadillidium vulgare* (now in Armadilloidea) has been attributed to an expansion of feeding habits (grazing and detritus feeding) in use of a wide variety of materials (Paris and Sikora, 1965).

The lack of a water-proof cutaneous layer, such as exists in insects and arachnids, has been noted repeatedly in reviews of water relations, respiration, and ecology of isopods (Edney, 1960, 1968; Warburg, 1968; Cloudsley-Thompson, 1977). Isopods are among the most common of terrestrial macroinvertebrates; a few species have adapted to truly xeric conditions, despite their apparent inability to retain water efficiently.

The complexity and diversity of these adaptations can be readily demonstrated in *Hemilepistus reaumurii,* a species inhabiting deserts of North Africa and the Middle East: (1) dorsal arching of the carapace reduces the amount of effective insolation; (2) individuals share communal, vertical burrows that are several centimeters deep; (3) several animals may associate to dig a single burrow; (4) this species is large for a land isopod, about 2.0 cm long; (5) it stands and runs with its body well above the ground, avoiding contact with the heated substrate; and (6) it exhibits crepuscular feeding activity (Vandel, 1943; Edney, 1968; Cloudsley-Thompson, 1977). Unlike many land isopods, *H. reaumurii* shows a positive photoreaction (except at high temperatures) and does not conglobate (Warburg, 1968). By contrast, desert species of *Porcellio* are strictly nocturnal and do not dig burrows (Edney, 1958); they show a negative photoreaction, except at high temperatures (Warburg, 1968). Some xeric species, such as *Venezillo arizonicus,*

TABLE IV

Characteristics of Terrestrial Isopods[a]

Preadaptations	Modifications in terrestrial species
STRUCTURAL CHARACTERISTICS	
Dorsoventral compression: stability and maneuverability on land Pereopods: strong for walking, climbing, and burrowing Pleopods: gas exchange and swimming; both rami modified as gills	Pleopods: exopodite may be modified as protective covering Capillary water transfer: uptake of water from substrate by uropods and transfer to pleopods in some supralittoral and mesic species; secondary loss in xeric species[3] Pseudotrachae: pleopods modified as invaginated system of tubes in some terrestrial species (Porcellionidae and Armadillidiidae)[1]
PHYSIOLOGICAL CHARACTERISTICS	
Development: direct (epimorphic) to an immature adult; retention and protection of eggs in brood pouch; internal fertilization Ecdysis: separate for anterior and for posterior portions of animal; results in partial retention of mobility	Hygroreceptors: may exist in some species (evidence inconclusive, based on morphology and behavior)[6] Development: ? reduction in number of eggs/brood Desiccation: resistance to further desiccation may increase due to changes in cutaneous permeability[4] Thermoregulation: evaporative cooling in littoral, supralittoral, and mesic forms; not used by most xeric forms[2] Permeability: increased waterproofing of nonpleopodal areas; oxygen transfer across integument decreases with increased terrestriality[1] Excretion: may or may not be reduction in intake or metabolism of proteins; ammonia is major form of waste product[1,5]
ECOLOGICAL AND BEHAVIORAL CHARACTERISTICS	
Phototropism: usually negative[2] Thigmotaxis: positive, strong[7] Gregariousness: in some species[8]	Phototropism: usually negative; positive in forest-dwellers like *A. vulgare* and desert species like *H. reaumurii*[2]

(continued)

TABLE IV *(Continued)*

Preadaptations	Modifications in terrestrial species
Feeding: consumption of detritus or scavenging by most species	Conglobation: helps reduce water loss, seen in many littoral, mesic, and some xeric species[2] Water relations: avoidance of submergence by active migration in most species; relation of hygroreactions to habitat, temperature, and state of desiccation[2]

[a] References: (1) Edney, 1968; (2) Warburg, 1968; (3) Barnes, 1932; (4) Edney and Spencer, 1955; (5) Dresel and Moyle, 1950; (6) Jans and Ross, 1963; (7) Friedlander, 1965; (8) Kuenen and Nooteboom, 1963.

may be physiologically adapted to their environment, whereas many of the more diverse mesic forms, such as *A. vulgare,* appear to rely primarily on behavioral adaptations (Warburg, 1965). The importance of the interaction of these various factors is becoming clearer with recent studies on the communal social system of *H. reaumurii* (Linsenmair, 1972) and *H. aphganicus* (Schneider, 1970) and on seasonal adjustments of activity with respect to food, predators, and climatic conditions in *H. reaumurii* (Dubinsky *et al.,* 1979).

Conglobation, the ability to curl into a protective ball, is present in all major groups of land isopods, but not in all species of any group (Warburg, 1968). Conglobating species lose less water than do nonconglobating species, since most water is lost from the ventral respiratory organs (Warburg, 1965). The tegumental glands, which are present only in terrestrial species, are reduced in most conglobating species. Gorvett (1956) argued that they may have evolved as a protective device against spiders and other arthropods, and that conglobation has partially supplanted their defensive function. Other behavioral attributes of terrestrial species include the presence of a partially endogenous circadian rhythm of activity (Cloudsley-Thompson, 1952, 1956), marked thigmokinesis that is subordinate to a preference for high humidity (Friedlander, 1963, 1965), pronounced gregariousness that reduces water loss and that may be mediated by chemical attraction (Kuenen and Nooteboom, 1963), and activity responses to saturation deficit (Cloudsley-Thompson, 1956) or to actual relative humidity (Warburg, 1964).

The difficulty of sorting among these complex behavioral and physiological responses was noted by Edney (1968), as he commented on the problems of analyzing laboratory-based data. As in any organism, several physiological systems are responding simultaneously to a complex environment; thus, inferences drawn solely from one-dimensional studies of one or two parameters may be irrelevant or illogical with regard to the natural situation. Unfortunately, much older, as well as current, literature may be subject to this criticism. We second the call by Edney (1968) for more research, such as that by Den Boer (1961) on *P. scaber* and by Paris (1963) on *A. vulgare,* on isopods in natural microhabitats.

F. Decapoda: Caridea (Family Hippolytidae)

A semiterrestrial shrimp, *Merguia rhizophorae,* was observed by Holthuis (1959) in Suriname and described by Abele (1970) from a mangrove area on the Atlantic coast of Panama. Individuals were found above the high tide mark on damp sand, either with algal mats or under driftwood. Emerging at night, they were able to leap distances of 20–30 cm when disturbed. Although little specialization for terrestrial existence was evident, the females carried large eggs (0.9 mm in diameter compared with 0.28 mm for eggs of a related aquatic hippolytid species); this suggests an abbreviated interval of larval development. The eyes of this shrimp are large and contain a strong reflecting layer, a fact that is probably related to the organism's nocturnal habits. The shrimp supports its body on land with robust pereopods. This species is absent from nearby water and it apparently avoids water during escape responses. A related species, *M. oligodon,* was reported from intertidal mangrove roots in Malaysia (Rathbun, 1900).

G. Decapoda: Astacidea (Crayfishes)

Most species of crayfish are associated with permanent sources of freshwater; but some species are capable of withstanding prolonged intervals of drought, provided that a small amount of water remains in the living chambers of their burrows (Fig. 1). A number of species of Parastacidae, found in the Southern Hemisphere, are habitually encountered in dry soils, well above the level of the permanent water table. Many Australian burrowing crayfishes, or "yabbies," erect chimneys of mud pellets on the surface around the entrances of their burrows; the chimneys may be further raised during the wet season so that the burrow opening remains above the rising water level of the creek bed (Clark, 1936a,b; Suter and Richardson, 1977). North American astacids may also construct chimneys in swampy habitats or in areas of heavy vegetation (Crocker and Barr, 1968).

A correlation between communal association and dry substrates has been noted for species of *Engaeus*. In Australia, *E. victoriensis* inhabits community burrows in dry soils, and *E. fultoni* is a solitary resident of burrows in moist, swampy areas (Clark, 1936a); in Tasmania, *E. cisternarius* lives communally in dry soils, and *E. fossor* is a solitary resident of burrows in damp soils (Suter, 1977; Suter and Richardson, 1977). The number of eggs carried by comparably-sized females of the latter two species is inversely correlated with evidence of terrestrialness, a trend common to other crustacean groups (see Table VI).

H. Decapoda: Thalassinidea

The burrowing habits of *Thalassina anomala* in the Philippines are similar to those of terrestrial crayfish. A mound of earth is deposited at the burrow opening, forming a tower and entrance above the high tide mark (Pearse, 1911). The burrows extend to below the water table and may attain a length of 1 m or more (Fig. 1). *Thalassina* is nocturnal or crepuscular, and the burrows are usually sealed during the day, unless the weather is rainy or overcast. This species apparently stridulates, producing a sound similar to that of uncorking a bottle; but the functional attributes of this behavior remain to be investigated.

I. Decapoda: Anomura

COENOBITIDAE (LAND HERMIT CRABS)

Coenobitids have successfully adapted to arid insular and coastal habitats, apparently without the transitional step that involves a moist supralittoral or terrestrial environment (de Wilde, 1973). Like their aquatic relatives, the terrestrial species of hermit crabs occupy mollusk shells and other materials. In all probability, the protective advantages of the shell made it possible for hermit crabs to adapt to land directly from the intertidal zone. The shell has been characterized as a portable burrow (Reese, 1969), providing a source of water and refuge from predators, but with little restriction on movement. Availability of suitable shells is a potential limiting factor for hermit crabs (Reese, 1969), and competition among individuals can be intense (Hazlett, 1968), whether in the intertidal or terrestrial environment. *Coenobita brevimanus,* a widely-distributed species, occasionally uses pieces of coconut when gastropod shells are unavailable (Borradaile, 1903).

Two genera constitute the Coenobitidae: *Birgus,* containing the single species of robber, or coconut, crab, *B. latro* (L.); and *Coenobita,* a widespread tropical group of smaller extralittoral species. Unlike *Coenobita, Birgus* retains a shell only during the glaucothoë stage and first 2 years as a small crab, apparently as protection against desiccation (Reese, 1968). When it abandons the use of shells, *Birgus* is still smaller in size than is an adult *Coenobita,* which retains a shell throughout its life. In the laboratory,

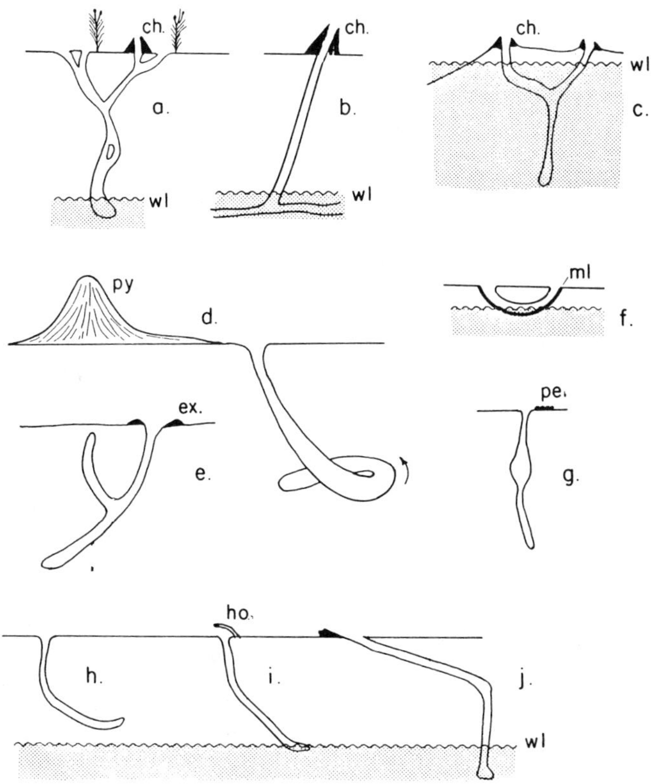

Fig. 1. Burrow structures. (a) *Engaeus fossor* (after Suter and Richardson, 1977). (b) *Thalassina anomala* (after Pearse, 1911). (c) *Cambarus fodiens*[1] (after Crocker and Barr, 1968). (d) *Ocypode śaratan* (after Linsenmair, 1967). (e) *Ocypode quadrata* (L W. Powers, unpublished). (f) *Macrophthalmus depressus* (after Silas and Sankarankutty, 1967). (g) *Scopimera pilula* (after Silas and Sankarankutty, 1967). (h) *Uca panacea* (after Powers, 1975a). (i) *Uca pugilator* (L. W. Powers, unpublished observations). (j) *Uca pugnax* (L. W. Powers, unpublished observations). Abbreviations: ch, chimney; ex, excavated material; hd, hood; ml, mud lining; pe, pellets; py, pyramid; wl, level of water.

small individuals of *Birgus* that lack shells are strongly nocturnal; they dig deep burrows, into which they retreat during the day. By contrast, individuals of *Coenobita* without shells are active throughout the day and night and dig burrows only prior to molting. Differences in shell use and activity may decrease competition between the two genera where they are sympatric (Reese, 1969).

Birgus latro is the largest terrestrial crustacean, reaching a carapace length of 60 cm (Wiens, 1962). Its ability to climb trees has been well-documented (e.g., Gibson-Hill, 1947; Johnson, 1965). but tales of this crab cutting down

[1]Now known as *Fallicambarus fodiens*.

intact coconuts have generally been discounted. Courtship behavior generally lacks precopulatory displays and postcopulatory association; this is in agreement with the solitary and highly agonistic nature of this species (Helfman, 1977a). In addition to scavenging, *Birgus* may prey actively on other crabs, such as *Cardisoma* (Harms, cited in Johnson, 1965).

Generalizations regarding the biology of *Coenobita* are difficult to maintain, even for a single species. Two ecological forms of *C. clypeatus* (Herbst) were described on Curaçao (de Wilde, 1973): a coastal population of small-sized individuals competes with other decapods for limited food resources, whereas an inland population has access to fresh water, shady trees, and adequate food. The inland population must migrate overland to spawn. Shells may not be as readily available inland, thus setting a limit on density and size-structure of this population, but competition for shells with other decapods is absent. Ecdysis mainly occurs while the crab remains within the shell, so that much water is retained; yet de Wilde was unable to determine whether shell water is transferred during exchange of shells or replaced immediately thereafter. Differences between populations of different islands may also be marked: *C. clypeatus* on Curaçao avoids permanently humid places and rarely visits vegetation; in Bimini it forages in the branches of small shrubs located in exposed, dry areas. In Jamaica, the same species lives in open, very arid habitats, remaining under large mangrove leaves during midday; this is similar to their habits in Florida. Females were never observed to immerse themselves when spawning on Curaçao, but did so on Saba and other islands (de Wilde, 1973). Diurnal digging behavior in *C. rugosus* is a function of the ratio of seawater to sand: digging decreases as the ratio increases, until water is present in sufficient amounts to be absorbed directly from the substrate (Vannini, 1975). Migrational movements of *C. rugosus* in Somalia between sand dunes and inland vegetation during the rainy season assume a daily periodicity; during the dry season, movements occur between dunes and the exposed beach flat at night, but assume a tidal periodicity as well (Vannini, 1976a). Once again we stress the need for long, well-integrated field studies of organisms that reveal the diversity of activities and, one may hope, the underlying factors responsible for their expression.

J. Decapoda: Brachyura (True Crabs)

Brachyurans are present in many types of marine, freshwater, and terrestrial habitats, including the intertidal zone of rocky shores, sandy beaches, estuarine marshes, and mangrove swamps; in and along rivers, lakes, and ponds; in freshwater caves (Hartnoll, 1964); in tanks of bromeliad plants of upland rain forests (Hartnoll, 1964; Abele, 1972); on mountain slopes hundreds of meters above sea level (Johnson, 1965; Abele and Means, 1977); and on arid plains of Australia (Williams, 1980). As with Isopoda and Amphipoda, various genera of Brachyura have independently adapted to land via diverse ecological routes, resulting in unique combinations of physiological and behavioral features. Some species of Parathelphusidae and sesarmid Grapsidae (*Geosesarma* group) utilize direct development from egg to immature adult crab (Johnson, 1965), but most terrestrial crabs require an accessible source of water in which to spawn and mature. The powerful appendages and compact body of true crabs are well suited for rapid locomotion on land, as well as for climbing, burrowing, and manipulating objects.

Whereas anomurans retain their microhabitat by carrying a mollusk shell, brachyurans create an extension of their aquatic environment by excavating and inhabiting a burrow or by utilizing available crevices. For various species, burrows and crevices offer protection from predators and from unfavorable climatic factors; a source of water or high humidity; and a site at which to court, mate, and incubate eggs. The subject of burrows and burrowing is treated in more detail later in this chapter.

For most features of brachyurans, it is difficult to generalize as to cause and effect. Yet in the Ocypodidae and Gecarcinidae, closely related families that are composed entirely of intertidal and terrestrial species, one can make some statements regarding the coevolution of various adaptive features.

Among the Ocypodidae, visual capability is enhanced by elongation of eyestalks, especially in some species of *Ocypode* (ghost crabs), a genus of fast-running, highly aggressive facultative predators. Flexion of the eyestalks into horizontal grooves situated on the anterofrontal margin of the carapace occurs as the crab enters its burrow. The visual performance of *Uca* (fiddler crabs), a widely distributed genus of small ocypodids, has been documented according to physiological and behavioral criteria, including measurements of orientation based on horizontal features, polarized light, and position of the sun (Herrnkind, 1968a, 1972), and on discrimination of light intensity and color (Hyatt, 1974, 1975).

Color pattern in some gecarcinids (e.g., young *Cardisoma guanhumi*, most species of *Gecarcinus*) can be very striking. Yet these features may function as camouflage or disruptive coloration rather than as social signals. Little work has been done on ecological and behavioral aspects of color change, except for the blanching responses of *Uca* at high temperatures (e.g., Smith and Miller, 1973).

Whether a population is active diurnally or nocturnally, or both, may depend upon (1) preadaptations resulting from evolutionary history of the species (phylogenetic inertia), (2) physiological tolerance to daytime environmental stresses, (3) presence of significant predators, (4) habits of prey or location of food, and (5) social interactions related to resource partitioning and reproduction.

In lieu of generalization, we have chosen to describe some of the diverse functional adaptations that have been documented in terrestrial crustaceans.

IV. FUNCTIONAL ADAPTATIONS OF THE INDIVIDUAL

This section summarizes morphological, physiological, and behavioral attributes that relate to physical limiting factors, with emphasis on adaptations occurring in anomuran and brachyuran crabs. This subject is discussed in detail and for additional groups of crustaceans elsewhere in this treatise.

A. Locomotion and Related Processes

Terrestrial anomuran and brachyuran crabs generally utilize the same locomotory structures as do their aquatic relatives. Burrowing and climbing occur both on land and in the water; however, there is no terrestrial equivalent for swimming that can augment movement for feeding, for escape from predators, or for entry into a dispersive environment of currents and wave motion. Unlike spiders that utilize silk strands for gliding and the many insects that have evolved active flight, crustaceans have not taken advantage of the air as a dispersive medium.

For details regarding structure of appendages and the associated musculature, a comparative study by Manton (1977) on the evolution of limbs and locomotory mechanisms in various arthropods may be consulted; special problems pertaining to terrestrial crustaceans, however, received minimal treatment in Manton's work. Differences in morphological preadaptations between isopods and amphipods with regard to limb flexure and body stance have been reviewed by Edney (1960).

1. PATTERNS OF MOVEMENT ON LAND

Locomotion on land is best and most diversely illustrated by the Brachyura; examples include the rapid running of *Ocypode* (Milne and Milne, 1946), the agile movements of *Sesarma* and *Aratus* (Warner, 1969), and the versatile climbing abilities of numerous species of crabs (von Hagen, 1977). *Ocypode ceratophthalmus* has been clocked at speeds of 1.8 m/sec when running for short distances on a sandy beach (up to 2.3 m/sec on the wooden deck of a ship) (Hafemann and Hubbard, 1969); such speeds permit this crab to prey upon lizards, crabs, and other highly mobile animals on open, exposed shorelines, yet to escape from large avian and mammalian predators. The body of a ghost crab weighs little relative to its size; this is due to a lightweight endophragmal skeleton, an extremely thin carapace, and sizeable branchial chambers that are filled largely with air. In addition to its rapid speed, *Ocypode cordimana* can use a pouncing "jump-bounce" in capturing live, moving prey (Vannini, 1976b). Differences in speed of two species of ghost crab on Fanning Atoll in the Pacific may correspond to differences in habitat: *O. cordimana,* which lives in supralittoral areas where vegetation offers concealment but may impede locomotion, is relatively slow and docile; *O. ceratophthalmus,* which is intertidal, is fast-moving and highly aggressive (Fellows, 1973). The supralittoral Hawaiian species *O. pallidula* is similar in its movements to *O. cordimana.* Both *O. pallidula* and *O. cordimana* have shorter legs than has *O. ceratophthalmus.*

Use of the fifth pereopods (fourth walking legs) as tactile sensors and as grasping appendages is described for intertidal xanthid crabs by Knudsen (1960). When entering a small tunnel or crevice, a xanthid crab holds these legs dorsally over the carapace to contact the ceiling of the enclosed space. Extension of all legs facilitates immobilization during heavy wave action or resistance to dislodgement by predators.

Climbing activities are prevalent in several genera of crabs, many species of which, as opportunity permits, supplement scavenging on the surface with arboreal feeding (von Hagen, 1977). *Gecarcinus lateralis* often climbs shrubs and trees, sometimes to feed (Bliss *et al.,* 1978). *Uca pugnax,* normally a substrate deposit feeder, frequently climbs stalks of *Spartina* and other marsh plants to feed on portions of the stems and to scrape algae from the surface of the plant (L. W. Powers, unpublished observations). Many grapsids are commonly found on trees, including mangrove trees (e.g., *Aratus pisonii*), and on man-made wooden structures (e.g., *Sesarma cinereum*) (Warner, 1969; von Hagen, 1977). The bromeliad crab, *Metopaulias depressus,* is phytotelmatic throughout its entire life cycle, although adult crabs may leave the "tank" of the bromeliad to feed. This Jamaican crab is restricted to large saxicolous and epiphytic bromeliads at altitudes above 250 m (von Hagen, 1977). Bodily orientation of climbing crabs and various motivational factors and preadaptations for climbing are reviewed by von Hagen (1977).

2. BURROWING

For many terrestrial crustaceans, burrowing is an important activity; for numerous species, it may be the most critical adaptation to land. Desert isopods (*Hemilepistus reaumurii*) dig and occupy burrows as family groups (Linsenmair, 1972). Most terrestrial anomurans carry a gastropod shell, which serves as a refuge and source of water; but the thalassinid, *Thalassina anomala* digs a burrow to the water level (Fig. 1). Among Brachyura, grapsids generally hide in crevices or crawl under rocks, logs, or leaf litter when seeking refuge; some species of grapsids use the burrows of other species of crabs. *Sesarma reticulatum,* for example, often enlarges the burrows of *Uca pugilator* or *Uca pugnax,* sometimes displacing an occupant while doing so (L. W. Powers, unpublished observations). In general, however, burrowing appears to be a poorly developed form of activity among grapsids, possibly because they have adapted to land through the rocky intertidal zone.

Gecarcinids and ocypodids are proficient burrowers; most species inhabit substrates of soft mud or sand, either inland or near water. Gecarcinids dig relatively unspecialized tunnels. In the case of *Cardisoma guanhumi,* the burrows descend to the water table (Herreid and Gifford, 1963); this is not so for burrows of *Gecarcinus lateralis,* which the crab digs in loose, dry, sandy soil (Bliss *et al.,* 1978; Bliss, 1979).

Burrowing is most diverse and specialized in ocypodids. On a sandy beach, an intertidal species, *Dotilla mictyroides,* may occupy up to five temporary burrows within the period of a single low tide; excavation in such wet, sandy substrates is rapid, and the burrow is shallow (Tweedie, 1950). Related species of *Dotilla,* as well as *Scopimera,* which reside in higher zones of the intertidal, excavate more permanent burrows in firmer sands and muds (Fielder, 1970; Hartnoll, 1973).

Sexually mature males of *Ocypode saratan* dig spiral burrows (Fig. 1), different from those of females and sexually immature males (Linsenmair, 1967). Mounds of sand (Sandpyramides) in front of the burrows appear to serve as location signals for sexually receptive females. Similar "copulation" burrows have been described for *O. ceratophthalmus,* but in this species both males and females have a pile of sand near the entrance of their burrow (Hughes, 1973). Males of several species of *Uca* often build over their burrow entrance various structures, to which a number of functions have been hypothesized. Igloo-shaped domes, or "hoods," may function to partition available space and decrease territorial aggression in dense populations of *Uca musica terpsichores* (Zucker, 1974); in other species it may serve primarily to attract receptive females to the actively courting male (von Hagen, 1968). Females of *Uca pugnax* are attracted to the towerlike "chimneys" of males; many of these females posture in rigid, extended poses near the top of the chimneys, then undergo ecdysis within a few days (K. Sprague and L. W. Powers, unpublished observations). The "chimneys" of astacid crayfish (Crocker and Barr, 1968) and "castles" of *Cardisoma carnifex* (Silas and Sankarankutty, 1960) serve mainly to protect the burrow from inundation and exposure to aquatic predators. Of course, these burrow-related structures may serve any number of functions in a particular species or habitat, and more research is required to further delineate their functional significance in specific situations. Table V summarizes some structures associated with decapod burrows.

TABLE V

Burrow-Associated Structures in Decapods

Structure	Species	Functions	References
Plug: At burrow entrance, level with surface or at some distance below surface; made of pellets or single mass of substrate	*Thalassina;* species of *Gecarcinus, Cardisoma, Scopimera, Uca, Dotilla,* and *Ocypode*	Seals burrow from approaching tide, daylight, or extremes of temperature; protects burrow occupant from conspecifics or predators; sequester mated females	Pearse, 1911; Crane, 1975; Gifford, 1962; Tweedie, 1950; Cowles, 1908; Fellows, 1973; Silas and Sankarankutty, 1960, 1967; Bliss *et al.*, 1978; Christy, 1978
Hood or shelter: Over burrow entrance; made of excavation pellets; in form of cowl or partial cowl	Some *Uca* species: *beebei, latimanus, leptodactyla, minax, pugilator, pugnax, musica terpsichores*	?Microclimate modification; aid for burrow location or homing; focus and aid to courtship and territorial displays; decrease density of crabs	von Hagen, 1968; Zucker, 1974, 1978; Crane, 1975; Powers and Cole, 1976
Castle and chimney: Tower-like extensions of burrow above ground level; made of pellets or slabs of substrate	*Thalassina;* some astacid and parastacid crayfish; *Cardisoma carnifex; Barytelphusa guerini*	Protect burrow entrance from inundation, usually built in moist or intertidal habitats or among heavy vegetation; over burrow in dry season	Pearse, 1911; McCann, 1938; Silas and Sankarankutty, 1960; Crocker and Barr, 1968; Suter and Richardson, 1977
Lip and arena: Modified area around burrow at surface; lip is raised edge of entrance rim; arena is flattened substrate surface around burrow entrance	Some species of *Uca:* particularly sanddwelling species: *panacea, pugilator*	Seen only around burrows of displaying males; possible passive accretion due to movements of crab while legs contact rim; arena due to trampling by male; ? help	Crane, 1975; Powers, 1975a

(continued)

TABLE V *(Continued)*

Structure	Species	Functions	References
		female locate burrow	
Pyramid: Sand piled near entrance of burrow, forms cone; trenchlike pathway from pyramid to burrow; burrow spirals downward	*Ocypode saratan* and *O. ceratophthalmus*	Spacing mechanisms for colony; signal function for sexually receptive females and aggressive, intruding males; acts as stable display signal	Linsenmair, 1967; Lighter, 1974
Igloo: Pile of excavation pellets immediately over burrow entrance, in wet intertidal sand	*Dotilla mictyroides* and *Scopimera inflata*	Results from excavation of new burrows by *Dotilla* and from feeding over old burrows by *Scopimera*	Tweedie, 1950; Fielder, 1970

Most features of burrows are not constant for a species, a population, or an individual. *Ocypode* digs deeper burrows in winter than in summer (Milne and Milne, 1946; Utashiro and Horii, 1965); in many species of *Uca,* depth of burrow depends on depth of the water table (Crane, 1975; Powers, 1975b). Plugging of the entrance of a burrow is seen in species of diverse phylogeny and habitat. Certain species may plug the entrance during the day (Silas and Sankarankutty, 1960), with an approaching tide (Tweedie, 1950; Fielder, 1970), with the onset of winter (Bliss *et al.*, 1978), with onset of the dry season (Gifford, 1962), or in response to agonistic social interactions (Powers, 1975a). The topographic form of a burrow may vary with sex and size (Linsenmair, 1967; Fellows, 1973), or with other factors, e.g., the variety of unbranched, Y-shaped, and complex burrows of *Ocypode* (Hill and Hunter, 1973; Vannini, 1980).

3. PERIODS OF ACTIVITY

With increasing distance from the intertidal zone, crustaceans usually exhibit a predominant activity pattern based on light and dark intervals, rather than changes in water level (Barnwell, 1968; Palmer, 1973). Many species are nocturnal or crepuscular, taking advantage of lower temperatures and increased ground moisture in the evening, at night, or at dawn. Fiddler crabs (*Uca*) are well-known as diurnally active animals of the marsh, but most species avoid the hottest surface temperatures at midday. Many temperate and subtropical species of *Uca* are nocturnal as well as diurnal in their surface activity. These species possess advanced forms of social behavior and ecological adaptation (see Crane, 1975, for review of genus); nocturnal courtship, by means of auditory signals, may be compensation for the shorter mating season imposed by nonequatorial climates (Salmon, 1965).

Although *Ocypode* is often nocturnal, some species are primarily diurnal (Linsenmair, 1967); some populations may be either nocturnal or diurnal, depending upon local conditions, such as exposure to wave action on the beach, interactions with predators or prey, and disturbance by human beings (Jones, 1972; Vannini, 1976b). For example, *O. ceratophthalmus* is nocturnal in its activity on western, sheltered shores of Inhaca Island and diurnal on eastern, exposed shores (Hughes, 1966); on Aldabra Atoll, just the reverse situation occurs (Farrow, 1971). In the laboratory under constant conditions, the activity of this same species is very irregular (Jones, 1972). Thus, it is possible that the activity of an opportunistic scavenger–predator such as *Ocypode* may occur more in response to environmental factors, such as local sources of food and potential predators, than to a cycle of tides or light/dark intervals.

In contrast to ocypodids, in which diurnal activity predominates, gecarcinids are primarily nocturnal (Palmer, 1971). In the genus *Gecarcinus,* the eyestalks are comparable in length to those of most other crabs (not elongated as in ocypodids), and complex or prolonged social exchanges are noticeably lacking. On overcast days, during rain showers, and in densely shaded areas, *G. lateralis* may be diurnally active; but generally, in its normally dry environment of exposed dunes and sandy grasslands, this crab is nocturnal (Henning and Klaassen, 1973; Bliss *et al.*, 1978; Bliss, 1979). Even among individuals within diurnally active populations of *G. lateralis,* behavioral interactions are limited; they consist mainly of threatening postures of low intensity and mutual avoidance (L. W. Powers, unpublished observations).

Cardisoma guanhumi is also primarily nocturnal, but this gecarcinid tends to dig its burrows in the shade of palm trees or of certain shrubs; its crepuscular periods of maximum activity may start well before dusk and terminate well after dawn (Henning and Klaassen, 1973; Henning, 1975a). In addition, within the shaded biotope, very young individuals are frequently active during the daytime, particularly if there is a cooling breeze; heavy rain showers usually bring small and large crabs out in great numbers (D. E. Bliss, unpublished observations).

B. Procurement of Food

Among crustaceans, most mechanisms for obtaining food in aquatic environments are applicable on land; such mechanisms include deposit feeding (e.g., astacids, coenobitids, ocypodids), algal scraping (isopods, grapsids of the rocky shores), chewing on vascular plants (gecarcinids, grapsids of the mangroves), scavenging (coenobitids, ocypodids), and active predation (ocypodids). Some species are highly specialized: *Ligia oceanica,* an intertidal isopod, feeds mainly on the alga *Fucus vesiculosus* (Nicholls, 1931); a species of *Coenobita* on Aldabra Island feeds primarily on feces of a giant tortoise (Grubb, 1971); one species of isopod lives in ant nests and eats the feces of ants but not dead ants (Marshall and Orr, 1960).

Many terrestrial species of decapod crustaceans are opportunistic and can utilize a wide range of food items. *Ocypode quadrata* preys on mollusks (*Donax variabilis*), mole crabs (*Emerita talpoida*), stranded jellyfishes, dead fishes, beached whales, bird's eggs, hatchling sea turtles, lizards, other terrestrial crabs (including smaller individuals of its own species), and various arthropods; *O. quadrata* also scavenges through drift material on the beaches, to consume seaweed and other organic debris (Koepcke and Koepcke, 1953; Fales, 1976; Wolcott, 1978; L. W. Powers, unpublished observations; see also Vannini, 1976b, on *O. kuhlii*). *Ocypode quadrata* drags many objects to its burrow; but, if the object is too large to be moved, one or several crabs may dig new burrows next to the object or underneath it.

Deposit and flotation feeders, such as many intertidal ocypodids, may have adapted to specific substrates; their mouthparts have become specialized for removing microorganisms and organic films from particles of a particular grain size and adherence (Miller, 1961; Fielder, 1970; Hartnoll, 1973). For example, *Uca pugilator* is found predominantly on sandy substrates, while *U. pugnax* occurs in muddy areas; although overlap between these species often occurs in habitats of mixed mud and sand, each species shows a higher efficiency of food extraction on its own substrate. In *Uca,* variations in structure of mouthparts and in functional mechanisms of feeding were described by Miller (1961) and Altevogt (1976). Differences in feeding efficiency attributable to sexual dimorphism have been investigated by Valiela *et al.* (1974): males feed at lower rates because the enlarged (major) cheliped is not used, but they may compensate by possessing a somewhat enlarged minor cheliped.

C. Water Relations

For a terrestrial crustacean, as for all other terrestrial animals, it is of crucial importance that loss of water and salts not exceed the animal's capacity to replenish these substances. Adaptations designed to reduce loss of water and facilitate its uptake and storage exist; these adaptations are discussed here. For comparison of the specific adaptations of terrestrial crustaceans with those of other terrestrial arthropods (exclusive of insects), see Mantel (1979). Osmotic and ionic regulation in terrestrial crustaceans was reviewed briefly by Bliss (1968) and in detail by Mantel (1979); osmotic and ionic regulation in all crustaceans, including terrestrial forms, is discussed in Chapter 2, Volume 5 of this treatise and Chapter 4 of this volume. The subject of osmotic and ionic regulation is not discussed in this present chapter.

1. ADAPTATIONS TO REDUCE LOSS OF WATER

In living organisms, loss of water can be reduced by: (1) decrease in permeability of the integument, (2) decrease in the area of exposed respiratory surfaces, (3) increase in retention of water during excretion of waste products, and (4) decrease in exposure to desiccating environments. In one way or another, terrestrial crustaceans exploit all of these means of minimizing the amount of water lost from their body.

Many terrestrial crustaceans lose water more rapidly than do other arthropods (Edney, 1977), probably because of differences in their cuticular waterproofing mechanisms. Yet, in terrestrial species of brachyuran crabs, rates of water loss are only about one-half of rates in intertidal species; aquatic species lose some three to five times as much water as do terrestrial species (Herreid, 1969b). Most of the water is lost through the integument, rather than specifically from the respiratory surfaces (Herreid, 1969a). Despite this, for several species of crabs, the number of gills and the total area of respiratory surface generally decrease with increasing terrestrialness of habitat (Gray, 1957; Bliss, 1968).

The land crab *Cardisoma guanhumi* can tolerate high levels of ammonia in the hemolymph; yet large, white, extracellular accumulations of uric acid (up to 16% of dry body weight) can often be seen in its body (Gifford, 1968). Apparently, in this species, waste products of nitrogen metabolism can assume, at least in part, this insoluble form, which is less toxic than is ammonia and does not require copious amounts of water for its removal. In three other species of land crabs, *Ocypode cordimana, Gecarcinus lateralis,* and *Cardisoma crassum,* similar but less extensive accumulations of uric acid have been observed (Rao, 1968; D. M. Skinner, C. A. Mason, D. B. Bright[2]). Uric acid may be lacking in *C. crassum* in areas where ground water is substantial; uric acid may be present in areas where ground water fluctuates in amount. Thus, it would appear that the amount of water available to a terrestrial crab may determine the ratio of ammonia to uric acid present in the body (Bliss and Mantel, 1968).

In terrestrial hermit crabs, loss of water is reduced by the operculum-like closure of the major and minor chelae, as well as the close fit of a properly sized shell (Harms, 1929). Weakened individuals of *Coenobita clypeatus*

[2]Participants in symposium discussion. *Am. Zool.* **8,** 542–543 (1968).

show before death a sudden increase in loss of water that has been attributed to the crab's inability at that time to close the shell completely (de Wilde, 1973). Although *Birgus latro* foregoes a shell as it matures, an increase in calcification of the abdominal integument and a decrease in the surface-to-volume ratio of the body help the adult individual to minimize water loss (Harms, 1932). In addition, *B. latro* can secrete into the branchial chambers a material that forms a mucuslike fluid, capable of resisting evaporation (Harms, 1932).

Exposure may be especially critical when an organism undergoes ecdysis. Loss of water may be lessened if ecdysis takes place during the night, when the relative humidity is higher (e.g., *Grapsus grapsus* in Bermuda), or within a burrow (e.g., *Uca panacea* in Texas) (L. W. Powers, unpublished observations). Isopods undergo ecdysis in separate anterior and posterior stages; this leads to a slight increase in transpiration (Edney, 1977), but probably decreases vulnerability to predators.

2. ADAPTATIONS FOR INTAKE OF WATER

Because retention of water in terrestrial crustaceans is always incomplete, these animals must periodically replenish their stores of water. To accomplish this, the animals must (1) have access to appropriate sources of water, and (2) be able to transfer the water from the source into their body.

a. Access to Water. Water may be available in a number of forms, from atmospheric moisture and dew to bodies of standing water and moisture within the substrate (interstitial water). Water is also available in sources of food, either directly, or indirectly as a result of metabolism. Salinity can vary from that of hypersaline lagoons and evaporating tide pools to that of normal sea water and brackish waters of estuaries and bays to that of fresh water. Different species of terrestrial crustaceans may gain access to water in the same habitat by entirely different means; a single species may employ several techniques, according to its other activities and the type of water that is available.

Ocypodid crabs typically live either between the tides or above, but close to the high tide line; they venture only short distances inland, if at all. For example, *Uca pugilator* is found on intertidal sandy beaches and in marshes; it often immerses itself in seawater as it carries on flotation feeding at the water's edge (Miller, 1961). Some individuals may feed at some distance from standing water; these crabs either enter their burrow periodically to immerse themselves in water at the base of the burrow, or they obtain water from the damp substrate by pressing their body to the surface (see below). Although *Ocypode quadrata* can utilize either source of water, this crab enters the water less frequently than does *Uca*. *Ocypode quadrata* inhabits sandy flats and foredunes of seaside beaches; its burrows may terminate in moist sand, but standing water is rarely present. The crab obtains most of its water by pressing its body onto the damp substrate and, aided by tufts of setae, conducting water from the substrate surface to the branchial chambers.

Substrates containing high percentages of clay and organic materials with high water-binding capacities offer greater resistance to capillary transfer of water; thus, species of *Uca* on mud flats may require periodic immersion to obtain water. The inland distributional limits of various species of *Uca* on different substrates appear to confirm such a relationship (see Crane, 1975, for a review of the genus).

The hermit crab *Coenobita clypeatus* obtains most of its water either with food or by drinking, even at times using water carried in the shell for drinking. Whether shell water can enter the crab's body by permeation through the soft integument of the abdomen is not known. Differences in the type and amount of available water may affect the activity budget of individuals. In Curaçao, inland populations of *C. clypeatus* require only brief contact with standing water, which may be fresh or slightly brackish, and they carry large volumes of water in their shells. Coastal populations require frequent contact with seawater and spend greater amounts of time engaged in water uptake; they carry little or no water in the shell (de Wilde, 1973).

As indicated earlier (Section IV,A), all species of the genus *Cardisoma* construct burrows only in low-lying areas where the crabs can dig down to ground water. Although typically this water is low in salinity, some populations of *C. guanhumi* have, within their burrows, water ranging from slightly brackish to the salinity of seawater (Herreid and Gifford, 1963; Taissoun, 1974).

For another genus of gecarcinid, water is not always so accessible. *Gecarcinus lateralis* constructs its burrows in loose, dry, sandy soil and is dependent on outside sources of water; these comprise rain water, dew, and water present in plant tissues that regularly serve as its food or in animal tissues that the crab occasionally eats. *Gecarcinus planatus,* a crab that is extremely abundant on Clipperton Island (a coral atoll in the Pacific), also burrows in dry terrain or hides under whatever shelter is available; its principal source of water appears to be the brackish waters of the lagoon (salinity of 3–5 ppt at the surface; 14–17 ppt at the bottom). The crabs immerse themselves in water of up to 20–50 cm in depth in order to feed on the luxuriant algae and aquatic plants that grow there; even when not feeding, crabs can occasionally be seen in waters of the lagoon (Ehrhardt and Niaussat, 1970). To our knowledge, this is the only species of the genus *Gecarcinus* that regularly enters a body of standing water—except for females of all species, which briefly enter the ocean to release their spawn.

b. Transfer of Water. Crabs of the families Ocypodidae, Gecarcinidae, Mictyridae, and Grapsidae are able to extract interstitial water from a sandy substrate and lift it to the branchial chambers; here it moistens the gills and, along with salt, can be actively taken into the hemolymph. With use of the scaphognathites, ocypodids can also drive water from the branchial chambers to the buccal cavity for flotation feeding (Miller, 1961). Some of the water may be swallowed; in experiments with transfer of a solution of dye (0.01 Rhodamine WT), not only did gills and branchial tufts of *Ocypode quadrata* become stained by the dye, but large amounts of the solution appeared in the crab's foregut (T. G. Wolcott, personal communication). It is likely that water and salts can enter the hemolymph through both gills and gut.

In order to extract interstitial water, ocypodids press against the sandy substrate tufts of setae that project from the coxa of the second and third walking legs. These setae are smooth, unbranched, and annulated over at least part of their length (Hartnoll, 1973). In *Uca pugilator,* the frequency of setal pressing against the substrate increases with rising temperatures and increasing locomotor activity; wandering males that engage in frequent combat and exploration of burrows take up water more often than do males that are feeding or defending burrows (L. W. Powers, unpublished observations). Hundreds of meters inland from open water and tidal inundation, the substrate on which *U. panacea* in Texas occurs is often so dry that the crabs must periodically enter their burrow to obtain water. Whether or not this is accomplished by setal pressing is unclear (Powers, 1975a).

During setal uptake of water from moist sand, capillary forces exerted by the bilateral tufts of setae are, in *Ocypode quadrata,* supplemented by negative pressures of up to 76 mm Hg below ambient pressure within the branchial chambers. These negative pressures, which apparently are produced by beating of the scaphognathites, can extract water from beach sand containing as little as 3–5% of water by weight (Wolcott, 1976, also personal communication).

The burrows of *O. quadrata* usually contain no free water, and the content of interstitial water in and around the burrow may be as low as 4% by weight (L. W. Powers, unpublished data). Thus, it would appear possible that the inland distribution of this species, which, according to Hay and Shore (1918), may extend up to 400 m from the coast, may be limited by the amount of moisture available in the substrate, as well as by such biotic factors as the type and density of vegetation and the availability of food (see Henning and Klaassen, 1973; Fimpel, 1975). Yet, according to T. G. Wolcott (personal communication), on barrier islands of North Carolina there is ample interstitial moisture in sandy soil at the center of the islands (up to 400 m from either shore); *O. quadrata* remains largely an animal of the beaches, probably because food is abundant there and no selective pressure to move inland exists. Small inland subpopulations, according to Wolcott, may represent subadults temporarily forced off the beach by intraspecific aggression (e.g., cannibalism).

Of three species of *Ocypode* occurring on sandy beaches along the east coast of India, *O. macrocera* and *O. platytarsis* live intertidally in substrates containing about 20% of water by weight, while *O. cordimana* lives up to 180 m inland from the beach in substrates with a water content of 8–12% (Rao, 1968). These values are well within the range of moisture content of sandy soils for which capillary forces and suction should suffice to lift interstital moisture. Rao (1968) has stated that all three species can maintain normal levels of hydration by setal uptake.

In *Gecarcinus lateralis,* external setae occur on the lateral edges of the first three abdominal segments; these setae lie in close proximity to the setae-fringed posterior extension of the pericardial sacs, and a setae-lined channel connects external setae with the pericardial sacs (Bliss, 1963). Free water reaching the external setae moves swiftly along the channel to the pericardial sacs and up over the convoluted surface of the sacs to the gills (Mason, 1970). According to T. G. Wolcott (personal communication), there is in *G. lateralis* evidence for bulk uptake of water from moist sand, such as occurs in *O. quadrata* when forces of suction are operative. Although both capillarity and suction are probably involved, it is unclear how advantageous this mechanism may be to *G. lateralis* in nature. Frequently, the amount of moisture in the soil surrounding the crab's burrow is so low that, according to T. G. Wolcott (personal communication), neither capillarity nor suction can function to extract interstitial water. It is more likely that these forces lift drops of dew or rainwater from the substrate to the branchial chambers. For further discussion of these possibilities, see Mason (1970), Wolcott (1976), and Bliss (1979).

Bulk uptake of interstitial water from beach sand can occur in another gecarcinid, *Cardisoma guanhumi* (T. G. Wolcott, personal communication). Furthermore, in *C. guanhumi,* as in *G. lateralis,* there is a setae-fringed posterior extension of each pericardial sac and, close by, setae on the anterior lateral borders of the abdomen (Bliss, 1963). Physiological and morphological evidence suggests that in *C. guanhumi* capillarity and suction can operate to draw interstitial water into the branchial chambers. For a different species, *C. carnifex,* Gross *et al.* (1966) showed that dehydrated individuals can rehydrate if they are in contact with sand dampened with fresh water or with seawater, or even if they are in contact with a piece of cloth dampened with seawater. The significance of this adaptation to this genus of crab is unclear, since the animal burrows down to the water table. Furthermore, since *C. guanhumi* would not be able to extract water from soil that is drier than beach sand or holds water more tightly (as do most soils), it is likely that the mechanism of bulk uptake is useful to the crab only to lift free water, such as dew and puddles of rain water, to the branchial chambers (T. G. Wolcott, personal communication). Yet this capacity to extract interstitial water from a sandy substrate may be advantageous during spawning and nonspawning migrations (see Gifford, 1962), when the crab may be away from its burrow for several days.

Diurnally active on intertidal sandy beaches, the Australian soldier crab, *Mictyris longicarpus,* has a double row of dense setae on the proximal border of the first abdominal segment and on the posterior margin of the carapace; transverse passageways situated behind the setae lead to the branchial chambers. The crab can use these setae to transport both interstitial water and films of water from tidal puddles (Hartnoll, 1973; Quinn, 1980).

3. STORAGE OF WATER

In terrestrial crustaceans, there exists a steady demand for water that can be used for maintenance of moist respiratory membranes, as well as for metabolism and, in some species, for evaporative cooling. Species with uncertain sources of water must presumably rely to some extent upon storage mechanisms; ecdysis and reproduction probably occur, at least partially, as water-dependent events. The xeric Australian crab *Holthuisana (Austrothelphusa) transversa,* remaining within its burrow during a prolonged drought, may maintain a normal water balance in part by absorbing water that condenses on permeable regions, such as the gills (Greenaway and MacMillen, 1978). Yet specific mechanisms of water storage may be involved in this crab's survival under such conditions; these require further investigation.

As ecdysis approaches, the needs of a terrestrial crustacean for water becomes more critical, since this fluid is required in order to stretch the new exoskeleton before postecdysial hardening takes place. The pericardial sacs tend to be larger in terrestrial species of crabs than in aquatic species; they are largest of all in *Gecarcinus lateralis* (Bliss, 1963). The large size facilitates storage of water during proecdysis, for these sacs can expand enormously as water enters. Yet experimental evidence indicates that the sacs do not themselves store water, but that they expand in response to an increasing internal hydrostatic pressure as water is taken into the hemolymph (Mason, 1970). In precise terms, therefore, proecdysial storage of water occurs within the hemolymph, although external evidence of this exists in the greatly swollen pericardial sacs.

In the laboratory, when *G. lateralis* in proecdysis is maintained on sand moistened with tap water, swelling of the pericardial sacs becomes more pronounced as ecdysis approaches. Should moist sand be replaced by dry sand, the pericardial sacs decrease in size, and proecdysis, as indicated by a regenerating limb, is delayed, or sometimes halted (Bliss and Boyer, 1964). Thus, the availability of water sufficient to maintain adequate stores within the hemolymph is crucial to the normal progress of proecdysis.

4. EXCESS OF WATER

During the winter rainy season, drowning is the major cause of mortality for the isopod *Armadillidium vulgare,* which escapes inundation by remaining near the surface of the ground both day and night (Paris, 1963). Isopods actively seek drier microhabitats and eliminate excess water by evaporation (den Boer, 1961).

Crabs that are adapted to the higher metabolic functions that are associated with an aerial environment may actively avoid prolonged submersion. Many agile climbing species of grapsids, such as *Metasesarma rubripes* and *Goniopsis cruentata,* move into the upper branches of mangrove trees during tidal flooding (von Hagen, 1977). *Sesarma cinereum* quickly emerges from water into which it is chased or placed; locomotor activity decreases to a low level in animals that are forced to remain submerged (L. W. Powers, unpublished observations). Although *Sesarma* and other semiterrestrial crabs do not drown when submerged, less oxygen is available to these active species. Heavy rains cause *Gecarcinus lateralis* to abandon its flooded burrow and seek shelter on or above the surface of the ground; temporarily, entire populations may thus be displaced from their customary living quarters (Bliss *et al.,* 1978).

D. Aerial Respiration

The availability of oxygen is greater and usually more constant in aerial environments than in aquatic habitats, which may turn anaerobic when circulation of water is limited or respiratory demands of living organisms are high. Terrestrial crustaceans must maintain moist respiratory surfaces; water balance and respiration are thus interdependent functions.

Among isopods, terrestrial adaptations for respiration include: (1) increased pleopodal surface area in the Oniscidae (Sílen, 1954); (2) decreased nonpleopodal gas exchange in proportion to pleopodal gas exchange (Edney, 1960); (3) presence of pseudotracheal tufts on pleopods of Porcellionidae and Armadillidiidae (Edney and Spencer, 1955); and (4) increase in waterproofing of nonpleopodal areas in such genera as *Porcellio* (Edney, 1968). Amphipods lack specific respiratory modifications for use on land (Hurley, 1968). Because they probably emerged onto land via moist leaf mold habitats, they have retained their basic gill structure. Their gills account for up to 60% of the total volume of gas exchange (Duncan, in Hurley, 1968).

Previous attempts to summarize respiratory measurements, such as those of oxygen consumption (e.g., Wolvekamp and Waterman, 1960), among distantly related decapods suggest the difficulties inherent in such comparisons. Lack of standardized methods and variability of respiratory rate with sex, size, metabolic condition, stage of the molting cycle, geographical origin, and state of acclimation are some of the confounding variables. Because of the complexity of evolutionary adaptation to divergent selective pressures, respiratory adaptations in land crabs are not likely to be correlated in a simple or direct relationship with habitat.

For the branchial chambers of terrestrial crabs, three patterns of development have been described by Diaz and Rodriguez (1977) as possible correlates of mobility, metabolic activity, and other life processes—but not of habitat. *Gecarcinus* and trichodactylids have large branchial chambers, with a simple lining and heavy vascularization. The metabolic requirements of these crabs are lower than those of the more active genus *Ocypode,* in which the branchial chambers are of intermediate size, but the branchial lining is folded so that the effective surface for gaseous exchange is greater. The third pattern, typical of the freshwater Pseudothelphusidae (exception: *Potamocarcinus mocinoi,* a cave-dwelling species from Mexico), consists of a complex branchial lining and a perforated "lung" structure in a relatively small epibranchial space. Diaz and Rodriguez discussed the possible relationships among such factors as robustness of carapace, mobility and activity, defense from predation, expansion of ovaries into the branchial chambers, and respiratory requirements, as seen in divergent groups of terrestrial brachyurans. Within a given family, such as the Pseudothelphusidae, it is easier to demonstrate a correlation between the complexity of the gill chambers and other terrestrial features (Rodriguez and Smalley, 1969). In crabs, reduction in the number and surface area of the gills in relation to increasing terrestrialness had previously been noted (Pearse, 1929; Gray, 1957).

The most advanced type of crustacean "lung" is found in the coconut crab, *Birgus latro.* This anomuran crab can survive in air with its gills removed (Harms, 1932); the crab drowns if submerged in water (as does *Ocypode:* see Gross, 1964). The efficiency of oxygen extraction in lungs of *Birgus* is low (2–8%: Cameron and Mecklenburg, 1973) compared with that of gills of aquatic crabs (20–70%: Larimer, 1961; Johansen *et al.,* 1970), but the low efficiency is apparently compensated by high rates of air flow generated by large scaphognathites. Other differences in respiratory processes separate *Birgus* from aquatic forms: acid–base readjustments of its hemolymph involve respiratory exchange rather than ionic mechanisms; the blood pressure of *Birgus* is two to four times higher (up to 50 mm Hg) than that of aquatic crabs; reversals of ventilation, common in aquatic crabs, are lacking in *Birgus;* and stoppages of ventilation in *Birgus* are infrequent (Cameron and Mecklenburg, 1973).

Rarely is lack of oxygen as much of a problem in aerial habitats as is accumulation of carbon dioxide (e.g., in a burrow); thus, most land crabs exhibit marked increases in ventilation in response to hypercapnia, but show little reaction to hypoxia. This differential response is seen in *Birgus* (Cameron and Mecklenburg, 1973) and in the brachyurans *Gecarcinus lateralis* and *Cardisoma guanhumi,* which, like *Birgus,* have a low efficiency of extraction of oxygen and a high rate of air flow (Standaert, 1970, as cited by Diaz and Rodriguez, 1977; Cameron, 1975). Interestingly, *Coenobita clypeatus,* which has other intermediate adaptations for a terrestrial existence, has respiratory responses that are similar to those of marine crabs; these are probably attributable to its use of mollusk shells to maintain an aquatic type of life-style (McMahon and Burggren, 1979).

The scaphognathites, which play a major role in circulating water through the branchial chambers of aquatic species and which create a gaseous respiratory current in gecarcinids and in *Birgus* (Cameron and Mecklenburg, 1973), do not beat during aerial respiration in *Holthuisiana transversa* (Greenaway and Taylor, 1976). In this Australian xeric species, inspiration and expiration are achieved by retraction and expansion of the membranous thoracic wall and roof of the branchial chambers, as the thoracic viscera move alternately to and fro. Ventilation is tidal, with the aperture of the base of each cheliped closed and both inspiration and expiration occurring via the prebranchial apertures (excurrent openings). Probably little or no air flows between the gills or the gill lamellae (Taylor and Greenaway, 1979); thus, there is undoubtably a greater dependence on the branchial epithelium for gaseous exchange than there is in gecarcinids and coenobitids. This mechanism in *H. transversa* is unique among land crabs thus far examined; its efficiency approaches that of vertebrate respiratory mechanisms.

E. Temperature Relations

It is difficult to differentiate the effects of heat load, water loss, and respiratory stress when evaluating upper lethal thermal limits of various organisms, particularly under field conditions. Part of the problem involves the imprecision with which many physical parameters are measured within the microhabitat of an organism. Ambient temperatures are often estimated from mesoclimatic data, such as those provided by local weather stations, or are based on field procedures that do not sample the actual microhabitat (for exceptions, see Edney, 1961; Smith and Miller, 1973; Powers and Cole, 1976). Relative humidity or saturation deficit is also difficult to determine. Thus, many data regarding thermal tolerance are derived from laboratory studies. A typical procedure involves the immersion of a crab in water for defined time intervals and at defined temperatures in order to determine its survival (e.g., Teal, 1958; Vernberg and Tashian, 1959).

Terrestrial crabs in the field use a variety of mechanisms to avoid prolonged exposure to lethal thermal conditions. Tropical species frequently approach or exceed the laboratory-determined upper limits for short intervals without noticeable harm (Edney, 1961; Crane, 1975). Evaporative cooling can lower the body temperature of *Uca pugilator* by 4°C in summer and 1°C in winter (Smith and Miller, 1973); but this mechanism operates most effectively when the relative humidity is low and convective currents are strong. Periodic replacement of water, as observed in ocypodids, must accompany any prolonged use of evaporative cooling. *Uca* alters its chromatophoric patterns in accordance with temperature, as well as with color of substrate and behavioral phase (Crane, 1944; Fingerman *et al.*, 1958; Smith and Miller, 1973). Behavioral thermoregulation plays a major role in reducing heat load in daytime-active crabs.

Many diurnal species shift their activities to cooler areas of their habitat. The burrow-retreat response of *Uca* increases in frequency as ambient temperatures rise in mangrove swamps and on exposed flats of barrier islands (Macintosh, 1978; Powers, 1975b). At midday, temperatures within burrows may be up to 15°C cooler than temperatures at the surface of the ground (Powers and Cole, 1976; Macintosh, 1978). Similar but less marked benefits are derived from the shade of vegetation and banks and from immersion in nearby standing water. Some tropical species may avoid midday activity: *Coenobita clypeatus* seeks shelter under palm fronds (de Wilde, 1973), and various species of intertidal *Uca* may remain in their burrows when low water occurs at midday (Crane, 1975). *Uca pugnax* displays a phototactic rhythm of diurnal periodicity: crabs are attracted to light when temperatures are moderate, such as in the morning and evening, but avoid light when temperatures are high (Palmer, 1964).

Lower thermal tolerances are critical for species of the temperate zones that are subject to seasonal extremes. In subtropical and temperate areas, some species of *Uca* with wide latitudinal ranges are restricted in social activities to intervals of climatic warmth, whereas the crabs may be perennially active in tropical areas (e.g., *U. rapax,* in Crane, 1975). Locomotory inhibition occurs between 11° and 12°C in *U. minax, U. pugnax,* and *U. pugilator* from temperate areas and between 13° and 14°C in subtropical *U. thayeri* and *U. rapax* (Miller and Vernberg, 1968). In eyestalkless *U. pugnax,* proecdysis is not initiated when the temperature is below 15°C and is delayed when the temperature is between 15° and 22°C (Passano, 1960). Larvae of *Uca* are more tolerant of low water temperatures than are adults (Vernberg and Costlow, 1966); thus, the adult population might be decimated by severe or prolonged freezing of the ground, but would be replaced by larval recruitment from southern populations.

Further information on temperature as a factor of importance to terrestrial crustaceans can be found in a review by Edney (1960) and in summaries of work on metabolism and acclimation by Vernberg and Vernberg (1968) and Miller and Vernberg (1968) (also in relevant chapters of this treatise).

F. Reproduction

We have selected three aspects of reproduction in crustaceans that pertain to terrestrial species: (1) relationship of mating to molting, (2) brood care in species that retain their larvae, and (3) larval dispersal in species that do not. In this treatise, physiological aspects of reproduction are discussed in Chapter 3 of Volume 9; ecological aspects in Chapter 5 of this volume.

1. RELATIONSHIP OF MATING TO MOLTING

Precopulatory associations in portunid and cancrid crabs are initiated by the release of chemicals from premolt females; males search actively and clasp receptive females, remaining with them and protecting them until ecdysis ensues and the new exoskeleton has hardened (Ryan, 1966; Eales, 1974). These crabs, as well as most other aquatic species of crabs, copulate immediately after the female has completed ecdysis, when the genital opercula are decalcified; but some aquatic species, notably xanthids (Knudsen, 1964; Hartnoll, 1969) and the burrowing corystid crab, *Corystes cassivelaunus* (Hartnoll, 1972), may copulate during brief intervals of opercular decalcification during intermolt.

Many semiterrestrial and terrestrial species of crabs, including gecarcinids, ocypodids, and most grapsids, mate while the female is hard (see Hartnoll, 1969, for list of species). Exceptions among the Grapsidae include *Sesarma bidentatum* (Hartnoll, 1964) and *Pachygrapsus crassipes* (Hiatt, 1948). About the latter, there may be some question, since both Hiatt (1948) and Bovjberg (1960) state that the female is vigorously active at the time of mating; therefore, these females must not be as soft at this phase of activity as are their aquatic counterparts.

In some terrestrial and semiterrestrial crabs (e.g., most species of *Uca:* Crane, 1975), mating involves complex courtship displays; in certain others, there is little preliminary association (e.g., the surface copulations of *Scopimera:* Yamaguchi *et al.*, 1979; of most sesarmids; and of some tropical species of *Uca*). Presumably, copulation in terrestrial species during inter-

TABLE VI

Reproduction in Terrestrial Crabs

Species	Number of eggs	Spawns/year	Incubation time	References
Coenobitidae				
Coenobita clypeatus	1,000–50,000		3–4 weeks	de Wilde, 1973
Parathelphusidae				
Barytelphusa guerini	100–300	1		Gangotri *et al.*, 1977
Gecarcinidae				
Cardisoma guanhumi	20,000–1,200,000	up to 5	16 days	Taissoun, 1974; Henning, 1975b
Grapsidae				
Aratus pisonii	3,000–10,000	5–6	16 days	Warner, 1967
Helice crassa		1–3	22–53 days	Nye, 1977
Hemigrapsus nudus	441–36,456	1		Knudsen, 1964
Hemigrapsus oregonensis	800–11,000+	2		Knudsen, 1964
Hemigrapsus penicillatus	8,000–100,000	5–6	35–40 days	Pillay and Ono, 1978
Metopaulias depressus	63–98	2	10–12 weeks	Hartnoll, 1964
Pachygrapsus crassipes	$\bar{x} = 50,000$	1–2	26–31 days	Hiatt, 1948
Sesarma cinereum		4–6	about 30 days	Seiple, 1979
Sesarma jarvisi	7–18			Abele and Means, 1977
Sesarma reticulatum		2–3	about 45 days	Seiple, 1979
Sesarma pictum	15,000–45,000	2–3	about 30 days	Pillay and Ono, 1978

molt permits greater mobility and enhances the potential for wider selection of a mate by females; but to establish this, there is need for more research on field populations (see Section V,A).

Copulation independent of ecdysis makes possible a larger number of broods within each cycle of intermolt, especially in older females that molt less frequently. Table VI compares some species with regard to reproductive activity. More field observations on closely related species in a range of habitats are needed in order to establish correlations between environmental and reproductive factors. For example, *Cardisoma guanhumi,* which inhabits low-lying areas and burrows down to the water table, exhibits some advanced terrestrial adaptations, but retains a high fecundity with large brood size and the ability to spawn several times in a season (Taissoun, 1974). Evolutionary compromises with regard to fecundity, mortality, and various ecological factors as they relate to terrestriality have been reviewed by Lindberg (1976), with *Pachygrapsus crassipes* as an example.

Among hermit crabs, *Birgus latro* mates while the female is hard, with minimal precopulatory and postcopulatory association (Helfman, 1977a). Although aquatic pagurids can transfer a spermatophore in less than 1 min (Hazlett, 1972), their courtship and that of *Coenobita perlatus* may last 1 hr or longer (Helfman, 1977a). Several field studies (e.g., Johnson, 1965; Helfman, 1977b) have noted the contrast between gregariousness of most hermit crabs and the solitary, highly agonistic behavior of *Birgus*. In this and many other features, *Coenobita* resembles more closely its aquatic, more distant relatives than it does its closer relative, *Birgus*. For instance, *Coenobita perlatus* probably carries enough water (3–4 ml) in its shell for the eggs to be fertilized in water, as they are in aquatic species; in the one mating of *Birgus* that has been documented, fertilization took place without the crab being immersed in water (Helfman, 1977a).

2. BROOD CARE

Parental care rarely occurs among Crustacea. Most eggs hatch in water, and the larvae and postlarvae risk high mortality until metamorphosis into first adult stage. Among amphibious crabs, metamorphosis entails emergence onto land during or shortly after the megalops stage (Herrnkind, 1968b; Klaassen, 1975).

Isopods and amphipods, whether aquatic or terrestrial, retain their young in a brood pouch and release them when they are in an advanced stage of development, thereby reducing their mortality. Larval stages are avoided altogether, but at the expense of dispersive marine stages that provide geographical expansion.

Reduction in larval life is known for some terrestrial brachyurans, particularly among families of freshwater crabs. *Holthuisana transversa* in northeastern Australia has no apparent larval stages outside of the egg. The female carries miniature adults on her ventrum for several days after they have emerged (Williams, 1980); this constitutes an obvious advantage for an inland xeric species with very limited access to standing water. Like potamonids and pseudothelphusids, montane species of *Sesarma* (e.g., *S. foxi* in southeast Asia) may also undergo direct development; but little is known of their reproductive habits (Johnson, 1965). Less drastic larval reduction may involve the loss of one or more zoeal stages, as in *Metopaulias depressus,* the bromeliad crab (Hartnoll, 1964), or the shortening of each zoeal stage. In contrast, terrestrial species, such as gecarcinids and ocypodids, that retain some marine chracteristics must return to the sea to release their young, sometimes enduring extensive spawning migrations overland to do so.

3. LARVAL DISPERSAL

Observations of *Coenobita clypeatus* in the laboratory indicate that the female can "shoot" her eggs into the water, thereby releasing her spawn without becoming immersed. In nature, likewise, females of this species deposit their eggs without direct contact with free water, either of tide pools or of the open sea. At low tide, the female hermit crab uses her fifth pereopods to pick small clusters of eggs from the egg mass on her pleopods, transfers the eggs to her maxillipeds and chelipeds, and then spreads the eggs on wet stones near the water line. Later, during high tide, the larvae are washed into the ocean. A female may deposit only a portion of her eggs, probably only those that are fully developed, on any given night; the remainder may be deposited on subsequent nights (de Wilde, 1973).

Cardisoma guanhumi in Florida may, in one or two nights, travel up to 5 or 6 km to the shore from inland burrowing sites in order to release spawn (Gifford, 1962). Release may occur in several stages, with the female seeking cover near the water's edge between immersions. Gifford's (1962) unsuccessful attempts to hatch eggs and rear larvae of *C. guanhumi* in water from a drainage ditch might appear to confirm the dependence of this crab on the sea for development. Yet Costlow and Bookhout (1968) demonstrated that *C. guanhumi* could develop up to the first crab stage in salinities of 15–45 ppt (ca. 48–128% seawater) provided the temperature was 25° to 30°C (at 20°C the few larvae that reached the megalops stage soon died). In Colombia, Henning (1975b) observed that while some females of *C. guanhumi* were releasing larvae in the sea, others simultaneously were releasing larvae in water of drainage ditches (salinity 1–2 ppt at the surface, 12 ppt at a depth of 50–80 cm); he subsequently collected both zoeas and megalopas from the drainage ditches.

Mass spawning migrations have been reported for *Gecarcoidea lalandei* on islands of the Pacific (Gibson-Hill, 1947; Johnson, 1965). In Colombia, both *C. guanhumi* and *Gecarcinus lateralis* undergo lunar and semilunar spawning migrations to the same beach, but the maxima of larval release do not coincide; *C. guanhumi* largely sheds its spawn before or at the full moon or new moon, while *G. lateralis* usually releases larvae 4–5 days later (Klaassen, 1975). Release of spawn always occurs at night, even in diurnal species such as those of *Uca* (Crane, 1975).

Recent studies have focused on temporal aspects of reproduction by relating both mating and larval release to lunar–tidal events. Female receptivity in several species of *Uca* appears to occur at monthly peaks, during the time of the full moon (von Hagen, 1962; Zucker, 1978); but it occurs at semimonthly intervals in other species of *Uca* (von Hagen, 1970; Christy, 1978). In most species of *Uca*, courtship activities of males occurs in semimonthly peaks (near the time of new moon and full moon), although in some species there are distinct monthly peaks (Zucker, 1978). Further observations by Zucker revealed that the monthly peaks shift to coincide with less extreme spring tides; release of spawn occurs about 2 weeks later, during the greater spring tides. Studies in the laboratory concerning synchrony of larval release and maturation confirm such a temporal pattern: *Uca pugnax* releases zoeas in peaks that correspond to lunar spring tides; development from zoea I through the megalopa then averages about 28 days (Wheeler, 1978).

Semilunar rhythms of larval release have been reported for *Sesarma intermedium* and *S. haematocheir* in Japan (Saigusa and Hidaka, 1978; Saigusa, 1980). Moonlight, rather than tidal inundation, is believed to be the responsible Zeitgeber[3] (Saigusa, 1980). In North Carolina, the number of gravid females of *S. cinereum* shows a distinct lunar periodicity, which becomes less pronounced as summer progresses (Seiple, 1979).

Uca pugilator in South Florida exhibits clear synchrony with lunar periodicity (4–5 days before spring tides) in its peaks of male courtship (waving of chela and construction of hood on burrow), in female reproductive activity (presumed by closing of burrow and cessation of feeding at surface), and in larval release (Christy, 1978). Although other workers had noted the lunar cyclicity of male courtship activity and had attributed it to availability of receptive females (Feest, 1969; von Hagen, 1970), Christy hypothesized that larval recruitment and dispersal may be the determining factors; this hypothesis is supported by known larval maturation times (Herrnkind, 1968b). This would be particularly adaptive for estuarine intertidal species, because maximal flushing by spring tides would carry larvae into the marine environment optimal for maturation and return them one lunar cycle later to emerge onto land either as megalops or as juvenile crabs.

[3]Zeitgeber: literally "time-giver," an environmental synchronizing factor.

G. Growth, Development, and Molting

The frequency of molting and, with some exceptions, the percentage increase in size at ecdysis (percentage molt increment) generally decrease with age in both terrestrial and aquatic species of decapod crustaceans (see Hartnoll, Chapter 3, Volume 2 of this treatise). Anecdysis, a long period devoid of molting, is frequently correlated with exposure of an animal to low temperatures. Consequently, anecdysis can be expected to be more prevalent in temperate terrestrial species than in tropical terrestrial and aquatic species of crustaceans; however, extensive comparative data regarding this are not available. Prolonged anecdysis may be expected in xeric species like *Holthuisana transversa,* which may remain underground for one to several years during severe droughts (MacMillen and Greenaway, 1978).

Terminal anecdysis is common among portunids and majids; whether it occurs among semiterrestrial and terrestrial crustaceans is open to question. Two studies have shown that the intertidal crabs *Pachygrapsus crassipes, P. transversus,* and *P. gracilis* do not undergo terminal anecdysis (Hiatt, 1948; Hartnoll, 1965). There is no evidence that either of the land crabs *Cardisoma guanhumi* (Henning, 1975b) or *Gecarcinus lateralis* (D. E. Bliss, unpublished data) has a terminal anecdysis. As Hartnoll (Chapter 3, Volume 2 of this treatise) has pointed out, growth in crustaceans can be limited with age either by a decreasing percentage molt increment, which, in decapods at least, is the prevailing mechanism; or by cessation of molting, i.e., by terminal anecdysis. These two mechanisms appear not to operate in the same animal: a species that undergoes terminal anecdysis does not usually exhibit a decreasing percentage molt increment prior to the final ecdysis (but growth may be truncated); and a species that shows a decreasing percentage molt increment with age does not appear to undergo terminal anecdysis. Hormonal mechanisms that affect terminal anecdysis are discussed by Skinner (Chapter 2, Volume 9 of this treatise).

Few, if any, absolute correlates exist between habitat and crustacean developmental parameters. In gecarcinids and coenobitids, which return to the sea to release their young, larval development is not abbreviated. Direct larval development, taking place entirely within the egg and thus of possible advantage to a terrestrial species, occurs among all species of freshwater crabs, not just among those that, as adults, spend large amounts of time on land (see Bliss, 1968). Furthermore, direct larval development occurs in several species of dromiids (Montgomery, 1922; Hale, 1925), as well as in the majid *Naxioides serpulifera* (Guérin) (Rathbun, 1914); and a reduction in the number of larval stages is common among majids (Hartnoll, 1965).

At ecdysis, the need of terrestrial crustaceans for obtaining water with which to stretch the new, soft exoskeleton can, for gecarcinids and coenobitids, further limit their distribution to coastal zones or islands, since the water must be suitable as regards salinity and ionic content. In the laboratory, *Gecarcinus lateralis* can survive for many months with distilled, demineralized water or with seawater of a range of salinities (2–30 ppt) made available in a small, shallow dish; but the crabs maintain their initial weight more readily when the available water has a salinity of about 15 ppt. Furthermore, the pericardial sacs of destalked individuals of *G. lateralis* become most swollen just prior to ecdysis and postecdysial survival is greatest when the salinity of the water is 15–23 ppt (Bliss *et al.*, 1966). In nature, *G. lateralis* in Bimini and Bermuda may be able to derive sufficient supplies of salt from wind-blown sea salt present in dew (Bliss, 1963; Wolcott, 1980). Similar sources of salt have been postulated for inland populations of *Coenobita clypeatus* on Curaçao (de Wilde, 1973).

For three species of *Ocypode,* differences in size and swelling of the pericardial sacs have been correlated with habitat (Rao, 1968). In *O. cordimana,* which lives up to 180 m inland from the seashore, the pericardial sacs are large. During proecdysis, because of setal uptake (see Section IV,C,2,b) and a subsequent intake of water that leads to an increase in hydrostatic pressure of the hemolymph, the pericardial sacs become swollen. In *O. macrocera* and *O. platytarsis,* which live intertidally, the pericardial sacs are relatively small; they swell only briefly during ecdysis as water enters the hemolymph. Comparative data on the many species of *Uca* are not available, but they would provide a more extensive test of correlation with habitat.

Differences exist between two gecarcinids as to the time at which and the extent to which the pericardial sacs swell. Ecdysis occurs within the burrow for both species. According to Henning (1975b), *Cardisoma guanhumi* closes the entrance of its burrow with wet mud brought to the surface from below, and the crab then descends to water at the base of the burrow; in this water the crab undergoes ecdysis. Pericardial sacs of *C. guanhumi* swell only slightly during ecdysis, as water enters. *Gecarcinus lateralis* also undergoes ecdysis within its burrow. In Bermuda during the month of March, Weitzman (1963) found both the crab and the exuvium within burrows. Yet, no free water is available in the burrow of *G. lateralis* in this habitat; usually the soil surrounding the burrow feels quite dry to the touch. Water needed for stretching the new soft shell at ecdysis is obtained elsewhere during proecdysis. By the time of ecdysis, in the laboratory at least, the pericardial sacs of *G. lateralis* can be quite swollen (Bliss and Boyer, 1964; Bliss *et al.*, 1966). Also in the laboratory, environmental conditions (darkness, solitude, adequate moisture, favorable temperatures) that prevail in a burrow tend to induce proecdysis; under these same conditions, pericardial sacs tend to swell to their maximal extent (Bliss and Boyer, 1964).

Particularly adaptive is the way in which the water taken into *G. lateralis* during proecdysis moves from the hemolymph into the lumen of the foregut at ecdysis. Here the water maintains turgidity of the thoracic viscera, including the midgut glands, until some hardening of the soft exoskeleton takes place. In this way, a smoothly rounded, characteristic profile in the hardened crab is ensured. It is believed that water in the foregut at ecdysis eventually returns to the hemolymph; movement of the water in both directions appears to be under neuroendocrine control (Bliss, 1968; 1979; Mantel, 1968; Mantel *et al.*, 1975).

For *C. guanhumi* and *G. lateralis,* differences exist as to the storage of calcium salts needed for hardening the new shell. In *G. lateralis,* two pairs of large gastroliths are formed prior to ecdysis, are dropped into the lumen of the foregut as the lining is shed during ecdysis, and are then dissolved by enzymes in fluids of the foregut, with the calcium salts quickly reentering the hemolymph for transport to the epidermal tissues. In addition, in the laboratory where no coral sand is available, *G. lateralis* eats its exuvium after ecdysis and further supplements its diet with egg shells that are provided (for further details, see Bliss, 1968).

As regards *C. guanhumi,* no reports have appeared on gastroliths large enough to serve as a postecdysial source of calcium salts. According to Henning (1975b), in this species calcium must be stored by the hemolymph and the midgut glands; additional calcium may come from the water of the burrow and from soil surrounding the burrow. Henning considers that the exuvium consumed by a crab after ecdysis can be an important source of calcium.

In a recent review, Graf (1978) pointed out that the degree to which crustaceans store calcium before ecdysis is related to their way of life. When the old exoskeleton is shed, a percentage of calcium in the body is lost: the supralittoral isopod *Ligia exotica* loses 25%, the supralittoral amphipod *Orchestia gammarella* 33%, the terrestrial isopod *Oniscus asellus* 40%, and the terrestrial isopod *Porcellio scaber* 51%; these values compare with 74% for the freshwater crayfish *Austropotamobius pallipes,* 77–89% for the freshwater amphipod *Gammarus pulex,* and 92% for the marine crab *Carcinus maenus* (Graf, 1978). Thus, supralittoral and terrestrial crustaceans retain far more calcium prior to ecdysis than do aquatic crustaceans.

In addition to the hemolymph, which serves temporarily for storage while calcium is in transit to and from the cuticle, the principal sites of storage of calcium are the midgut glands and special structures that include the gastroliths (some decapods), sternal plates (supralittoral and terrestrial isopods), and the posterior ceca (some amphipods) (Graf, 1978). In crustaceans in which these structures are well developed, the amount of calcium stored in the midgut glands before ecdysis is minimal.

Just as gecarcinids use a burrow for ecdysis, the terrestrial hermit crab *Coenobita clypeatus* uses a hiding place and its borrowed snail shell. The crab remains in hiding as much as possible and sheds its old exoskeleton largely while within the shell, which retains some water (de Wilde, 1973).

Among various species of the crabs *Uca, Sesarma,* and *Grapsus* that have been observed in the field (L. W. Powers, unpublished data), ecdysis occurs predominantly at night. In the laboratory, *Uca panacea* sheds its exoskeleton while immersed in water; whether shedding in nature takes place only in a burrow is unclear, since, in Texas, burrows of this species commonly contain little or no standing water (Powers, 1975a).

Detailed physiological studies of decapod crustaceans have necessarily been carried out on laboratory populations (see reviews by Bliss, 1968, 1979). Evolutionary relationships of molting to ecological factors are speculative for most species. Ecdysis in terrestrial species, like copulation, has not been observed often enough under natural conditions to make possible a comprehensive understanding of controlling processes and their manifestations. Clearly, more field studies are needed.

V. BEHAVIORAL INTERACTIONS

The importance of an organism's behavior for adaptation to land has been emphasized previously (Edney, 1968; Hurley, 1968). Space does not permit a review of the many manifestations of behavior that relate to life on land, so we are restricting our comments to aspects of social versus nonsocial intraspecific relationships and to competition among closely related species and with other arthropods.

A. Intraspecific Relationships

1. NONSOCIAL GREGARIOUSNESS

Various physical factors, detectable as gradients within microhabitats, can promote aggregations of animals by serving as negative or positive stimuli for repulsion or attraction, respectively. As ambient temperature increases, for example, dark, moist refuges become increasingly attractive to isopods (Warburg, 1968); land hermit crabs aggregate under palm fronds and other vegetation during midday (de Wilde, 1973). The extent of clumping is a function of both the intensity of the environmental stimuli and the quality and quantity of suitable refuges. Flooding or prolonged drought may result in migration of *Uca* to restricted portions of a habitat, with a resulting increase in observed density (Powers, 1975a).

Aggregations of animals may be a result of common attraction to particular features of a habitat or due to social interactions or both. Spawning migrations of *Coenobita clypeatus* (de Wilde, 1973), *Cardisoma guanhumi* (Gifford, 1962; Henning and Klaassen, 1973), and *Sesarma* spp. (Saigusa and Hidaka, 1978) are synchronized by lunar or tidal events, yet social facilitation may reinforce the environmental Zeitgebers. Likewise, the tidal rhythmicity of many activities that are observed in intertidal species (Palmer, 1973) can be regarded as nonsocial, whereas phases of waving and combat in *Uca,* also showing lunar cyclicity (von Hagen, 1970; Christy, 1978; Zucker, 1978), are clearly social interactions. Feeding aggregations of *Uca* are harder to classify: large groups of male and female *U. pugilator* travel onto the exposed sand or mud flats during low tide to feed. Movement by some crabs as they leave the burrow area stimulates movement by others, so that large feeding groups result; yet dispersal and individual movements can be seen, once the crabs reach the flats. Return to the burrow area also seems to involve social facilitation (Powers, unpublished observations).

Despite the aggressive nature of crabs, individuals of the same species may tolerate one another when in close proximity, if so required by environmental circumstances. Intertidal xanthids and grapsids are frequently found in dense groups at low tide, under rocks, and in crevices. *Uca,* which normally is the sole occupant of a burrow, will temporarily tolerate conspecifics seeking refuge from predators. All species must decrease their level of agonism to permit copulation, whether it be brief and perfunctory (e.g., *Birgus*) or prolonged and complex, as in some aquatic crabs (Hartnoll, 1969). As *Gecarcinus lateralis* feeds and moves about, it displays minimal social interactions, yet it maintains relatively close proximity, both during the day (Bliss *et al.,* 1978) and at night (L. W. Powers, unpublished observations). Two notable exceptions that show little toleration of conspecifics are the coconut crab, *B. latro* (Helfman, 1977b), and the various species of ghost crabs, *Ocypode.*

2. SOCIAL AGGREGATIONS AND COMMUNICATION

a. Courtship and Mating. Aspects of reproductive behavior have been previously reviewed by Schöne (1968), Wright (1968), and Hartnoll (1969), and in this treatise by Salmon and Hyatt (Chapter 1, Volume 7 of this treatise) and Sastry (Chapter 5 of this volume). Visual displays are most prominent in diurnal ocypodids, such as *Uca* spp. (Salmon, 1965; Crane, 1975; Hyatt, 1977), *Dotilla* spp. (Altevogt, 1957), *Scopimera globosa* (Yamaguchi *et al.,* 1979), *Hemiplax latifrons,* and *Heloecius cordiformis* (Griffin, 1968). Males signal to females by species-typical patterns of chela movements. Nocturnal species may utilize acoustic and tactile signals, as exemplified by the stridulatory and drumming sounds produced by *Ocypode* and *Uca,* respectively (Salmon, 1965; Horch, 1975). Little work has been reported on olfactory communication in land crabs, although some workers (e.g., Crane, 1975, for *Uca*) have speculated on its potential as a social factor. Courtship signaling may be provided by structural additions to the burrows of male ocypodids, such as the "Sandpyramides" of *Ocypode saratan* (Linsenmair, 1967) and the hoods or shelters of various species of *Uca* (Crane, 1975; Zucker, 1978).

Copulation may be preceded by elaborate, ritualized displays (e.g., those of *Uca*) or it may occur without apparent preliminaries, as in many grapsids (Broekhuysen, 1941; Knudsen, 1964). Males of *Sesarma reticulatum* approach and leap onto conspecific females (and occasionally, onto other males), grasp them across the carapace, and attempt copulation. Receptive females usually remain passive, whereas males and nonreceptive females struggle and dislodge the male (L. W. Powers, unpublished observations). Some tropical species of *Uca* engage in similar copulatory interactions on the surface, near the female's burrow (Crane, 1975). Other species rely on exchanges of signals between males and females, such as the circling movements of the chela of *Goniopsis cruentata* (Schöne and Schöne, 1963). Females of some fiddler crabs have specific displays to indicate rejection of a male's attempts at courtship (Zucker and Denny, 1979; Powers, 1979). Species with elaborate rituals of courtship perform them in close proximity to a burrow or other refuge, and copulation usually occurs within the burrow or refuge. Species that engage in rapid copulation and minimal courtship display can avoid prolonged exposure to predators even though they do not utilize refuges for mating. In summary, courtship repertoires of land crabs tend to be more complex than are those of aquatic species, and diurnal species generally possess the most advanced forms of display.

b. Agonistic Interactions. Territorial defense is best developed in diurnal ocypodids that occupy semipermanent burrows. In *Uca,* such a burrow may be an important resource as a refuge and an arena for courtship (Powers, 1975a; Greenspan, 1980). Since time and energy are required for excavation, defense of the burrow is not unexpected. However, abandonment of a successfully defended burrow is not uncommon; burrows acquired by victorious combatants ("aggressive wanderers" of Crane, 1975) may be occupied for only a brief interval, then abandoned (Powers, 1975a). A surplus of empty burrows exists in most *Uca* habitats; thus, the value of contesting for occupied burrows is not fully understood (however, see Christy, 1978; Greenspan, 1980).

TABLE VII

Studies of Agonistic Behavior in Terrestrial Crabs

Species	Subject of study	References
Coenobitidae		
Birgus latro	Threat displays	Helfman, 1977b
Mictyridae		
Mictyrus longicarpus	Burrow defense by males	Cameron, 1966
Gecarcinidae		
Cardisoma guanhumi	Territorial combat by males	Henning, 1975a
Gecarcinus lateralis	Agonistic acoustic signaling	Klaassen, 1973
Grapsidae		
Aratus pisoni	Dominance hierarchies; displays	Warner, 1970
Goniopsis cruentata	Threat displays	Warner, 1970
Helice crassa	Ritualized combat	Beer, 1959
Metapograpsus thukuhar	Dominance hierarchy	Alexander *et al.*, 1973
Ocypodidae		
Dotilla mictyroides	Aggression by wandering males	Tweedie, 1950
Heloecius cordiformis	Burrow defense by males	Griffin, 1968
Hemiplax hirtipes	Male combat rituals	Beer, 1959
Hemiplax latifrons	Burrow defense by males	Griffin, 1968
Ilyoplax pusillus	Burrow defense in fixed territories	Ono, 1965
Macrophthalmus japonicus	Weak territorial defense	Ono, 1965
Ocypode ceratophthalmus	Aggressive encounters; territoriality	Hughes, 1966; Lighter, 1974
Ocypode kuhlii	Aggressive encounters	Jones, 1972
Deiratonotus cristatum	Weak territorial defense	Ono, 1965
Scopimera globosa	Burrow defense	Harada and Kawanabe, 1955; Ono, 1965
Scopimera inflata	Feeding territories, burrow defense	Fielder, 1970
Uca arcuata	Defense of fixed territories	Ono, 1965
Uca lactea	Defense of fixed territories	Ono, 1965
Uca maracoani	Intermale combat rituals	Crane, 1958
Uca musica terpsichores	Burrow defense; female display to reject courting males	Zucker and Denny, 1979; Zucker, 1974, 1977
Uca panacea	Dominance hierarchy; burrow defense	Powers, 1975a

TABLE VII *(Continued)*

Species	Subject of study	References
Uca pugilator	Quantitation of male combat; agonistic displays by females	Hyatt and Salmon, 1978, 1979; Powers, 1979
Uca pugnax	Quantitation of male combat; burrow defense	Hyatt and Salmon, 1978, 1979; Miller, 1961
Uca rapax	Intermale combat ritualization	Crane, 1967

Combat between males and territorial behavior have been described for a large number of terrestrial crabs (Table VII), but quantitative data have been relatively rare (see Hyatt and Salmon, 1978, 1979). In *Uca pugilator*, combat between a male resident of a burrow and a male wanderer frequently attracts other nearby residents (Powers, unpublished observations). This most often occurs when the intruding male is very aggressive and reaches into the resident's burrow with his large cheliped or when he attempts to dig the resident out. Neighboring males usually interrupt the attempted displacement and often engage the intruder, eventually forcing his withdrawal. Attention may then be directed to the original defending resident, as one or more neighbors briefly interact with threats or "chela grapples," then withdraw to their respective burrows. It is unlikely that such behavior represents communal defense. Yet, as a result, residents are less apt to be displaced from their burrows by larger or more aggressive males that intrude, and fighting is less intense and shorter in duration; this leaves more time for courtship and feeding. Females tend to avoid males that are actively engaged in combat; thus, interruption may be adaptive for nearby courting males. Male individuals of *Uca musica terpsichores* push sand into the burrows of other males that burrow too closely (Zucker, 1977), and males of other species of *Uca* may deliberately destroy the hoods of displaying conspecific males (Crane, 1975).

Nocturnal species generally have less complex agonistic behaviors, yet some may be very aggressive (e.g., species of *Ocypode*) while others are relatively passive (e.g., species of *Gecarcinus*). Agonistic encounters are more ritualized in terrestrial than in aquatic species of Brachyura. Serious injury rarely occurs in intermale combat of grapsids and ocypodids, whereas aquatic portunids and xanthids may inflict considerable damage on the carapace of an opponent or may cause the opponent to autotomize an appendage (Schöne, 1968).

c. Other Social Activities. Cooperative behavior, either as division of labor or as role specialization, is lacking in crustaceans. Minor exceptions include the protection offered to premolt females by males, prior to copulation, in some aquatic species of crabs (Hartnoll, 1969), and the passive parental care rendered by a few species of decapods and isopods that carry their newly-hatched young (e.g., the freshwater crab genus, *Kinsleya:* see Warner, 1977, p. 139). Although complex social behavior is apparently confined to decapods, at least one species of isopod lives in family groups, sharing burrows with overlapping generations (Linsenmair, 1972). A species of Japanese *Ligia* moves to and from intertidal feeding areas as a cohesive group (Ondo and Mori, 1956). The "trekking" behavior of soldier crabs, *Mictyris longicarpus,* appears to depend on social interactions (Cameron, 1966), although the significance and adaptive advantages of this behavior are not fully understood. *Uca* also displays a variety of group movements involving both sexes; these movements may only partially be related to feeding.

The behavioral repertoires of land crabs should be of interest to sociobiologists attempting to correlate environmental factors with the evolution of social systems. Fiddler crabs are a relatively recent group, and their social behavior and terrestrial adaptations appear to be coevolving rapidly (Crane, 1975). The social behavior of these crustaceans may be comparable to the primitive levels of sociality of semisocial insects and spiders (see Wilson, 1971); thus, these interactions may be precursors to true social behavior.

B. Interspecific Competition

Exploitation of a new environment involves physiological adjustments to physical factors and behavioral adaptations to the presence of other species: prey, predators, and organisms that compete directly or indirectly for various resources. Compared with many other arthropods, crustaceans exhibit relatively diverse feeding habits; many species are opportunistic and feed on whatever is available. Among the specialized crustaceans with narrower food habits, food materials are frequently abundant and competition is probably minimal. For example, intertidal deposit feeders (e.g., grapsid and ocypodid crabs) have a food source that is periodically refurbished and is virtually unlimited (Miller, 1961; Warner, 1969; Crane, 1975); leaf litter and algae are relatively abundant in those habitats that contain them (see Hurley, 1968, for a discussion of the leafmold environment).

Refuge space, be it in burrow, in crevice, or under rocks, is often shared with other animals. A variety of insect symbionts can be found in the burrows of many species of land crabs (Bright and Hogue, 1972), although opportunistic predators, such as *Uca* and *Ocypode,* will occasionally eat arthropods and worms that they encounter (Powers, 1975a).

A variety of animals prey on land crabs and other terrestrial crustaceans, but few seem to be specialized. Racoons, rails, egrets, and gulls make sporadic forays on colonies of fiddler crabs (L. W. Powers, unpublished observations); in Bermuda, night herons feed nocturnally on land crabs. The largely anecdotal literature on known and suspected predators cannot be reviewed here. The success of land crabs, anomuran or brachyuran, on many oceanic islands has been attributed to the lack of predation and competition by mammals, especially rats (Johnson, 1965); on many Pacific atolls, coconut crabs are the largest nonavian animals present (Wiens, 1962; Johnson, 1965). For most species in most habitats, however, the major factors limiting density and distribution appear to be mainly physical: i.e., the type of substrate and amount of water present, the availability of standing water or refuges of high humidity, and the presence of suitable vegetation for shade or food or both. Interactions with other species are usually sporadic and infrequent; mutual avoidance is the most common result of potential encounters.

VI. CONCLUSIONS

A. Transition from Sea to Land

As a group, crustaceans are generally ill-equipped for life on land. Most species exist by avoiding typically terrestrial conditions, either by remaining close to aquatic habitats or by occupying moist microhabitats. Unlike other arthropods, crustaceans have not fully exploited ecological opportunities with respect to food specialization; thus their niches are relatively broad within the nutritional constraints of their habitats. Exceptions exist, such as the scatophagous isopods and the bromeliad-dwelling grapsids noted above.

The routes of terrestrial invasion have been the subject of continuing discussion. Because of the greater variety and variability of freshwater habitats, which consequently provide greater opportunities for serial adaptations, Carter (1931) believed that the majority of terrestrial animals reached land via fresh water rather than via the marine supralittoral zone with its more abrupt transitions. This concept may not apply to amphipods and isopods; Hurley (1959, 1968) noted that terrestrial genera include supralittoral species, while freshwater genera are more closely related to marine forms than to supralittoral and terrestrial species. Pearse (1929) also supported the hypothesis of a route through the littoral zone, because of the relatively primitive features of genera like *Ligia* and *Halophiloscia*. Evidence from brachyurans shows considerable diversity in physiological adaptations, especially with regard to salt and water balance (Bliss, 1968, 1979; Mantel, 1979). Hartnoll (1971) proposed a scheme for the subgenus *Sesarma (Sesarma)* in Jamaica, in which a marine stock gave rise to freshwater forms that subsequently invaded land and also led to the specialized bromeliad genus, *Metopaulias*. A cave-dwelling species may have arisen from either freshwater or terrestrial ancestors. Ocypodids are clearly marine in origin, but some are specialized inhabitants of marine supralittoral zones, while others have penetrated estuaries and brackish water to reside in marshes with nearly fresh water (e.g., *Uca minax*). The larvae of such species must tolerate a broad range of salinities as they mature while moving seaward and then reverse their exposure on return to the marsh as megalops and young adults. Freshwater crabs and crayfish remain in their specific habitat throughout life; the former are well adapted to terrestrial existence, retain their young offspring, and represent ideal candidates as intermediates for truly terrestrial species. However, not all freshwater crabs are derived from estuarine-penetrating ancestors; physiological evidence from a species of *Sudanonautes* (family Potamonautidae) indicates that it may have been directly adapted from a marine environment (Lutz, 1969). Lutz thus proposes two groups: freshwater/land crabs and saltwater/land crabs, terms that have also been used by Bliss (1979) to discuss transitions in *Gecarcinus*.

B. Relative Success of Major Crustacean Groups

Preadaptations to land have been significant for all groups of Crustacea: isopods and amphipods retain their young and exhibit the same feeding and cryptozoic habits on land as in the water. Isopods and decapods are relatively mobile and adapted for running and climbing. Decapods exhibit great manueverability and rapid responses to potential threats. Few significant advances have been made by terrestrial crustaceans. Quantitative comparative studies are lacking, but the visual acuity of diurnal ocypodids is probably superior to that of aquatic crabs. Some decapods have, *de novo*, evolved the ability to retain postzoeal offspring (a few aquatic species release eggs with developmentally advanced zoeas, but they are unrelated to the terrestrial species), which provides independence from standing water for development. Amphipods are the most restricted of the three major groups, in terms of variety of habitat and ability to survive dehydrating conditions. Isopods inhabit some of the most xeric conditions, such as the Sahara desert (Cloudsley-Thompson, 1977); but they are still restricted to burrows and nocturnal activity. Decapods have an advantage in larger size and less exposed surface area-to-volume ratio, plus the ability to dig deeper burrows and reach ground water and areas of lower temperature.

All of these terrestrial species suffer to some degree in having inefficient capabilities for water retention; this is probably the single most important factor in limiting their success on land. Terrestrial hermit crabs that retain shells (*Coenobita*) have solved this problem by carrying water with them, but their maximal sizes are limited by the availability of suitable gastropod shells. *Birgus* is many times larger than *Coenobita* because of a more efficient epicuticle and respiratory system that allows it to abandon shells after 2 years of growth. Within each of the three major groups of terrestrial Crustacea, a number of different adaptive transitions and experiments are in progress, each successful to varying degrees in different habitats. The next ten-million years should be a challenging period for these animals and for their scientific observers.

ACKNOWLEDGEMENT

We appreciate the effort and time of Dr. Eric Edney and Dr. Linda Mantel in providing critical comments on our manuscript. The extensive resources of the library of the American Museum of Natural History were used frequently; we thank the staff for their help and patience. Very special thanks are due Dr. Thomas Wolcott for comments and for allowing us to include several of his unpublished observations. Finally, the senior author wishes to acknowledge the help, friendship, and countless contributions of his students: Doris Bull, Andrea Fochios, Jean Fruci, and Olivia Janovitz.

REFERENCES

Abele, L. G. (1970). Semi-terrestrial shrimp (*Merguia rhizophorae*). *Nature (London)* **226,** 661–662.

Abele, L. G. (1972). A note on the Brazilian bromeliad crabs (Crustacea, Grapsidae). *Arq. Cienc. Mar.* **12,** 123–126.

Abele, L. G., and Means, D. B. (1977). *Sesarma jarvisi* and *Sesarma cookei:* Montane, terrestrial crabs in Jamaica (Decapoda). *Crustaceana* **32,** 91–93.

Alexander, A. J., Caiger, K. M., and Smith, M. G. (1973). Preliminary investigations of the social relationships of a mud-flats crab, *Metopograpsus thukuar* (Owen). *S. Afr. J. Sci.* **69,** 87–89.

Altevogt, R. (1957). Beitrage zur Biologie und Ethologie von *Dotilla blanfordi* (Alcock) und *Dotilla myctiroides* (Milne-Edwards) (Crustacea Decapoda). *Z. Morphol. Oekol. Tiere* **46,** 369–388.

Altevogt, R. (1976). Feeding apparatus and feeding process in Indian fiddler crabs: a SEM-study. *In* German Scholars on India. *Contributions to Indian Studies,* **2,** 1–13.

Barnes, T. C. (1932). Salt requirements and space orientation of the littoral isopod *Ligia* in Bermuda. *Biol. Bull. (Woods Hole, Mass.)* **63,** 496–504.

Barnwell, F. H. (1968). The role of rhythmic systems in the adaptation of fiddler crabs to the intertidal zone. *Am. Zool.* **8,** 569–583.

Beer, C. G. (1959). Notes on the behaviour of two estuarine crab species. *Trans. R. Soc. N. Z.* **86,** 197–203.

Bishop, J. A. (1967). The zoogeography of the Australian freshwater decapod Crustacea. *In* "Australian Inland Waters and their Fauna" (A. H. Weatherley, ed.), pp. 107–122. Aust. Natl. Univ. Press, Canberra.

Bliss, D. E. (1963). The pericardial sacs of terrestrial Brachyura. *Mus. Comp. Zool., Spec. Publ.* pp. 59–78.

Bliss, D. E. (1968). Transition from water to land in decapod crustaceans. *Am. Zool.* **8,** 355–392.

Bliss, D. E. (1979). From sea to tree: Saga of a land crab. *Am. Zool.* **19,** 385–410.

Bliss, D. E., and Boyer, J. R. (1964). Environmental regulation of growth in the decapod crustacean *Gecarcinus lateralis. Gen. Comp. Endocrinol.* **4,** 15–41.

Bliss, D. E., and Mantel, L. H. (1968). Adaptations of crustaceans to land: A summary and analysis of new findings. *Am. Zool* **8,** 673–685.

Bliss, D. E., Wang, S. M. E., and Martinez, E. A. (1966). Water balance in the land crab, *Gecarcinus lateralis,* during the intermolt cycle. *Am. Zool.* **6,** 197–212.

Bliss, D. E., van Montfrans, J., van Montfrans, M., and Boyer, J. R. (1978). Behavior and growth of the land crab *Gecarcinus lateralis* (Fréminville) in southern Florida. *Bull. Am. Mus. Nat. Hist.* **160,** 111–152.

Borradaile, L. A. (1903). Land crustaceans. *In* "The Fauna and Geography of the Maldive and Laccadive Archipelagoes" (J. S. Gardiner, ed.), Vol. I, pp. 64–100. Cambridge Univ. Press, London and New York.

Bousfield, E. L. (1968). Discussion session I: Transition to land. *Am. Zool.* **8,** 393–395.

Bovjberg, R. V. (1960). Courtship behavior of the lined shore crab, *Pachygrapsus crassipes* Randall. *Pac. Sci.* **14,** 421–422.

Bright, D. B., and Hogue, C. L. (1972). A synopsis of the burrowing land crabs of the world and list of their arthropod symbionts and burrow associates. *Los Angeles County Mus. Contrib. Sci.* **220,** 1–58.

Broekhuysen, G. J., Jr. (1941). The life-history of *Cyclograpsus punctatus* M. Edw.: Breeding and growth. *Trans. R. Soc. S. Afr.* **28,** 331–366.

Cameron, A. M. (1966). Some aspects of the behaviour of the soldier crab, *Mictyris longicarpus. Pac. Sci.* **20,** 224–234.

Cameron, J. N. (1975). Aerial gas exchange in the terrestrial Brachyura *Gecarcinus lateralis* and *Cardisoma guanhumi. Comp. Biochem. Physiol.* **52A,** 129–134.

Cameron, J. N., and Mecklenburg, T. A. (1973). Aerial gas exchange in the coconut crab, *Birgus latro* with some notes on *Gecarcoidea lalandii. Respir. Physiol.* **19,** 245–261.

Carter, G. S. (1931). The fauna of the swamps of the Paraguayan Chaco in relation to its environment. II. Respiratory adaptations in the fishes. *J. Linn. Soc. London, Zool.* **37,** 327–368.

Christy, J. H. (1978). Adaptive significance of reproductive cycles in the fiddler crab *Uca pugilator:* A hypothesis. *Science* **199,** 453–455.

Clark, E. (1936a). Notes on the habits of land crayfishes. *Victoria Nat.* **53,** 65–68.

Clark, E. (1936b). The freshwater and land crayfishes of Australia. *Mem. Natl. Mus. Victoria, Melbourne* **10,** 5–58.

Cloudsley-Thompson, J. L. (1952). Studies in diurnal rhythms. II. Changes in the physiological responses of the woodlouse *Oniscus asellus* to environmental stimuli. *J. Exp. Biol.* **29,** 295–303.

Cloudsley-Thompson, J. L. (1956). Studies in diurnal rhythms. VII. Humidity responses and nocturnal activity in woodlice (Isopoda). *J. Exp. Biol.* **33,** 576–582.

Cloudsley-Thompson, J. L. (1977). "The Water and Temperature Relations of Woodlice." Meadowfield Press Ltd., Durham, England.

Costlow, J. D., Jr., and Bookhout, C. G. (1968). The effect of environmental factors on development of the land-crab, *Cardisoma guanhumi*. *Am. Zool.* **8,** 399–410.

Cowles, R. P. (1908). Habits, reactions and associations of *Ocypode arenaria*. *Pap. Tortugas Lab.* **2,** 1–41.

Crane, J. (1944). On the color changes of fiddler crabs (genus *Uca*) in the field. *Zoologica (N.Y.)* **29,** 161–168.

Crane, J. (1958). Aspects of social behavior in fiddler crabs, with special reference to *Uca maracoani* (Latreille). *Zoologica (N.Y.)* **43,** 113–130.

Crane, J. (1967). Combat and its ritualization in fiddler crabs (Ocypodidae) with special reference to *Uca rapax* (Smith). *Zoologica (N.Y.)* **52,** 49–76.

Crane, J. (1975). "Fiddler Crabs of the World. Ocypodidae: Genus Uca." Princeton Univ. Press, Princeton, New Jersey.

Crocker, D. W., and Barr, D. W. (1968). "Handbook of the Crayfishes of Ontario," Misc. Publ. R. Ontario Mus. Univ. of Toronto Press, Toronto.

Den Boer, P. J. (1961). The ecological significance of activity patterns in the woodlouse, *Porcellio scaber* Latr. (Isopoda). *Arch. Neerl. Zool.* **14,** 283–409.

de Wilde, P. A. W. J. (1973). On the ecology of *Coenobita clypeatus* in Curaçao with reference to reproduction, water economy, and osmoregulation in terrestrial hermit crabs. *Nat. Stud. Suriname Ned. Antil.* **76,** 1–138.

Diaz, H., and Rodriguez, G. (1977). The branchial chamber in terrestrial crabs: A comparative study. *Biol. Bull. (Woods Hole, Mass.)* **153,** 485–504.

Dresel, E. I. B., and Moyle, V. (1950). Nitrogenous excretion of amphipods and isopods. *J. Exp. Biol.* **27,** 210–225.

Dubinsky, Z., Steinberger, Y., and Shachak, M. (1979). The survival of the desert isopod *Hemilepistus reaumurii* (Audouin) in relation to temperature (Isopoda, Oniscoidea). *Crustaceana* **36,** 147–154.

Duncan, K. W. (1969). The ecology of two species of terrestrial Amphipoda (Crustacea: Family Talitridae) living in waste grassland. *Pedobiologia* **9,** 323–341.

Eales, A. J. (1974). Sex pheromone release in the shore crab *Carcinus maenas,* and the site of its release from females. *Mar. Behav. Physiol.* **2,** 345–355.

Edney, E. B. (1954). Woodlice and the land habitat. *Biol. Rev. Cambridge Philos. Soc.* **29,** 185–219.

Edney, E. B. (1958). The microclimate in which woodlice live. *Proc. Int. Congr. Entomol. 10th, 1956* Vol. 2, pp. 709–712.

Edney, E. B. (1960). Terrestrial adaptations. *In* "The Physiology of Crustacea" (T. H. Waterman, ed.), Vol. 1, pp. 367–393. Academic Press, New York.

Edney, E. B. (1961). The water and heat relationships of fiddler crabs (*Uca spp.*). *Trans. R. Soc. S. Afr.* **36,** 71–91.

Edney, E. B. (1968). Transition from water to land in isopod crustaceans. *Am. Zool.* **8,** 309–326.

Edney, E. B. (1977). "Water Balance in Land Arthropods," Ser. Zoophysiol. Ecol., Vol. 9, Springer-Verlag, Berlin and New York.

Edney, E. B., and Spencer, J. O. (1955). Cutaneous respiration in woodlice. *J. Exp. Biol.* **32,** 256–269.

Ehrhardt, J. P., and Niaussat, P. (1970). Ecologie et physiologie du brachyoure terrestre *Gecarcinus planatus* Stimpson (d'après le individus de l'atoll de Clipperton). *Bull. Soc. Zool. Fr.* **95,** 41–54.

Fales, R. R. (1976). Apparent predation on the mole crab *Emerita talpoida* (Say) by the ghost crab *Ocypode quadrata* (Fabricius). *Chesapeake Sci.* **17,** 65.

Farrow, G. E. (1971). Back-reef and lagoonal environments of Aldabra Atoll distinguished by their crustacean burrows. *Symp. Zool. Soc. London* **28,** 455–500.

Feest, J. (1969). Morphophysiologische Untersuchungen zur Ontogenese und Fortpflanzungsbiologie von *Uca annulipes* und *Uca triangularis* mit Vergleichsbefunden an *Ilyoplax gangetica*. *Forma Functio* **1,** 159–225.

Fellows, D. P. (1973). Behavioral ecology of the ghost crabs *Ocypode ceratophthalmus* and *Ocypode cordimana* at Fanning Atoll, Line Islands. *Hawaii Inst. Geophys. [Rep.] HIG* **HIG-73-13,** 219–241.

Fielder, D. R. (1970). The feeding behavior of the sand crab *Scopimera inflata* (Decapoda, Ocypodidae). *J. Zool.* **160,** 35–49.

Fimpel, E. (1975). Phänomene der Landadaptation bei terrestrischen und semiterristrichen Brachyura der brasilianischen Küste (Malacostraca, Decapoda). *Zool. Jahrb., Abt. Syst. (Oekol.), Geogr. Biol.* **102,** 173–214.

Fingerman, M., Lowe, M. E., and Mobberley, W. C., Jr. (1958). Environmental factors involved in setting the phases of activity of tidal rhythm of color change in the fiddler crabs *Uca pugilator* and *Uca minax*. *Limnol. Oceanogr.* **3,** 271–282.

Frey, D. G. (1980). The non-swimming chydorid Cladocera of wet forests, with descriptions of a new genus and two new species. *Int. Rev. Ges. Hydrobiol.* **65,** 613–641.

Friedlander, C. P. (1963). Thigmokinesis in woodlice. *Anim. Behav.* **12,** 164–174.

Friedlander, C. P. (1965). Aggregation in *Oniscus asellus* Linn. *Anim. Behav.* **13,** 342–346.

Gangotri, M. S., Vasantha, N., and Venkatachari, S. A. T. (1977). Sexual maturity and breeding behaviour in the freshwater crab, *Barytelphusa guerini* Milne Edwards. *Proc.—Indian Acad. Sci., Sect. B* **87,** 195–201.

Gibson-Hill, C. A. (1947). Field notes on the terrestrial crabs. *Bull. Raffles Mus.* **18,** 43–53.

Gifford, C. A. (1962). Some observations on the general biology of the land crab, *Cardisoma guanhumi* (Latreille), in South Florida. *Biol. Bull. (Woods Hole, Mass.)* **123,** 207–223.

Gifford, C. A. (1968). Accumulation of uric acid in the land crab, *Cardisoma guanhumi*. *Am. Zool.* **8,** 521–528.

Gorvett, H. (1956). Tegumental glands and terrestrial life in woodlice. *Proc. Zool. Soc. London* **126,** 291–314.

Graf, F. (1978). Les sources de calcium pour les crustacés venant de muer. *Arch. Zool. Exp. Gen.* **119,** 143–161.

Gray, I. E. (1957). A comparative study of the gill area of crabs. *Biol. Bull. (Woods Hole, Mass.)* **112,** 34–42.

Greenaway, P., and MacMillen, R. E. (1978). Salt and water balance in the terrestrial phase of the inland crab *Holthuisana (Austrothel phusa) transversa* Martens (Parathelphusoidea: Sundathelphusidae). *Physiol. Zool.* **51,** 217–229.

Greenaway, P., and Taylor, H. H. (1976). Aerial gas exchange in Australian arid-zone crab, *Parathelphusa transversa* Von Martens. *Nature (London)* **262,** 711–713.

Greenspan, B. (1980). Male size and reproductive success in the communal courtship system of the fiddler crab, *Uca rapax*. *Anim. Behav.* **28,** 387–392.

Griffin, D. J. G. (1968). Social and maintenance behaviour in two Australian ocypodid crabs (Crustacea: Brachyura). *J. Zool.* **156,** 291–305.

Gross, W. J. (1964). Trends in water and salt regulation among aquatic and amphibious crabs. *Biol. Bull. (Woods Hole, Mass.)* **127,** 447–466.

Gross, W. J., Lasiewski, R. C., Dennis, M., and Rudy, P., Jr. (1966). Salt and water balance in selected crabs of Madagascar. *Comp. Biochem. Physiol.* **27,** 641–660.

Grubb, P. (1971). Ecology of terrestrial decapod crustaceans on Aldabra. *Phil. Trans. R. Soc., Ser. B* **260,** 411–416.

Hafemann, D. H., and Hubbard, J. L. (1969). On the rapid running of ghost crabs (*Ocypode ceratophthalma*). *J. Exp. Zool.* **170,** 25–31.

Hale, H. M. (1925). The development of two Australian sponge crabs. *Proc. Linn. Soc. N. S. W.* **50,** 405–413.
Harada, E., and Kawanabe, H. (1955). The behaviour of the sand-crab, *Scopimera globasa* de Haan, with special reference to the problems of coaction between individuals. *Jpn. J. Ecol.* **4,** 162–165.
Harding, J. P. (1953). The first known example of a terrestrial ostracod, *Mesocypris terrestris* sp. nov. *Ann. Natal Mus.* **12,** 359–365.
Harms, J. W. (1929). Die realisation von Genen und die Consecutive Adaptation. I. Phasen in der Differenzierung der Anlagenkomplexe und die Frage der Landtierwerdung. *Z. Wiss. Zool.* **133,** 211–397.
Harms, J. W. (1932). Die realisation von Genen und die Consecutive Adaptation. II. *Birgus latro* L. als Landkrebs und seine Beziehungen zu den Coenobiten. *Z. Wiss. Zool.* **140,**167–290.
Hartnoll, R. G. (1964). The freshwater grapsid crabs of Jamaica. *Proc. Linn. Soc. London* **175,** 145–169.
Hartnoll, R. G. (1965). Notes on the marine grapsid crabs of Jamaica. *Proc. Linn. Soc. London* **176,** 113–147.
Hartnoll, R. G. (1969). Mating in the Brachyura. *Crustaceana* **16,** 161–181.
Hartnoll, R. G. (1971). *Sesarma cookei* n.sp., a grapsid crab from Jamaica (Decapoda, Brachyura). *Crustaceana* **20,** 257–264.
Hartnoll, R. G. (1972). The biology of the burrowing crab, *Corystes cassivelaunus. Bijd. Dierk.* **42,** 139–155.
Hartnoll, R. G. (1973). Factors affecting the distribution and behaviour of the crab *Dotilla fenestrata* on East African shores. *Estuarine Coastal Mar. Sci.* **1,** 137–152.
Hay, W. P., and Shore, C. A. (1918). The decapod crustaceans of Beaufort, N.C., and surrounding region. *Fish. Bull.* **35,** 369–475.
Hazlett, B. A. (1968). Effects of crowding on the agonistic behavior of the hermit crab, *Pagurus bernhardus. Ecology* **49,** 573–574.
Hazlett, B. A. (1972). Shell fighting and sexual behavior in the hermit crab genera *Paguristes* and *Calcinus,* with comments on *Pagurus. Bull. Mar. Sci.* **22,** 806–823.
Helfman, G. S. (1977a). Copulatory behavior of the coconut or robber crab *Birgus latro* (L.) (Decapoda, Anomura, Paguridea, Coenobitidae). *Crustaceana* **33,** 198–202.
Helfman, G. S. (1977b). Agonistic behaviour of the coconut crab, *Birgus latro* (L.). *Z. Tierpsychol.* **43,** 425–438.
Henning, H. G. (1975a). Ökologische, ethologische und sinnesphysio-logische Untersuchungen an der Landkrabbe *Cardisoma guanhumi* Latreille (Decapoda, Brachyura) in Nordkolumbien. *Forma Functio* **8,** 253–304.
Henning, H. G. (1975b). Kampf-, Fortpflanzungs- und Häutungverhalten; Wachstum und Geschlechtsreife von *Cardisoma guanhumi* Latreille. *Forma Functio* **8,** 463–510.
Henning, H. G., and Klaassen, F. (1973). Dekapode Crustaceen auf der Isla de Salamanca (Atlantik-Küste, Kolumbien). *Mitt. Inst. Colombo-Aleman Invest. Cient.* **7,** 63–84.
Herreid, C. F., II (1969a). Water loss of crabs from different habitats. *Comp. Biochem. Physiol.* **28,** 829–839.
Herreid, C. F., II (1969b). Integument permeability of crabs and adaptation to land. *Comp. Biochem. Physiol.* **29,** 423–429.
Herreid, C. F., and Gifford, C. A. (1963). The burrow habitat of the land crab, *Cardisoma guanhumi* (Latreille). *Ecology* **44,** 773–775.
Herrnkind, W. F. (1968a). Adaptive visually-directed orientation in *Uca pugilator. Am. Zool.* **8,** 585–598.
Herrnkind, W. F. (1968b). The breeding of *Uca pugilator* (Bosc) and mass rearing of the larvae with comments on the behavior of the larval and early crab stages (Brachyura, Ocypodidae). *Crustaceana, Suppl.* **2,** 214–224.

Herrnkind, W. F. (1972). Orientation of shore-living arthropods, especially the sand fiddler crab. *In* "Behavior of Marine Animals" (H. E. Winn and B. L. Olla, eds.), Vol. I, pp. 1–59. Plenum, New York.

Hiatt, R. W. (1948). The biology of the lined shore crab, *Pachygrapsus crassipes* Randall. *Pac. Sci.* **2,** 135–213.

Hill, G. W., and Hunter, R. E. (1973). Burrows of the ghost crab *Ocypode quadrata* (Fabricius) on the barrier islands, south-central Texas coast. *J. Sediment. Petrol.* **43,** 24–30.

Holthuis, L. B. (1959). The Crustacea Decapoda of Suriname (Dutch Guiana). *Zool. Verh., Leiden* **44,** 1–296.

Horch, K. (1975). The acoustic behavior of the ghost crab *Ocypode cordimana* Latreille, 1818 (Decapoda, Brachyura). *Crustaceana* **29,** 193–205.

Hughes, D. A. (1966). Behavioural and ecological investigations of the crab *Ocypode ceratophthalmus* (Crustacea: Ocypodidae). *J. Zool.* **150,** 129–143.

Hughes, D. A. (1973). On mating and the "copulation burrows" of crabs of the genus *Ocypode* (Decapoda, Brachyura). *Crustaceana* **24,** 72–76.

Hurley, D. E. (1959). Notes on the ecology and environmental adaptations of the terrestrial Amphipoda. *Pac. Sci.* **13,** 107–129.

Hurley, D. E. (1968). Transition from water to land in amphipod crustaceans. *Am. Zool.* **8,** 327–353.

Hyatt, G. W. (1974). Behavioural evidence for light intensity discrimination by the fiddler crab *Uca pugilator* (Brachyura Ocypodidae). *Anim. Behav.* **22,** 796–801.

Hyatt, G. W. (1975). Physiological and behavioral evidence for colour discrimination by fiddler crabs (Brachyura, Ocypodidae, genus *Uca*). *In* "Physiological Ecology of Estuarine Organisms" (F. J. Vernberg, ed.). Univ. of South Carolina Press, Columbia.

Hyatt, G. W. (1977). Field observations of size-dependent changes in waving display and other behavior in the fiddler crab, *Uca pugilator* (Brachyura, Ocypodidae). *Mar. Behav. Physiol.* **4,** 283–292.

Hyatt, G. W., and Salmon, M. (1978). Combat in the fiddler crab *Uca pugilator* and *U. pugnax:* A quantitative descriptive analysis. *Behaviour* **65,** 182–211.

Hyatt, G. W., and Salmon, M. (1979). Comparative statistical and information analysis of combat in the fiddler crabs, *Uca pugilator* and *U. pugnax. Behaviour* **68,** 1–23.

Jans, D., and Ross, K. F. A. (1963). A histological study of the peripheral receptors in the thorax of land isopods, with special reference to the location of possible hygroreceptors. *Q. J. Microsc. Sci.* [N. S.] **104,** 337–350.

Johansen, K., Lenfant, C., and Mecklenburg, T. A. (1970). Respiration in the crab, *Cancer magister. Z. Vergl. Physiol.* **70,** 1–19.

Johnson, D. S. (1965). Land crabs. *J. Malaya Branch R. Asiat. Soc.* **38,** 43–66.

Jones, D. A. (1972). Aspects of the ecology and behaviour of *Ocypode ceratophthalmus* (Pallas) and *O. kuhlii* de Haan (Crustacea: Ocypodidae). *J. Exp. Mar. Biol. Ecol.* **8,** 31–43.

Klaassen, F. (1973). Stridulation und Komminikation durch Substratschall bei *Gecarcinus lateralis* (Crustacea Decapoda). *J. Comp. Physiol.* **83,** 73–79.

Klaassen, F. (1975). Ökologische und ethologische Untersuchungen zur Fortpflanzungsbiologie von *Gecarcinus lateralis* (Decapoda, Brachyura). *Forma Functio* **8,** 101–174.

Knudsen, J. W. (1960). Reproduction, life history and larval ecology of the California Xanthidae, the pebble crabs. *Pac. Sci.* **14,** 3–17.

Knudsen, J. W. (1964). Observations of the reproductive cycles and ecology of the common Brachyura and crab-like Anomura of Puget Sound, Washington. *Pac. Sci.* **18,** 3–33.

Koepcke, H. W., and Koepcke, M. (1953). Contribucion al conocimiento de la forma de vida de *Ocypode gaudichaudii* Milne Edwards & Lucas (Decapoda, Crust.). *Publ. Mus. Hist. Nat. (Lima), Ser. A., Zool.* **13,** 1–46.

Kuenen, D. J., and Nooteboom, H. P. (1963). Olfactory orientation in some land isopods (Oniscoidea, Crustacea). *Entomol. Exp. Appl.* **6,** 133–142.

Larimer, J. L. (1961). Measurement of ventilation volume in decapod crustaceans. *Physiol. Zool.* **34,** 158–166.

Lighter, F. J. (1974). A note on a behavioral spacing mechanism of the ghost crab *Ocypode ceratophthalmus* (Pallas) Decapoda, family Ocypodidae). *Crustaceana* **27,** 312–314.

Lindberg, W. J. (1976). Reproductive ecology of a shore crab: an evolutionary analysis. *Biologist* **58,** 61–66.

Linsenmair, K. E. (1967). Konstruktion und Signalfunktion der Sandpyramide der Reiterkrabbe *Ocypode saratan* Försk. (Decapoda Brachyura Ocypodidae). *Z. Tierpsychol.* **24,** 403–456.

Linsenmair, K. E. (1972). The importance of family-specific "badges" for the cohesion of families in the social desert woodlouse, *Hemilepistus reaumurii. Z. Tierpsychol.* **31,** 131–162.

Lutz, P. L. (1969). Salt and water balance in the West African fresh-water/land crab *Sudanonautes africanus africanus* and the effects of desiccation. *Comp. Biochem. Physiol.* **30,** 469–480.

McCann, C. (1938). Notes on the common land crab, *Paratelphusa (Barytelphusa) guerini* of Salsette Island. *J. Bombay Nat. Hist. Soc.* **39,** 531–542.

Macintosh, D. J. (1978). Some responses of tropical mangrove fiddler crabs (*Uca* spp.) to high environmental temperatures. *In* "Physiology and Behaviour of Marine Organisms" (D. S. McLusky and A. J. Berry, eds.), pp. 49–56. Pergamon, Oxford.

McIntyre, R. J. (1954). A common sand-hopper. M.Sc. Thesis, Canterbury University, Christchurch, New Zealand.

McMahon, B. R., and Burggren, W. W. (1979). Respiration and adaptation to the terrestrial habitat in the land hermit crab *Coenobita clypeatus. J. Exp. Biol.* **79,** 265–281.

MacMillen, R. E., and Greenaway, P. (1978). Adjustments of energy and water metabolism to drought in an Australian arid-zone crab. *Physiol. Zool.* **51,** 230–240.

Mantel, L. H. (1968). The foregut of *Gecarcinus lateralis* as an organ of salt and water balance. *Am. Zool.* **8,** 433–442.

Mantel, L. H. (1979). Terrestrial invertebrates other than insects. *In* "Comparative Physiology of Osmoregulation in Animals" (G. M. O. Maloiy), Vol 1, pp. 175–218. Academic Press, New York.

Mantel, L. H., Bliss, D. E., Sheehan, S. W., and Martinez, E. A. (1975). Physiology of hemolymph, gut fluid, and hepatopancreas of the land crab *Gecarcinus lateralis* (Fréminville) in various neuroendocrine states. *Comp. Biochem. Physiol.* **51A,** 663–671.

Manton, S. M. (1977). "The Arthropoda: Habits, Functional Morphology, and Evolution." Oxford Univ. Press (Claredon), London and New York.

Marshall, S. M., and Orr, A. P. (1960). Feeding and nutrition. *In* "The Physiology of Crustacea" (T. H. Waterman, ed.), Vol. 1, pp. 227–258. Academic Press, New York.

Mason, C. A. (1970). Function of the pericardial sacs during the molt cycle in the land crab *Gecarcinus lateralis. J. Exp. Zool.* **174,** 381–390.

Miller, D. C. (1961). The feeding mechanism of fiddler crabs, with ecological considerations of feeding adaptations. *Zoologica (N.Y.)* **46,** 89–100.

Miller, D. C., and Vernberg, F. J. (1968). Some thermal requirements of fiddler crabs of the temperate and tropical zones and their influence on geographic distribution. *Am. Zool* **8,** 459–469.

Miller, M. A. (1938). Comparative ecological studies of the terrestrial isopod Crustacea of the San Francisco Bay region. *Univ. Calif., Berkeley, Publ. Zool.* **43,** 113–142.

Milne, L. J., and Milne, M. J. (1946). Notes on the behavior of the ghost crab. *Am. Nat.* **80,** 362–380.

Montgomery, S. K. (1922). Direct development in a dromiid crab. *Proc. Zool. Soc. London* **1922,** 193–196.
Nicholls, A. G. (1931). Studies on *Ligia oceanica*. II. The processes of feeding, digestion and absorption with a description of the structure of the foregut. *J. Mar. Biol. Assoc. U.K.* **17,** 675–708.
Nye, P. A. (1977). Reproduction, growth, and distribution of the grapsid crab *Helice crassa* (Dana, 1851) in the southern part of New Zealand. *Crustaceana* **33,** 75–89.
Ondo, Y., and Mori, S. (1956). Periodic behavior of the shore isopod *Megaligia exotica* (Roux). I. Observations under natural conditions. *Jpn. J. Ecol.* **5,** 161–167.
Ono, Y. ((1965). On the ecological distribution of ocypoid crabs in the estuary. *Mem. Fac. Sci., Kyushu Univ., Ser. E.* **4,** 1–60.
Palmer, J. D. (1964). A persistent, light-preference rhythm in the fiddler crab, *Uca pugnax* and its possible adaptive significance. *Am. Nat.* **98,** 431–434.
Palmer, J. D. (1971). Comparative studies of circadian locomotory rhythms in four species of terrestrial crabs. *Am. Midl. Nat.* **85,** 97–107.
Palmer, J. D. (1973). Tidal rhythms: The clock control of the rhythmic physiology of marine organisms. *Biol. Rev. Cambridge Philos. Soc.* **48,** 377–418.
Paris, O. H. (1963). The ecology of *Armadillidium vulgare* (Isopoda: Oniscoidea) in California grassland: Food, enemies and weather. *Ecol. Monogr.* **33,** 1–22.
Paris, O. H., and Sikora, A. (1965). Radiotracer demonstration of isopod herbivory. *Ecology* **46,** 729–734.
Passano, L. M. (1960). Low temperature blockage of molting in *Uca pugnax*. *Biol. Bull. (Woods Hole, Mass.)* **118,** 129–136.
Pearse, A. S. (1911). On the habits of *Thalassina anomala* (Herbst). *Philipp. J. Sci.* **6,** 213–217.
Pearse, A. S. (1929). Observations on certain littoral and terrestrial animals at Tortugas, Florida, with several references to migrations from marine to terrestrial habitats. *Pap. Tortugas Lab.* **26,** 205–223.
Pillay, K. K., and Ono, Y. (1978). The breeding cycles of two species of grapsid crabs (Crustacea: Decapoda) from the north coast of Kyushu, Japan. *Mar. Biol. (Berlin)* **45,** 237–248.
Powers, L. W. (1975a). The fiddler crab burrow: A study in behavioral ecology. Doctoral Dissertation, University of Texas, Austin.
Powers, L. W. (1975b). Fiddler crabs in a nontidal environment. *Contrib. Mar. Sci.* **19,** 67–78.
Powers, L. W. (1979). Agonistic behavior in female fiddler crabs. *Uca pugnax* and *U. pugilator*. *Am. Zool.* **19,** 913.
Powers, L. W., and Cole, J. F. (1976). Temperature variation in fiddler crab microhabitats. *J. Exp. Mar. Biol. Ecol.* **21,** 141–158.
Quinn, R. H. (1980). Mechanisms for obtaining water for flotation feeding in the soldier crab, *Mictyris longicarpus* Latreille, 1806. *J. Exp. Mar. Biol. Ecol.* **43,** 49–60.
Rao, K. R. (1968). The pericardial sacs of *Ocypode* in relation to the conservation of water, molting and behaviour. *Am. Zool.* **8,** 561–567.
Rathbun, M. J. (1900). Results of the Branner-Agassiz Expedition to Brazil. 1. The decapod and stomatopod Crustacea. *Proc. Wash. Acad. Sci.* **1900,** 133–156.
Rathbun, M. J. (1914). Stalk-eyed crustaceans collected at the Monte Bello Islands. *Proc. Zool. Soc. London* pp. 653–664.
Reese, E. S. (1968). Shell use: An adaptation for emigration from the sea by the coconut crab. *Science* **161,** 385–386.
Reese, E. S. (1969). Behavioral adaptations of intertidal hermit crabs. *Am. Zool.* **9,** 343–355.
Rodriguez, G., and Smalley, A. E. (1969). Los cangrejos de agua dulce de Mexico de la familia Pseudothelphusidae (Crustacea, Brachyura). *An. Inst. Biol., Univ. Nac. Auton. Mex., Ser. Cienc. Mar. Limnol.* **40,** 69–112.

Ryan, E. P. (1966). Pheromone: Evidence in a decapod crustacean. *Science* **151,** 340–341.

Saigusa, M. (1980). Entrainment of a semilunar rhythm by a simulated moonlight cycle in the terrestrial crab, *Sesarma haematocheir. Oecologia* **46,** 38–44.

Saigusa, M., and Hidaka, T. (1978). Semilunar rhythm in the zoea-release activity of the land crabs *Sesarma. Oecologia* **37,** 163–176.

Salmon, M. (1965). Waving display and sound production in the courtship behavior of *Uca pugilator,* with comparisons to *U. minax* and *U. pugnax. Zoologica (N.Y.)* **50,** 123–150.

Schmalfuss, H., and Ferrara, F. (1978). Terrestrial isopods from West Africa. 2. Families Tylidae, Ligiidae, Trichoniscidae, Styloniscidae, Rhyscotidae, Halophiloscidae, Philosciidae, Platyarthridae, Trachelipidae, Procellionidae, Armadillidiidae. *Monit. Zool. Ital.* **NS 11,** Suppl, 15–97.

Schneider, P. (1970). Lebensweise und soziales Verhalten der Wustenassel *Hemilepistus aphganicus* Borutzky 1958. *Z. Tierpsychol.* **29,** 121–133.

Schöne, H. (1968). Agonistic and sexual display in aquatic and semi-terrestrial brachyuran crabs. *Am. Zool.* **8,** 641–654.

Schöne, H., and Schöne, H. (1963). Balz und andere Verhaltensweisen der Mangrovekrabbe *Goniopsis cruentata* Latr. und das Winkverhalten der eulitoralen Brachyuren. *Z. Tierpsychol.* **20,** 642–656.

Seiple, W. (1979). Distribution, habitat preferences and breeding periods in the crustaceans *Sesarma cinereum* and *S. reticulatum* (Brachyura: Decapoda: Grapsidae). *Mar. Biol. (Berlin)* **52,** 77–86.

Silas, E. G., and Sankarankutty, C. (1960). On the castle building habit of the crab *Cardisoma carnifex* (Herbst) (Family Geocarcinidae), of the Andaman Islands. *J. Mar. Biol. Assoc. India* **2,** 237–240.

Silas, E. G., and Sankarankutty, C. (1967). Field investigations on the shore crabs of the Gulf of Mannar and Palk Bay, with special reference to the ecology and behaviour of the pellet crab *Scopimera proxima* Kemp. *Proc. Symp. Crustacea Ser.* **2,** 1008–1025.

Sílen, L. (1954). On the circulatory system of the Isopoda Oniscoidea. *Acta Zool. (Stockholm)* **35,** 11–70.

Smith, W. K., and Miller, P. C. (1973). The thermal ecology of two South Florida fiddler crabs: *Uca rapax* Smith and *U. pugilator* Bosc. *Physiol. Zool.* **46,** 186–207.

Standaert, T. A. (1970). Aerial and aquatic respiration of the land crab *Cardisoma guanhumi.* Doctoral Dissertation, Duke University, Durham, North Carolina.

Suter, P. J. (1977). The biology of two species of *Engaeus* (Decapoda: Parastacidae) in Tasmania. II. Life history and larval development, with particular reference to *E. cisternarius. Aust. J. Mar. Freshwater Res.* **28,** 85–93.

Suter, P. J., and Richardson, A. M. M. (1977). The biology of two species of *Engaeus* (Decapoda: parasticidae) in Tasmania. III. Habitat, food, associated fauna and distribution. *Aust. J. Mar. Freshwater Res.* **28,** 95–103.

Taissoun, E. (1974). El cangrejo de Tierra *Cardisoma guanhumi* (Latreille) en Venezuela. *Bol. Cent. Invest. Biol.* **10,** 5–50.

Taylor, H. H., and Greenaway, P. (1979). The structure of the gills and lungs of the arid-zone crab, *Holthuisana (Australothelphusa) transversa* (Brachyura: Sundathelphusidae), including observations on arterial vessels within the gills. *J. Zool.* **189,** 359–384.

Teal, J. M. (1958). Distribution of fiddler crabs in Georgia salt marshes. *Ecology* **39,** 185–193.

Tweedie, M. W. F. (1950). Notes on grapsoid crabs from the Raffles Museum. 2. On the habits of three ocypodid crabs. *Bull. Raffles Mus.* **23,** 310–324.

Utashiro, T., and Morii, Y. (1965). Some knowledge of *Ocypoda stimpsoni* Ortmann and on its burrows. Biological study of "Lebensspuren." VI. *Takeda Branch Fac. Educ., Niigata Univ., Res. Rep.* **9,** 121–141.

Valiela, I., Babiec, D. F., Atherton, W., Seitzinger, S., and Krebs, C. (1974). Some consequences of sexual dimorphism: Feeding in male and female fiddler crabs, *Uca pugnax* (Smith). *Biol. Bull. (Woods Hole, Mass.)* **147,** 652–660.

Vandel, A. (1943). Essai sur l'origine, l'évolution et al classification des Oniscoidea (Isopodes terrestres). *Bull. Biol. Fr. Belge., Suppl.* **30,** 1–143.

Vandel, A. (1946). Crustacéa Isopodes terrestres épigés et cavernicoles du Portugal. *An. Fac. Cienc. Porto* **30,** 135–427.

Vandel, A. (1954). La répartition geóographique des Oniscoidea (Crustaces Isopodes terrestres). *Bull. Biol. Fr. Belge* **74,** 221–272.

Vandel, A. (1964). De l'emploi les appareils respiratoires pour l'établissement d'une classification rationnelle des Isopodes terrestres (Oniscoidea). *Bull. Soc. Zool. Fr.* **89,** 730–736.

Vandel, A. (1965). Sur l'existence d'oniscoides très primitifs menant une vie aquatique et sur le polyphylétisme des isopodes terrestres. *Ann. Speleol.* **20,** 489–518.

Vandel, A. (1973). Les isopodes terrestres de l'Australie. Etude systématique et biogéographique. *Mem. Mus. Natl. Hist. Nat., Ser. A. (Paris)* [N.S.] **82,** 1–171.

Vannini, M. (1975). Researches on the coast of Somalia. The shore and the dune of Sar Uanle. 5. Description and rhythmicity of digging behaviour in *Coenobita rugosus* Milne Edwards. *Monit. Zool. Ital.* **NS 6,** Suppl. 233–242.

Vannini, M. (1976a). Researches on the coast of Somalia. The shore and the dune of Sar Uanle. 7. Field observations on the periodical transdunal migrations of the hermit crab *Coenobita rugosus* Milne Edwards. *Monit. Zool. Ital.* **NS 7,** Suppl., 145–185.

Vannini, M. (1976b). Researches on the coast of Somalia. The shore and the dune of Sar Uanle. 10. Sandy beach decapods. *Monit. Zool. Ital.* **NS 8,** Suppl., 255–286.

Vannini, M. (1980). Researches on the coast of Somalia. The shore and the dune of Sar Uanle. 27. Burrows and digging behaviour in *Ocypode* and other crabs (Crustacea Brachyura). *Monit. Zool. Ital.* **NS 13,** Suppl., 11–44.

Vernberg, F. J., and Costlow, J. D., Jr. (1966). Studies on the physiological variation between tropical and temperate-zone fiddler crabs of the genus *Uca*. IV. Oxygen consumption of larvae and young crabs reared in the laboratory. *Physiol. Zool.* **39,** 36–52.

Vernberg, F. J., and Tashian, R. E. (1959). Studies on the physiological variation between tropical and temperate zone fiddler crabs of the genus *Uca*. I. Thermal death limits and distribution. *Ecology* **40,** 589–593.

Vernberg, W. B. and Vernberg, F. J. (1968). Physiological diversity in metabolism in marine and terrestrial Crustacea. *Am. Zool.* **8,** 449–458.

von Hagen, H. O. (1962). Freilandstudien zur Sexual- und Fortpflanzungsbiologie von *Uca tangeri* in Andalusien. *Z. Morphol. Oekol. Tiere* **51,** 611–725.

von Hagen, H. O. (1968). Studienen peruanischen Winterkrabben (*Uca*). *Zool. Jahrb., Abt. Syst. (Oekol.), Geogr. Tiere* **95,** 395–468.

von Hagen, H. O. (1970). Anpassungen an das spezielle Gezeitzonen-Niveau bei Ocypodiden (Decapoda, Brachyura). *Forma Functio* **2,** 361–413.

von Hagen, H. O. (1977). The tree-climbing crabs of Trinidad. *Stud. Fauna Curaçao Carib. Islands* **54,** 25–59.

Warburg, M. R. (1964). The response of isopods towards temperature, humidity and light. *Anim. Behav.* **12,** 175–186.

Warburg, M. R. (1965). Water relation and internal body temperature of isopods from mesic and xeric habitats. *Physiol. Zool.* **37,** 99–109.

Warburg, M. R. (1968). Behavioral adaptations of terrestrial isopods. *Am. Zool.* **8,** 545–559.

Warner, G. F. (1967). The life history of the mangrove tree crab, *Aratus pisoni*. *J. Zool.* **153,** 321–335.

Warner, G. F. (1969). The occurrence and distribution of crabs in a Jamaican mangrove swamp. *J. Anim. Ecol.* **38,** 379–389.
Warner, G. F. (1970). Behaviour of two species of grapsid crab during intraspecific encounters. *Behaviour* **36,** 9–19.
Warner, G. F. (1977). "The Biology of Crabs." Elek Ltd., London.
Weitzman, M. C. (1963). The biology of the tropical land crab, *Gecarcinus lateralis* (Fréminville). Doctoral Dissertation, Albert Einstein College of Medicine, Yeshiva University, New York.
Wheeler, D. E. (1978). Semilunar hatching periodicity in the mud fiddler crab *Uca pugnax* (Smith). *Estuaries* **1,** 268–269.
Wiens, H. J. (1962). "Atoll Environment and Ecology." Yale Univ. Press, New Haven, Connecticut.
Wieser, W. (1967). Conquering terra firma: The copper problem from the isopod's point of view. *Helgol. Wiss. Meeresunters.* **15,** 282–293.
Wildish, D. J. (1979). Reproductive consequences of the terrestrial habit in *Orchestia* (Crustacea: Amphipoda). *In. J. Invertebr. Reprod.* **1,** 9–20.
Williams, W. D. (1980). "Australian Freshwater Life," 2nd ed. Macmillan, South Melbourne, Australia.
Williamson, D. I. (1951). On the mating and breeding of some semiterrestrial amphipods. *Rep. Dove Mar. Lab., Ser. 3* **12,** 49–62.
Wilson, E. O. (1971). "The Insect Societies." Belknap Press, Cambridge, Massachusetts.
Wolcott, T. G. (1976). Uptake of soil capillary water by ghost crabs. *Nature (London)* **264,** 756–757.
Wolcott, T. G. (1978). Ecological role of ghost crabs, *Ocypode quadrata* (Fabricius) on an ocean beach: Scavengers or predators. *J. Exp. Mar. Biol. Ecol.* **31,** 67–82.
Wolcott, T. G. (1980). Salt availability and range limitation in land crabs. *Am. Zool.* **20,** 925.
Wolvekamp, H. P., and Waterman, T. H. (1960). Respiration. *In* "The Physiology of Crustacea" (T. H. Waterman, ed.), Vol. 1, pp. 36–100. Academic Press, New York.
Wright, H. O. (1968). Visual displays in brachyuran crabs: Field and laboratory studies. *Am. Zool.* **8,** 655–665.
Yamaguchi, T., Noguchi, Y., and Ogawara, N. (1979). Studies of the courtship behavior and copulation of the sand bubbler crab, *Scopimera globosa*. *Publ. Amakusa Mar. Biol. Lab., Kyushu Univ.* **5,** 1–44.
Zucker, N. (1974). Shelter building as a means of reducing territory size in the fiddler crab, *Uca terpsichores* (Crustacea: Ocypodidae). *Am. Midl. Nat.* **91,** 224–236.
Zucker, N. (1977). Neighbor dislodgement and burrow-filling activities by male *Uca musica terpsichores:* A spacing mechanism. *Mar. Biol. (Berlin)* **41,** 281–286.
Zucker, N. (1978). Monthly reproductive cycles in three sympatric hood-building tropical fiddler crabs (genus *Uca*). *Biol. Bull. (Woods Hole, Mass.)* **155,** 410–424.
Zucker, N., and Denny, R. (1979). Interspecific communication in fiddler crabs: Preliminary report of a female rejection display directed toward courting heterospecific males. *Z. Tierpsychol.* **50,** 9–17.

7

Freshwater Adaptations

WINONA B. VERNBERG and F. JOHN VERNBERG

I. INTRODUCTION

The adaptations of crustaceans to a terrestrial mode of existence is discussed in a previous chapter; this chapter focuses on their adaptations to the nonmarine aquatic environment. Crustaceans have successfully invaded a wide variety of aquatic environments, ranging from hot springs to polar

THE BIOLOGY OF CRUSTACEA, VOL. 8

ISBN 0-12-106408-5

ponds and from large lakes to streams; and while many of the same abiotic factors influencing marine and brackish water crustaceans also affect freshwater forms, some unique conditions do occur. The most obvious is the marked decrease in salinity and the complex interaction of this with other factors. Other differences are associated with the temporary, isolated nature of some freshwater ponds and streams in contrast to the enormity and permanence of the world's oceans and the continuity of water masses of the various seas, bays, and estuaries. The freshwater community structure in adjacent but separated river systems may vary widely; however, many groups of smaller-sized organisms, including crustaceans, have evolved striking methods of dispersal resulting in their widespread distribution into widely separated aquatic systems. Unlike the conditions in the ocean, where oxygen and carbon dioxide concentrations are typically stable, these factors vary markedly in freshwater habitats and may limit the distribution of crustaceans.

II. CHARACTERISTICS OF FRESHWATER HABITATS

Before beginning a discussion about the functional adaptations of freshwater organisms, let us consider some of the characteristics of the freshwater environment. Fresh waters are distributed over the surface of the land in a series of different systems that can generally be divided into two groups based on the activity or rate of movement of the water: standing quiet (lentic) water, which includes lakes, ponds, swamps, and marshes; and flowing water (lotic), which are streams and rivers. Each of these presents a varied set of environmental characteristics.

Lakes and ponds are formed in many different ways and can be natural or artificial. Both change with time and are ultimately transformed into wetlands or meadows. Many different definitions have been offered to distinguish between lakes and ponds, but for our purposes we will define a lake as a large expanse of open water, and ponds as smaller bodies of water often filled with plant growth. Physical characteristics that animals encounter in standing water environments are quite different. The temperature of a lake varies with season and depth; even in the tropics some thermal stratification is present in the deeper lakes. Most temperate zone lakes exhibit an annual temperature cycle and a pronounced thermal stratification. Gradations also occur in the amount of oxygen and light in different parts of the lake. The depth of light penetration depends upon the turbidity of the water, and, generally, the stratification of oxygen more or less parallels water temperatures during the summer. Oxygen concentrations are highest at the surface in the warmer months, but during the spring and fall turnovers the deeper regions of the lake are reoxygenated. If the lake does not freeze over during the winter, the oxygen content will remain relatively constant; however, if ice forms, the amount of oxygen beneath it may be seriously depleted. In contrast, environmental conditions, i.e., temperature, oxygen, and light, are almost uniform in a pond. Since ponds often are temporary, they offer only transient existence for the animals living there. Some of these temporary ponds will recur regularly, and some quite irregularly, depending on the precipitation in the area. Both types, however, support a unique form of life. To live in a temporary pond, an organism must be able to reproduce very rapidly when conditions are favorable and also have some means of surviving in the absence of water.

With the exception of some swamps that develop on the flood plains of southern North America, bogs, swamps, and marshes develop during the late stages in lake succession. Bogs are present in humid areas and generally develop where the drainage is blocked and little organic matter can be carried away. The combination of anaerobic conditions and cold temperatures slow down biological activity. The organic debris partially decomposes to form peat, which releases humic acid; as a result the water in a bog is always acidic. Few plants and animals are able to tolerate these acidic conditions, but those that do may be present in large numbers. The marshes are wetlands characterized by grasslike vegetation, including sedges, cattails, and bull rush. Smith (1966) defines marshes essentially as wet prairies because they consist of low, poorly drained land that has standing water for several months of the year. Such marshes support a much richer fauna than do bogs. Swamps support large trees and shrubs and offer a wide variety of both aquatic and semiaquatic habitats, which in turn support a rich variety of animals.

A stream has been defined by Reid and Wood (1976) as a "mass of water with load moving in a more or less definite pattern and following the course of least resistance toward a lower elevation." Streams may begin as outlets of ponds or lakes or they may arise from seepage areas or springs. The chief characteristic of a stream is that the water is continuously moving, and as it flows from the source in the direction of least resistance, it cuts a channel by eroding small furrows that go into gullies. The greater the volume of water and the steeper the slope, the faster the land is eroded away. As the water moves along carrying a load of errosional debris, the channel is cut deeper and wider. Thus, the character of the stream changes as it flows from the source to the mouth. Up in the hills of the source, the head water streams are small and shallow; the current is swift, and the waters are cold. As the water moves downstream, the current slows and the water spreads more widely and is deeper. By the time the water reaches the lowlands, the current is very slow and the bottom is covered with silt and mud. Streams offer a diverse range of habitats, and species near the source of a stream differ markedly from those farther down the stream. Unlike lakes, streams usually produce a turbulence to prevent the stratification that is characteristic of lakes. Furthermore, the oxygen content in unpolluted streams generally is higher than in lakes. Turbidity, however, often is greater, especially during certain seasons. One of the most important environmental features that an organism encounters in streams is a continuous flow of water, and stream animals have evolved many morphological adaptations to this characteristic.

III. EVOLUTIONARY TRENDS

The various orders of Crustacea show rather remarkable differences in colonization patterns. For example, the Cephalocarida, Malacostraca, and Cirripedia are primarily marine; species of the Branchiura, Copepoda, and Ostracoda can be found in marine, brackish, and freshwater habitats; and the Branchiopoda are found chiefly in fresh water (Remane and Schlieper, 1971).

Recently Potts and Durning (1980) reviewed the interactions of physiological and morphological adaptations of crustaceans in relation to the evolution of branchiopods. Although some of the earliest fossil branchiopods are found in marine rocks, freshwater fossil species appear in the Devonian, and branchiopods have lived exclusively in freshwater for most of their existence. The early branchiopods probably were common in most freshwater habitats, but with the evolution of fast-swimming predators, these species were forced to retreat to where their predators could not survive. The formation of resistant eggs, coupled with a short life cycle, enabled some species to inhabit temporary ponds, and Potts and Durning have speculated that the branchiopods radiated out from these temporary ponds to other predator-free habitats. Thus, one species of Anostraca lives in pools of pure water formed by melting snow, another inhabits alkaline lakes where the pH may be 9.8, a third species is found in hot springs at temperatures up to 37°C, and another one can survive in highly saline pools.

Until the Oligocene period when the cladocerans first appeared, only primitive and conservative orders of branchiopods were found in freshwater. The Cladocera show a remarkable range of functional adaptations which have enabled them to invade most freshwater habitats. Their feeding habits cover the range of pelagic filter feeders to detritus feeders, scrapers, and carnivores. They are able to utilize parthenogenesis to ensure an increased rate of reproduction in a habitat where predation is high, and, in most species, the young are retained in a brood chamber and there are no free swimming larval stages. In addition to radiating into most freshwater habitats, several cladoceran genera returned to the oceans, where today they are ubiquitous. Like their freshwater ancestors, these oceanic species maintain a low blood osmoconcentration. Hence, to survive, these species have developed the ability to osmoregulate. In one species, *Evadne nordmanni,* the site of salt secretion is the neck organ (Haptorgan); the structure of this organ has some resemblance to the salt secretory cells found in the epipodites of *Artemia*. In *Artemia,* and possibly in cladocerans, there is a high permeability to sodium, but a low permeability to chloride ions; the chloride ions are actively extruded to maintain low blood concentrations. Other species that inhabit coastal and estuarine waters, such as those found in the Caspian Sea, have evolved into an exceptionally wide variety of forms, possibly because competition is reduced as a result of an unusual salinity regime.

IV. SALINITY

In some groups of crustaceans, species have invaded freshwater habits either permanently or temporarily. Some well-known examples include the blue crab (*Callinectes sapidus*), which periodically penetrates river systems; the Chinese wool-handed crab *(Eriocheir sinensis),* which spends most of its adult life in rivers, but returns to the sea to breed; the freshwater prawn *Macrobrachium rosenbergii,* which is endogenous to fresh and brackish waters; and the amphipod *Gammarus duebeni,* which lives in either marine or freshwater systems. In contrast, some of the freshwater crabs cannot withstand oceanic seawater. In general, a salinity of 5–8‰ acts as a critical boundary between freshwater and marine organisms (Remane, 1934; Khlebovich, 1969; Gilles and Pequeux, Chapter 4 of this volume).[1] This generality is based on the distribution of many groups of organisms and the responses of certain physiological processes (survival, growth, locomotion, cellular electrochemical properties, and ionic and osmotic regulation). A recent example illustrates this biogeographical trend. Regulation of river flow in the northwestern Black Sea area altered the hydrology of lagoons and estuarine areas associated with these rivers so that the freshwater amphipod *Dikerogammarus haemobaphes,* living in the Dneper River was exposed to saline waters. This species survived the changing environment, and Ponomareva (1975) determined that it is capable of adapting and living normally in a salinity range of 0.35–5‰. Some freshwater crabs can tolerate even higher salinities. An example is that of *Potamon niloticus,* which dies in 50% sea water, and *Pseudothelphusa americana jouyi* and *Metopaulias depressus,* which can survive salinities reaching 80% sea water (Thompson, 1970).

[1]The symbol ‰ indicates parts per thousand (ppt).

Although the blood osmoconcentration in freshwater crabs is significantly above that of the ambient fluid, it is less than that of osmoregulating marine crabs. The body surface of these freshwater animals is less permeable to passive loss of Na^+ than is that of marine and estuarine crabs. Other crustaceans, such as crayfish and some amphipods, are limited to freshwater or very dilute solutions and cannot live in salt water. Characteristically these animals have a low blood concentration of ions and produce a dilute urine by reabsorption of the filtered plasma solute from a relatively water-impermeable region in the kidney (the distal tubular segment). Details of ionic and osmotic regulation are found in Chapter 4 of this volume and in Chapter 2, Volume 5 of this treatise; for a general review see Gilles (1979).

Populations of the same species living in fresh water may exhibit different osmoregulatory capabilities from those found in brackish or sea water. Two studies serve to illustrate this point. Populations of the *Saduria entomon* from the Baltic Sea could not be acclimated to the freshwater environment of Scandinavian lakes, where another population of this species was found. Functional differences between freshwater and marine populations help explain this biogeographical observation. The permeability of freshwater forms was significantly less than that of seawater populations, and the active transport mechanism of lake forms had a greater ability to remove sodium from the ambient milieu (Croghan and Lockwood, 1968). Two populations of the freshwater crab *Potamon edulis* from Malta were found to be more sensitive to increased salinity than were animals found on the European continent (Harris and Micallef, 1971), and the tissues of Malta crabs could not tolerate increased hemolymph concentrations as well as could animals from the continent.

Dorgelo (1977) compared the adaptive responses of three species of gammarid amphipods from marine, brackish, and freshwater habitats to different salinity–temperature combinations. The freshwater species *(Gammarus fossarum)* showed the lowest level of hyperregulation of blood sodium. The potentially lethal blood levels observed at higher salinites might explain geographical limitations of this species. In contrast, no ecological implications of the influence of temperature on the level of regulation were apparent.

When subjecting freshwater-acclimated prawns *(Macrobrachium rosenbergii)* to a salinity of 24‰ for 48–72 hr, Armstrong *et al.* (1981) reported the following functional changes: ammonia production sharply decreased, blood Na^+ increased, but not to the level of the ambient concentration, and both blood ammonia and protein concentrations decreased. The authors suggested that ammonia was removed from the environment by the prawn as a result of the reversal of normal Na^+/NH_4^+ exchange, where blood Na^+ is hyporegulated and exogenous NH_4^+ is used as a counter-ion.

V. TEMPERATURE

A. Lethal Thermal Limits

Freshwater crustaceans occupy a wide variety of habitats which vary significantly in terms of their thermal extremes. Crustaceans living in hot springs regularly survive in waters of 45°–48°C (*Thermosbaena mirabilis*) or 55°C (*Thermobathynella adami*) (Capart, 1951), while animals in polar ponds survive subzero temperatures. As has been demonstrated in other animal groups, there is a general relationship between the distribution of animals and their lethal thermal limits: animals from cold water habitats tend to have lower lethal thermal limits than warm-water forms (Vernberg, 1962).

In recent years, the thermal lethal limits of many freshwater organisms have been reported, especially as part of environmental impact studies associated with siting of power plants. For example, the thermal tolerance of *Lirceus brachyurus,* an isopod living in Buffalo Creek, Tennessee, U.S.A., is labile and increases with increasing acclimation temperature (Cheper, 1980). Although the upper lethal temperatures were lower than those reported for other isopod species, the lower lethal temperatures were similar to those of other isopod species and were related to ice mineralization of the creek water. The laboratory findings on thermal limits correspond to maximum and minimum water temperatures recorded in this creek, demonstrating a close relationship between laboratory and field observations.

Many studies on the effects of temperature on longevity have involved organisms exposed to a constant temperature; however, animals typically live in a thermally fluctuating environment. Smyly (1980) found that the longevity of the freshwater copepod *Acanthocyclops viridis* varied depending on whether it was exposed to constant or alternating temperatures. Fluctuating thermal regimes shortened the expected life span of adult females but lengthened that of adult males. He concluded that the number of changes in temperature may not be responsible for the shortened life span of females, but repeated interruptions by varying temperature may have disrupted some physiological function involved in the acclimation process. In contrast, the increased life span of adult males may indicate that males acclimate faster than females, or that fluctuating temperature influences some metabolic function found only in females, i.e., the ovarian cycle. These data have added significance in that earlier estimates of the life span of various copepods, which were based on observations made entirely under constant thermal regimes, may be incorrect.

Although the effect of lethal temperature on individuals of a species has ecological significance, other physiological responses, such as reproduction, also are of importance to the continuity of a species. One example of this point is the recent review of Sarviro (1980), in which data from various sources on populations of the amphipod *Gammarus lacustris* from 14 lakes in the Soviet Union were analyzed. The lower thermal limit of this species in Lake Baikal is 0°C; animals acclimated to warm-water reservoirs have an uppler limit approaching 28°–30°C. However, the development of eggs is possible only within the range of 7°–23°C. Higher temperatures (23°–27°C) increase the length of embryogenesis, while low temperatures (4°–6°C) inhibit development. Based on these data, the zone of tolerance for the species is about 7°–23°C. Data on 29 biotic characteristics of different populations of *G. lacustris* were compared, and marked geographical differences were noted, which frequently could be correlated with the temperature regime of the lake. However, there are no reported data which indicate whether these differences reflect genetic differences or whether observed differences are environmentally-induced acclimation effects.

B. Temperature and Oxygen

Although data on the lethal effect of one factor are helpful in understanding the physiological ecology of organisms, the environment consists of many factors. Hence, studies of multiple factor effects on organisms more nearly simulate natural conditions.

An earlier study by Sprague (1963) serves as an excellent example of multiple factor effects (high temperature and low oxygen) on freshwater crustaceans from different habitats. The four species studied by Sprague were: (1) *Asellus intermedius,* an isopod, which can withstand water that is very warm or drastically reduced in oxygen content; (2) *Hyalella azteca,* the most widely distributed amphipod in North America, which can withstand high temperature, but is less tolerant than *A. intermedius* of hypoxic conditions; (3) *Gammarus fasciatus,* an amphipod, which is characteristic of warm, slow moving waters; (4) *G. pseudolimnaeus,* an amphipod, which is restricted to cool water. Both species of *Gammarus* live in water in which they experience only mild deoxygenation. In the laboratory, resistance was determined for animals acclimated to different temperatures and oxygen concentrations.

Although differences in heat resistance between species were observed, these were not as great as might be expected based on habitat differences, and temperature resistance seemed greater than necessary for survival in nature. In contrast, resistance times of the four species to low oxygen concentration were highly correlated with field observations: *A. intermedius, H. azteca,* and *G. fasciatus* live in habitats where dissolved oxygen comes close to their lethal limits, whereas *A. intermedius* tolerates hypoxic conditions.

This study illustrates the need to use both laboratory and field data to understand the physiological ecology of organisms, since temperature and salinity exhibit different roles in determining the distribution of crustaceans.

C. Adaptive Mechanisms

Although it is not the intent of this chapter to emphasize the physiological and biochemical mechanisms of adaptation to temperature stress, five examples are cited to demonstrate significant lines of research on this subject that have ecological significance.

1. As temperature decreases, the concentration of unsaturated fatty acids tends to increase, an observation reported by a number of investigators (see review of Hazel and Prosser, 1974). Recently, Farkas (1979) found that fats of freshwater planktonic crustaceans overwintering in an active state are richer in long chain polyunsaturated fatty acids than are the fats of species spending the winter as resting eggs. Exposure of warm-acclimated animals to 5°C in the laboratory resulted in a marked increase in the level of phospholipid decosahexaenoic acid (an unsaturated fatty acid). Since the animals received no extra food during the experiment, it would appear that the source of this fatty acid was some internal precursor. Farkas believes that the ability to adapt membrane fluidity to ambient temperature could be important for survival at low temperatures.

2. There is a correlation between the death of an organism and the inactivation of certain cellular processes. Cossins and Bowler (1976) demonstrated a close correspondence between the heat death temperatures of the crayfish *Austropotamobius pallipes* and the inactivation temperatures for several membrane-bound enzymes of muscle.

3. As temperature changes, the locomotor behavior of an organism may be altered. After acclimating crayfish *(Astacus astacus)* to different temperatures (5° and 20°C) and measuring their neural and motor function, Kivivuori (1980) found that both cold and heat resistance increased in the following order: righting reflex, walking, scaphognathite beat, escape reflex, and motor functions (spontaneous firing). The author suggested that in most cases, temperature acclimation shifted the resistance limits in an adaptive manner.

4. The response of organisms to thermal changes may be integrated by hormonal control. Nagabhushanam *et al.* (1979) reported changes in the brain and thoracic ganglion neurosecretory cells of the freshwater prawn *Macrobrachium kistnensis* after acclimation to either 13° or 31°C. After warm acclimation, the neurosecretory material (NSM) from "A" cells of the brain was depleted whereas accumulation was noted following cold stress. "B" cells remained unaffected. Both "A" and "B" cells of the thoracic

ganglion appeared to have an increase in NSM following both cold and warm stress. The authors postulated that these neurosecretory responses are important in maintaining homeostasis during periods of thermal stress.

5. The mechanisms of heat tolerance in the tropical freshwater crab *Barytelphusa cunicularis* were examined by Diwan and Nagabhushanam (1976). Acclimation to low temperature (15°C) lowered the upper lethal temperature and acclimation to high temperature (31°–32°C) increased resistance to high temperature. Certain chemical changes in response to thermal acclimation were found with increasing thermal acclimation temperature: (1) increased water content of the whole animal and of the midgut gland; (2) decrease in protein and fat content; and (3) increase in glycogen and blood glucose. These findings are in agreement with temperature acclimation response patterns observed in other groups of animals from marine and terrestrial habitats.

VI. DROUGHT

In response to adverse conditions, some crustaceans inhabiting temporary ponds form resistant eggs. Observations on the hatching of these resistant eggs show a variety of responses. Some will only hatch after being exposed to drying (*Streptocephalus seali,* a fairy shrimp: Moore, 1967). Eggs of certain anostracans hatch only after exposure to drying and/or freezing. Some species have two types of eggs, one which requires drying before hatching and a second which does not (Sam and Krishanaswamy, 1979). Resistant eggs have the capacity to hydrate, and the rate of hydration is dependent on the ionic content and osmolarity of the medium. Morris (1971) reported that in *Artemia salina* cysts hydrated in distilled water, but no hydration was observed in a 15% NaCl solution. However, the undried eggs in some other species required a hypoosmotic medium for the hydration and bursting of the outer shell. Sam and Krishnaswamy (1979) reported that in the undried eggs of tropical fairy shrimp, *Streptocephalus dichotomus,* embryonic development up to the breaking stage is completed before the onset of the dormant period. However, hatching is inhibited by high osmotic pressure of the medium, and nonhydrated eggs hatch and begin a new generation only following rains, which tend to dilute pond water.

The freshwater crab *Barytelphusa cunicularis* is able to survive drought conditions by a unique morphological adaptation, which permits water stored in the branchial chambers to be recycled over the gills (Diwan, 1975). On the other hand, some crustaceans, such as *Artemia salina,* survive adverse environments by forming cysts. Of physiological–ecological interest, it is hydration, not the salinity of ambient water, that initiates respiratory activity of the cysts (Boulton and Huggins, 1977).

If ambient conditions become too stressful, organisms may use behavioral mechanisms to seek less stressful sites. For example, the amphipod *Crangonyx antennatus* lives in pools and streams in caves. During periods of drought when the pools dry up, members of this species are able to burrow into the substratum and thus survive desiccation and predation from terrestrial animals. Holsinger and Dickson (1977) observed this response both in the field and in simulated laboratory experiments. The ability to burrow even when drought conditions are not experienced may afford this species protection from potential aquatic predators.

VII. GENETIC VARIATION

The genetic basis for differential adaptations of geographically separated populations of one species to various combinations of environmental factors is as yet poorly understood. Whether or not a population will adapt or become extinct in a particular freshwater system is dependent on the rapidity and severity with which the stress is introduced in relation to the inbuilt capacity of the population for genetic change. Fraser *et al.* (1978) studied this question using samples of *Asellus aquaticus* collected from three sites in the River Trent basin (England), which are subjected to different levels of lead pollution. They found that populations from unpolluted waters are highly sensitive to lead but that tolerances develop following exposure over time. Whether this population difference is due to selection of resistant animals or to genetic change in response to lead is as yet unknown.

Aston and Milner (1980) also examined the response of populations of a crustacean *(Asellus aquaticus)* found in the River Trent that were subjected to different combinations of environmental factors; in their study the environmental parameters were temperature and organic pollution. The populations from the region of highest organic pollution had the following characteristics: (1) increased population density, and (2) increased mean length of males, immature females, and nonovigerous females. Aston and Milner suggested that these population differences were probably the result of the greater availability of food (degraded sewage and bacteria) in the organically enriched habitats. Population characteristics and distribution appeared to be correlated with the thermal characteristics of the power plants, since the population from the warmest of the three heated sites had smaller males and ovigerous females and may have had shorter life cycles. No apparent genetic differences in populations were reported.

VIII. BIOENERGETICS

In recent years, an increased interest in the various aspects of bioenergetics of freshwater organisms has taken place. Energy budgets have been estimated for some populations by means of the following equation (Ricker, 1968; Petrusewicz and Macfadyen, 1970; Lei and Armitage, 1980):

$$C = P + R + F + U + D + F$$

where C is the energy ingested by the population; P, the total energy produced by the population as the sum of somatic growth, egg production, and exuvia; R, energy lost through respiration by the population; F, energy leaving the population as feces; U, energy lost as products of nitrogen metabolism; D, energy digested or absorbed.

Knowledge gained from bioenergetic studies is useful not only in understanding adaptive mechanisms of organisms to various environmental factors, but also in determining functional dynamics of communities. To construct an energy budget, numerous physiological parameters must be measured under conditions which simulate field conditions. In one sense, energy budget studies utilize the data of many separate physiological studies to present a broader integrative approach to adaptation at the population or community level. Although a detailed review of the bioenergetics of freshwater organisms is beyond the scope of this chapter, two examples will illustrate this research area: (1) the study of Lei and Armitage (1980), in which they estimated the energetic budget of a wild population of *Daphnia ambigua* inhabiting a pond in Kansas, U.S.A.; and (2) the study of Wulff (1980), who investigated the energetics of *Daphnia magna* from a rock pool in Sweden.

Lei and Armitage (1980) constructed monthly energy budgets for a 13-month period, expressed in calories per cubic meter. Energy intake ranged from 220.1 cal to 79.8 kcal; highest values were observed during the colder months when population densities were highest. A similar seasonal trend in production was noted, with values ranging from 26 cal to 10.2 kcal. Respiratory loss varied seasonally, but did not show peak activity in the same months as food intake production; this suggests that when high levels of food are available, more of the consumed energy is allocated to production. Energy flow is low during the summer and fall, but high in the winter and spring. The production and assimilation by *Daphnia* accounted for 9.3 and 23.6%, respectively, of the gross primary production of phytoplankton. Unexpectedly, Lei and Armitage found that a substantial amount of energy was lost as exuvia (25% of the annual production). The annual gross production efficiency (10.9–11.3%) is equivalent to the ecological efficiency and is within the range postulated by Slobodkin (1962) and Engelmann (1966) for other ecological systems.

Based on a 5-month period, Wulff (1980) calculated an energy budget for *Daphnia magna,* the dominant animals in a rock pool habitat. He developed a numerical model that describes an energy budget in terms of variation in weight-specific growth, reproduction, molting, feeding, and respiration rates in relation to temperature and food concentrations. Using this model, he was able to predict the chances for population survival and utilization of the available energy when exposed to varying food concentrations. Wulff reported that varying food concentrations had the greatest influence on energetics, even more so than varying temperature.

IX. RESPIRATION

Freshwater crustaceans are confronted with a wide variety of habitats, each of which presents the organism with a different set of environmental conditions. One functional indicator of an organism's ability to successfully meet these stresses in this array of habitats is respiration. Hence, knowledge of the mechanisms of uptake and utilization of oxygen affords an insight into how organisms are adapted to different habitats. Although oxygen uptake is a good indicator of energy utilization by organisms, in recent years emphasis is being placed on the important role of anaerobic pathways in bioenergetics. Chapter 1 in this volume deals with selected studies on the general topics of respiratory adaptations of all crustaceans; freshwater organisms will be emphasized in this chapter.

A number of factors influence the rate of oxygen consumption. An increase in temperature typically increases the oxygen uptake rate, but there are exceptions. For example, over the range of temperatures that copepodites and adults of the copepod *Mesocyclops brasiliensis* normally experience in a Venezuelan lake, the metabolic rate is unchanged ($Q_{10} = 1$) (Epp and Lewis, 1979). Also, the metabolic–temperature response can change depending on the acclimation temperature: acclimation to higher temperatures shifts the position of the M–T curve to the right. Such effects of acclimation have been reported in the cladoceran *Daphnia ambigua* (Armitage and Lei, 1979) and in the tropical freshwater crab *Parathelphusa hydrodromus* (Kotaiah and Rajabai, 1972).

Temperature–metabolic interaction can influence the population size of lake species, as illustrated by the work of Gophen (1976) on *Ceriodaphnia reticulata,* a cladoceran. Over the temperature range of 15°–22°C, food ingestion and ammonia excretion rates increased more rapidly than over a range of 22°–27°C. In contrast, the metabolic rate increased more rapidly over the higher thermal range; this change produced an energy deficit as a result of both reduction of body size and a decrease in egg production. Since Lake Kinneret, Israel, has summer temperatures of 27°–28°C, Gophen speculated that this metabolic–temperature response pattern at elevated temperatures explains the progressive decline of the population during the summer. Seasonal differences in metabolic rates also have been reported (Lei and Armitage, 1980), as well as differences in body size and stage of development (Green and Chapman, 1977; Ambore and Venkatachari, 1978; Laybourn-Parry and Strachan, 1980). Pollutants may influence metabolic rates, but not in a predictable manner (Sigmon, 1979).

The concentration of oxygen may influence the metabolic rate of freshwater crustaceans. Two main types of response have been noted: oxyconforming, where the rate varies with ambient oxygen concentration, and oxyregulating, where the rate is independent of external concentration (Maloeuf, 1937; Vernberg, 1972). For example, a relationship between the concentration of oxygen in the water and the respiratory response of the crawfish (*Pacifastacus leniusculus*) was reported by Moshiri *et al.* (1970). This species, which lives in water of low temperature and high oxygen content, showed no evidence of metabolic regulation over a wide range of oxygen tensions.

Correlations between oxygen consumption and habitat have been noted for freshwater crustaceans in a number of studies. For example, isopods and crayfish living in streams characterized by rapidly flowing water generally have higher metabolic rates than organisms inhabiting slow-moving waters (Wolky, 1934; Fox and Baldes, 1935), and cave-inhabiting crustaceans have lower oxygen consumption rates than surface-dwelling forms of amphipods (Derouet, 1953; Troiani, 1954) and crayfish (Burbanck *et al.,* 1948). A similar response was reported for the respiration rate, ATP turnover, and adenylate concentrations of excised gills of surface and cave species of crayfish (Dickson and Franz, 1980). This response pattern may be an adaptation to the reduced food supply in caves. The amphipod *Gammarus fossarum* is better adapted metabolically to live in habitats characterized by running water and low temperature than *G. pulex,* which is more adaptable to variable biotopes, especially at temperatures above 15°C (Roux, 1975). At reduced temperatures, *G. fossarum* has a competitive advantage over *G. pulex*. As a generalization it can be stated that metabolically conforming organisms typically are found in well-oxygenated waters, whereas animals from waters that have fluctuating oxygen levels show some degree of metabolic regulation (Vernberg and Vernberg, 1970).

Generally, species from higher latitudes have higher metabolic rates than those from lower latitudes (see Vernberg, Chapter 1 of this volume). A recent example can be found in the study by Smith *et al.* (1978) on populations of the copepod *Diaptomus dorsalis;* these authors reported that oxygen consumption rates of a Texas population was lower than that of congeners from northern populations at similar temperatures. Exceptions to this generalization have been observed, however. Epp and Lewis (1979) found that a tropical copepod (*Mesocyclops brasiliensis*) from Venezuela had a higher metabolic rate than most, but not all, temperate zone species. Another study in which results differed from expected patterns was that by Ostapenya *et al.* (1969). The purpose of their investigations was to determine the influence of low temperature on metabolism in planktonic copepods living at high latitudes in the U.S.S.R. The metabolic rate of winter populations of the dominant copepod *Diaptomus graciloides,* as expected, was influenced by temperature, but the response pattern did not fit Krogh's predicted "normal curve." Instead, Q_{10} values of 5.1 and 1.8 were reported for the thermal ranges of 0.5°–2.5°C and 2.5°–3.9°C, respectively, versus Krogh's curve predicted value of 10.9. The thermal range reported in Lake Naroch during the winter ranges from 0° to 4°C.

Some freshwater species, such as *Macrobrachium rosenbergii,* migrate to brackish water habitats during part of their life cycle. Nelson *et al.* (1977) found that salinity–temperature combinations significantly influence the metabolic rate of juvenile *Macrobrachium,* which in turn affects their distribution. The juvenile stage develops in estuarine waters and migrates into freshwater habitats. Nelson *et al.* suggested that the metabolic stress of high temperature and salinity was instrumental in initiating this migratory behavior of the juvenile stages. Thus, physiological responses are an important component in the niche selection of animals.

X. FEEDING

The food eaten by animals provides the source of energy needed to sustain the organism's metabolic machinery. Since energy is such a fundamental requirement, it is not surprising that there is a continuing interest in the dietary preferences and requirements of aquatic animals. In many species, dietary preferences change with the stage of the life cycle. In observing the diet of natural and cultured populations of *Mesocyclops leuckarti,* a dominant copepod in Lake Kinneret, Israel, Gophen (1977) found that nauplii and early copepodid stages were herbivorous, whereas later stages were carnivorous and selectively preyed on certain species. In another study, the feeding of two coexisting congeneric copepods from Lake Michigan, *Diaptomus sicilis* and *D. ashlandi,* was observed over a 10-month period (Bowers, 1980). *Diaptomus sicilis* preferred ingested particles in the largest size class, while *D. ashlandi,* the small-sized species, fed only on the smallest sized particles of chlorophyll *a*. During the summer, the highest feeding rates of *D. ashlandi* occurred in the epilimnion and that of *D. sicilis* in the metalimnion, where large particles were most prevalent. These observations confirm an earlier generality proposed by Hutchinson (1951) that coexisting and possibly competing calanoid copepods are often separated by size and selection differences in feeding behavior.

Feeding is not only necessary for growth and maintenance of an individual organism, but it is necessary for reproduction. To illustrate this point, Lampert (1978) demonstrated that egg production in a *Daphnia* (*longispina* group) population in Lake Constance, Germany, was restricted if the food concentration was less than 0.2 mg C/liter, but at 0.7 mg C/liter egg production reached a plateau. Based on field data, he concluded that *Daphnia* in Lake Constance seem to be food limited during most of the year.

There are many questions that must be answered before it is possible to state why crustaceans are able to inhabit specialized environments. One of the questions revolves around the adaptive processes at the cellular level, particularly the absorptive and secretory capabilities of the gastrointestinal tract. Brick and Ahearn (1978) summarized earlier studies on absorption of various dietary components by the gastrointestinal tract of various crustaceans and presented new data on the freshwater shrimp *Macrobrachium rosenbergii*. Although amino acid influx in the intestine of this species is similar to that reported for a large variety of cell types in other species, in addition, the carrier process has a high affinity for very low lysine concentrations.

Other studies have emphasized predator–prey interactions in freshwater systems. For example, Kerfoot (1977), investigated the adaptive significance of invertebrate predation on zooplankton communities. He found that one result of copepod *Epischura nevadensis* predation on zooplankton is that the abundance of the cladoceran *Bosmina longirostris* is reduced and larger-bodied species are favored. Hence, in a lake without vertebrate planktivores, but dominated by abundant invertebrate predators, selective removal of small-sized species could result in a community composed of large-sized species less susceptible to invertebrate predation. In another study on the effect of invertebrate predation on the zooplankton community, O'Brien and Vinyard (1978) found that the presence of backswimmers, *Anisops* sp., influenced the abundance of two varieties of *Daphnia carinata* in ponds located in India. *Anisops* fed more readily on *D. carinata* forma *typica* than var. *cephalata*. Apparently var. *cephalata* is better able to evade capture; however, in ponds where backswimmers are absent, forma *typica* has a competitive advantage in that its survivorship and rate of increase is higher and its feeding rate is more rapid.

Feeding efficiency of freshwater crustaceans may be influenced by a number of environmental factors. Kring and O'Brien (1976) demonstrated that when the oxygen concentration dropped below 3 mg 0_2/liter, the filtering rate of *Daphnia pulex* decreased sharply. However, after acclimation to low oxygen conditions the filtering rate gradually increased, and eventually it exceeded that reported under normal oxygen concentrations. a response probably correlated with increased production of hemoglobin under hypoxic conditions.

Food intake is also influenced by temperature, but not always in a predictable manner. The cladoceran, *Ceriodaphnia reticulata,* living in Lake Kinneret, Israel, increases its rate of food intake more rapidly over the thermal range of 15°–22°C than at a higher range of 22°–27°C (Gophen, 1976). In another cladoceran, *Daphnia ambigua,* the filtering rates increased with increasing body size and were significantly affected by environmental temperature (Armitage and Lei, 1979). Acclimation to cold shifted the filtering rate–temperature curve to the left, indicating that animals from colder environments filter feed at a faster rate at a given temperature than do warm-acclimated individuals. A temperature near 12°C separates warm- and cold-acclimatization processes (Armitage and Lei, 1979). Earlier workers showed various other factors influenced filtering rates (food concentration: Ryther, 1954; light intensity and quality: Buikema, 1973; pH: Ivanova and Kelkowski, 1972).

XI. REPRODUCTION

Although many groups of Crustacea successfully live in fresh water, not all have become reproductively independent of marine waters. For example, the portunid crab, *Callinectes sapidus,* and some species of fiddler crabs, *Uca* sp., may live in fresh water, but their larvae will survive only in saline waters. In contrast, many crustaceans utilize the freshwater habitat during some phase of development.

The initiation of reproductive cycles in freshwater crustaceans is dependent on many of the same environmental factors as are estuarine and marine species: temperature, daylength, and food availabiility. In many temperate zone species, daylength is often the main cue for the initiation and termination of the reproductive cycle. In a study of the freshwater amphipod, *Hyalella azteca* from southern Manitoba, DeMarch (1977) reported that daylength was the critical factor in both the initiation and termination of the reproductive cycle. A 12L–12D photoperiod is the cue which precedes warmer temperatures in the spring; in the fall this photoperiod precedes impending undesirable cooler temperatures. *Hyalella azteca,* however, apparently does not respond in the same manner to photoperiod over its entire geographical range. In experimental studies on a population of *H. azteca* from Oregon, Strong (1972) demonstrated that these animals showed no response to photoperiod. Instead, DeMarch (1977) has suggested that the photoperiodic response only occurs where populations of these amphipods experience extremes in daylength.

Tadini and Valentino (1969) demonstrated that light influences the reproductive period in the isopod *Asellus aquaticus,* and that in populations from western and central Europe, the presence of a nonbreeding period in winter is genetically determined. They suggested that the light stimulus in autumn is timed so that breeding stops before temperature falls, since low temperature is lethal to eggs and young.

In the isopod *Asellus aquaticus,* breeding is closely correlated with water temperature (Økland, 1978). This species, which is found over a wide geographic range in Europe, shows three main patterns of reproductive cycling. In southern Europe, where water temperatures are generally above 4°C, *A. aquaticus* reproduces throughout the year (Tadini and Valentino, 1969). In southern regions of France, populations of *A. aquaticus* breed only in winter (Vandel, 1926). Populations of these isopods from Norway breed only during the spring and summer months when temperatures reach 7°–8°C.

In the prawn *Macrobrachium malcolmsonii,* the breeding cycle is influenced by both temperature and rainfall (Rajyalakshmi, 1980). With rainfall, temperatures decrease and oxygen levels in the water rise. In many animals, combined conditions act as a stimulus for breeding activity. In the Madras population of *Paratelphusa hydrodromus,* rainfall is essential for release of larvae, and their release is synchronized with the northeast monsoon during the coldest month of the year. Initiation of the breeding cycle occurs when temperatures are high and during the dry season (Pillia, 1981).

A close connection between food supply and egg production is apparent in many animals. A good example are the cyclopoid copepods inhabiting temperate zone oligotrophic lakes. The breeding cycle of these organisms invariably coincides with spring and autumn phytoplankton peaks (Nilssen, 1980).

In general, both marine decapod crustaceans and those freshwater species which return to the sea to reproduce bear a relatively greater number of small-sized eggs than do crustaceans reproducing in freshwater. Pillia (1982) has summarized data from various sources and reported that in marine species the number of eggs varies from 3,000 to 10,000,000 per female, and the size of an egg varies from 0.2 to 1.8 mm; these figures for freshwater species are 10–800 and 0.2–2.8, respectively. Although there are exceptions to all generalities, eggs of marine crustaceans usually develop within a relatively short period of time, and few develop into adults. In contrast, many of the freshwater crabs incubate the eggs for longer periods of time, and as a result, older and better developed individuals are released into the environment, and survival rates are relatively high.

Although branchiopods originated in seawater, they are best represented in fresh water. In reviewing the evolution of these crustaceans, Potts and Durning (1980) suggested certain reproductive and developmental adaptations which favor life away from the world's oceans. Initially they were common in many freshwater habitats, but as more predators appeared, especially fishes, branchiopods were forced into habitats which excluded fishes, such as temporary ponds. The ability to form resistant eggs and to have short life cycles allowed branchiopods to survive the extreme environments.

The cladocerans have evolved other reproductive ploys to facilitate rapid generation time, such as the adaption of parthenogenesis and a brood chamber. Generally males are only found during periods of peak population densities, frequently preceding a period of adversity; however, in some species males have never been reported. Parthenogenetic species are preadapted to two kinds of niches: (1) persistent, low-density niches, and (2) fluctuating niches with unpredictable recurrences of low-density and high-density conditions. A detailed review of the adaptive significance of parthenogenesis in various groups of organisms was published by Gerritsen (1980). Many species of cladocerans form brood chambers in which young develop and leave in an advanced stage. Some brood chambers are highly developed, and the young are nourished by a placenta-like organ called the "Nahrboden." In a few cases, embryos of the third generation are developed before the birth of the second.

Temperature significantly influences the development of preadult stages of freshwater crustaceans. Over the range of 16°–28°C, growth of postembryonic stages of the branchiuran *Argulus foliaceus* increases with temperature, although the relation is not linear (Schlüter, 1979). Two species of diaptomid copepods from Bluff Lake, Nova Scotia, had similar egg development times at temperatures from 14° to 23°C, but the times at lower temperatures were significantly different (Cooley, 1978). One species, *Diaptomus minutus,* is better adapted for winter temperatures than the more warm-adapted *Aglaodiaptomus spatulocrenatus.* In an extensive paper, Sarvala (1979) reported on the effects of temperature on the development times of egg, nauplius, and copepodite stages of seven species of freshwater benthic copepods from Finland. The rate of development and the magnitude of thermal effect varied with the different species. However, some generalities were evident. With increasing temperature, development time decreased up to a point; beyond a certain temperature elevated temperatures inhibited development. All of the harpacticoid species developed more slowly than previously studied copepods from planktonic, littoral, or benthic habitats, but faster than subterranean animals.

Økland (1978) reported that the incubation time for the isopod *Asellus aquaticus* from a eutrophic lake in Norway was 23–25 days, corresponding to about 300 degree-days above 4°C. No growth occurred from November to March–April when the water temperature reached 3°–5°C.

In laboratory studies, DeMarch (1978) compared the effects of constant and variable temperature on the growth of an amphipod *Hyalella azteca*. As expected, the time to reach maturity and the embryonic development time were inversely correlated with temperature when animals were reared in constant temperatures. Over the thermal range of 10°–23°C, larger animals were produced at the lower temperatures; 18°C was the critical temperature separating large and small animals. Exposure to fluctuating thermal regimes showed a marked influence on growth patterns. The largest animals produced were those transferred from 10° to 23°C, while those transferred from 23° to 20°C were small. These responses were most noticeable in organisms transferred in an early stage of development. DeMarch (1978) suggested that these laboratory results could explain the seasonal cycles in size observed in natural populations.

Welton and Clarke (1980) observed the reproduction and growth of the amphipod *Gammarus pulex* in the laboratory at temperatures ranging from 5° to 20°C to provide data on production and population dynamics of field populations. The best equation to describe the relationship between brood development and temperature is the power law:

$$\log_e D = a + b \log_e^t$$

where $a = 5.752$ and $b = 0.962$, D is days, and t is temperature. The predicted longevity of field populations is 17–23 months for females and 2.5 years for males. These experiments were conducted at constant temperatures; it is possible that cyclic temperatures might alter developmental patterns.

Two species of copepods living in Bluff Lake, Nova Scotia, responded differently to temperature. One species, *Aglaodiaptomus spatulocrenatus,* did not produce eggs when the water temperature dropped below 10°C, while *Diaptomus minutus* appeared to be adapted to a winter existence (Cooley, 1978). Thus, both species could differentially exploit the resources of this lake on a temporal basis.

The migration of the freshwater decapod *Atyaephyra desmarestii* throughout northern and eastern Europe from the Mediterranean countries is correlated with its capacity to limit growth to periods of the year when the temperature is greater than 10°C (Descouturelle, 1979). This finding is based on a 12-month study in which laboratory temperatures simulated those in the field.

XII. BUOYANCY

Animals living in freshwater have a greater problem in keeping afloat than those inhabiting seawater. Numerous flotation mechanisms have evolved in both marine and freshwater species, including increased body water content, skeletal reduction, gas floats, changed body shape, and body projections that increase the body surface. In general, it appears that relatively more nannoplankton is present in fresh water than in sea water. Also, large-sized plankton are less plentiful in freshwater systems. Since viscosity of water increases with decreasing temperature, it has been suggested that cyclomorphosis in certain forms (e.g., cladocerans) might be related to buoyancy during various seasons. Recently, Hebert (1978) studied cyclomorphosis in natural populations of *Daphnia cephalata* in Australian ponds. He reported that head length to carapace length ratio (H/C ratio) is directly dependent on food intake, temperature, and the presence of weed growth in these ponds, but is inversely dependent on pond area, which suggests that turbulence is also negatively associated with head size. Hebert observed that the adaptive significance of this correlation might be based on the role of the helmet in gas exchange, because hemolymph circulates through the laminate helmet. Increased gas exchange would be adaptive, as the oxygen concentration in these senescent ponds is likely to be very low during the night because of plant respiration.

XIII. HYDROSTATIC PRESSURE

Freshwater crustaceans have invaded large lakes and occupy various vertically distributed habitats, including the deep water zone. Recently, Brauer *et al.* (1980) investigated the adaptive mechanisms which enable gammarids to withstand the high hydrostatic pressures found in Lake Baikal in Eastern Siberia. The lake is an outstanding site for studies of evolutionary biology because it is geologically old and has a diverse fauna that shows a high degree of endemism. Because of a horizontal and vertical circulation pattern, which maintains a high oxygen content throughout the entire water column, the fauna of Lake Baikal extends to all depths in this deepest of all lakes. Brauer *et al.* (1980) subjected nine species of gammarids (five shallow water species, living up to a capture depth of 50 m, and four species from depths ranging from 935 to 1400 m) to increasing hydrostatic pressures. Nine distinguishable response stages reflecting successive phases of hyperactivity, convulsions, and immobilization were described. The abyssal gammarids exhibited greater threshold pressures for the initiation of convulsions than did the shallow water gammarids, and these values averaged near the maximum pressure existing in the lake. In contrast, the threshold pressures for hyperactivity and immobilization were not only significantly higher than can possibly be encountered in Lake Baikal, but they differed little between deep and shallow water species. Brauer *et al.* speculated that adaptation to high pressure is an essential characteristic of deep water gammarids in Lake Baikal. The response of these organisms suggested that the adaptive mechanism probably involves changes in the properties of the central nervous system.

XIV. BEHAVIORAL ADAPTATIONS

Since animals depend on light for orientation, diurnal migration, and synchrony of rhythmic activities, light reception is probably the most important sensory modality in the exploration of the environment. Response to light is generally considered to be one of the most important factors to enable free-swimming organisms favorable placement in the water column. One animal, the freshwater cladoceran, *Daphnia,* depends on both photokinetic and phototactic responses in its migration (Harris and Mason, 1956). The migratory movement is independent of the direction of light and is entirely photokinetic. That is, light stimulates locomotion at a certain intensity irrespective of the direction of light. At high intensities, however, their response is phototactic and the light source governs the direction of locomotion. When light intensities are reduced, *Daphnia* moves toward the light source; when light intensities increase beyond a certain point, the animals move away from it. Blinded animals will respond photokinetically, but not phototactically. For directional responses, compound eyes are necessary; dermal responses are necessary for the kinetic ones.

There is also evidence that the endogenous activity changes are linked with high photosensitivity, explaining migratory patterns in some species. In a study on the calanoid copepod, *Pseudodiaptomus hessei,* in a subtropical lake in southern Africa, Hart and Allanson (1976) demonstrated that all stages in the life cycle showed a pronounced diel migration. The landlocked freshwater lake, Lake Sibaya, has a maximum depth of 40 m, and a mean depth of 13 m, with seasonal temperatures ranging from 18° to 28°C. Permanent stratification does not exist. Nighttime vertical migration of adult *P. hessei* appeared to be related to lunar conditions; that is, on bright moonlit nights these copepods occurred in deeper parts of the lake, while on dark, moonless nights they were found near the surface. During the daylight hours, adults occurred near the bottom of the lake. Observed differences in distribution of the larval stages suggest that as the instars age, they become increasingly photosensitive. Thus, any given instar lives by day at the depth at which it is just able to perceive light. Diurnal distribution patterns of all stages of *P. hessei* can be associated to changes in light intensity.

Changing light conditions also have been shown to be a factor in diurnal drift rates of stream dwelling organisms. The drift rate of the crustacean *Gammarus limnaeus* was found to increase sharply after sunset and fall sharply at sunrise (Waters, 1962).

Seasonal migrations of lake dwelling crustaceans have been linked to behavioral responses to decreasing light and decreasing temperature, as illustrated by the crayfish, *Pacifastacus leniusculus,* in Lake Tahoe (Flint, 1977). During summer and autumn, crayfish are found in shallow waters, but in autumn, as water temperature and solar radiation decrease, the adult population migrates into deeper waters. This migration is adaptive in that it enables the species to avoid the winter storms, which are particularly destructive to animals living in shallow waters.

Despite the fact that cave-dwelling crayfish have greatly reduced visual structures, many species have the capability of responding to light. In the crayfish, *Cambarus setosus,* it has been established that the lamina ganglionaris and the retinal elements are essentially nonexistent (Fingerman *et al.,* 1964). Yet, when the head of this animal is illuminated, it exhibits a locomotor response (Larimer, 1966). Because this response is present even when the eye sockets are removed, either a dermal light sense or a central nervous system receptor must be present. In a series of studies on both surface- and cave-dwelling species of crayfish, *Procambarus* (Mellon *et al.,* 1976; Mellon, 1977; Mellon and Lorton, 1977), it was reported that in a cave-dwelling species, *P. erythrops,* neither the eye muscles nor the motor neurons exhibited any growth abnormalities, but all of the eye motor structures in this species were very much reduced in size. In a subsequent paper, Mellon and Lnenicka (1980) compared the electrical and anatomical parameters of fibers in two eye muscles of surface and cave-dwelling crayfish. Although the muscles of the cave-dwelling species contained fewer fibers of smaller diameter and had a smaller cross-sectional area, the current voltage relationships were similar in both surface and cave-dwelling species. The authors speculated that the value of the ocular motor system to *P. erythrops* may be in spacial orientation.

Behavioral responses play a major role in Crustacea living in temporary pools and creeks. A good illustration of such adaptations can be found in the work of Styron and Burbanck (1967) and Styron (1968) on the asellid isopod, *Lirceus fontinalis,* in which the observed behavior patterns are associated with moisture, flowing water, and temperature. Under favorable conditions these organisms are photonegative, only rarely venturing from the underside of rocks. As the water in a temporary pool recedes through evaporation, the isopods maintain their preferred location by a turning reaction which keeps them in the receding water. As the water continues to recede, they aggregate in a torpid condition in or on top of moist sand or rock. In the creeks, which offer a more stable environment than the ponds, the animals show a rheotactic response to rapidly flowing water, which enables the isopods to move down the creek and repopulate the temporary pools following drought and high temperature. This rheotactic response, however, is negated when water temperatures are in excess of 26°C.

Stream-dwelling animals often show behavioral responses which keep them from being washed into unfavorable environments. Studies on stream dwelling populations of the amphipod *Crangonyx antennatus* have shown that these animals demonstrate a strong thigmotactic cryptic behavior pattern (Dickson, 1977). At all times they maintain contact with the substratum and are usually in the areas of rocks and spaces between benthic gravel. Populations of *C. antennatus* living in pools did not display the same behavior, but instead moved freely over the entire accessible area both by walking on the substrate and by swimming. The hypothesis that these differences represent behavioral adaptations to specific environments was supported by laboratory studies in which lower wash out rates of stream dwelling *C. antennatus* were observed in comparison with pool-inhabiting organisms.

XV. SUMMARY

Freshwater crustaceans occupy many diverse habitats and have evolved a wide range of adaptations enabling them to meet the unique environmental stresses which they confront. Although an increasing interest in studying the functional responses of these various crustaceans has resulted in a marked upsurge in published papers, it is apparent that many topics have only been addressed in a preliminary manner. With the increasing pressures for use of freshwater resources, to say that there is a pressing need to more fully understand the adaptive capacity of freshwater organisms is an understatement. Studies in recent years have sought to fill this information void, especially investigation of multiple environmental interactions which more closely simulate the animals' natural environments. Future investigations should focus on comparative data concerning the adaptive mechanisms of both freshwater and marine crustaceans. This effort is needed because of a tendency for scientists studying either freshwater or marine crustaceans to overlook the research findings of the other. The result is often a needless duplication of efforts or a lack of general understanding of basic mechanisms. For example, are the functional responses to temperature of freshwater species different from those of marine organisms?

The future for research on mechanisms of adaptation is very promising and undoubtedly major advances will be made through a combination of ecological and physiological–biochemical approaches.

REFERENCES

Ambore, N. E., and Venkatachari, S. A. T. (1978). Respiratory metabolism in relation to body size, sex and gill area of the freshwater crab *Barytelphusa guerini* Milne Edwards. *Indian J. Exp. Biol.* **16,** 465–472.

Armstrong, D. A., Strange, K., Corwe, J., Knight, A., and Simmons, M. (1981). High salinity acclimation by the prawn *Macrobrachium rosenbergii:* uptake of exogenous ammonia and changes in endogenous nitrogen compounds. *Biol. Bull. (Woods Hole, Mass.)* **160,** 349–365.

Armitage K. B., and Lei, C. H. (1979). Temperature acclimatization in the filtering rates and oxygen consumption of *Daphnia ambigua* Scourfield. *Comp. Biochem. Physiol.* **62A,** 807–812.

Aston, R. J., and Milner, A. G. P. (1980). A comparison of populations of the iopod *Asellus aquaticus* above and below power stations in organically polluted reaches of the River Trent. *Fresh. Biol.* **10,** 1–14.

Bowers, J. A. (1980). Feeding habits of *Diaptomus ashlandi* and *Diaptomus sicilis* in Lake Michigan. *Int. Rev. Ges. Hydrobiol.* **65,** 259–267.

Boulton, A. P., and Huggins, A. K. (1977). Biochemical changes occurring during morphogenesis of the brine shrimp *Artemia salina* and the effect of alterations in salinity. *Comp. Biochem. Physiol.* **57A,** 17–22.

Brauer, R. W., Bekman, M. Y., Keyser, J. B., Nesbitt, D. L., Sidelev, G. N., and Wright, S. L. (1980). Adaptation to high hydrostatic pressures of abyssal gammarids from Lake Baikal in Eastern Siberia. *Comp. Biochem. Physiol.* **65A,** 109–117.

Brick, R. W., and Ahern, G. A. (1978). Lysine transport across the mucosal border of the perfused midgut in the freshwater shrimp, *Macrobrachium rosenbergii. J. Comp. Physiol.* **124,** 169–179.

Buikema, A. L., Jr. (1973). Filtering rate of the cladoceran, *Daphnia pulex,* as a function of body size, light and acclimation. *Hydrobiologia* **41,** 515–527.

Burbanck, W. D., Edwards, J. P., and Burbanck, M. P. (1948). Toleration of lowered oxygen tension by cave and stream crayfish. *Ecology* **29,** 360–367.

Capart, A. (1951). *Thermobathynella adami,* gen. et spec. nov., Anaspidacé du Congo belge. *Inst. R. Sci. Nat. Belg. Bull.* **27,** 1–4.

Cheper, N. J. (1980). Thermal tolerance of the isopod *Lirceus brachyurus* (Crustacea: Isopoda). *Am. Midl. Nat.* **104,** 312–318.

Cooley, J. M. (1978). The effect of temperature on the development of diapausing and subitaneous eggs in several freshwater copepods. *Crustaceana* **35,** 27–34.

Cossins, A. R., and Bowler, K. (1976). Resistance adaptation of the freshwater crayfish and thermal inactivation of membrane-bound enzymes. *J. Comp. Physiol.* **111,** 15–24.

Croghan, P. C., and Lockwood, A. P. M. (1968). Ionic regulation of the Baltic and fresh-water races of the isopod *Mesidotea (Saduria) entomon* (L.). *J. Exp. Biol.* **48,** 141–158.

DeMarch, B. G. E. (1977). The effects of photoperiod and temperature on the induction and termination of reproductive resting stage in the freshwater amphipod *Hyalella azteca* (Saussure). *Can. J. Zool.* **55,** 1595–1600.

DeMarch, B. G. E. (1978). The effects of constant and variable temperatures on the size, gorwth, and reproduction of the freshwater amphipod *Hyalella azteca* (Saussure). *Can. J. Zool.* **56,** 1801–1806.

Derouet, L. (1953). Étude comparée du métabolisme respiratoire chez certaines espèces de crustacés cavernicoles et épigés. *Notes Biospéléologiques* **8,** 103–109.

Descouturelle, G. (1979). Existence de seuils dans la regulation thermique du cycle d'intermue chez la crevette D'eau douce *Atyaephyra desmarestii desmarestii* (Millet, 1831) Decapoda, Caridea). *Crustaceana* **5,** 167–175.

Dickson, G. W. (1977). Behavioral adaptation of the troglobitic amphipod crustacean *Crangonyx antennatus* to stream habitats. *Hydrobiologia* **56,** 17–20.

Dickson, G. W., and Franz, R. (1980). Respiration rates, ATP turnover and adenylate energy charge in excised gills of surface and cave crayfish. *Comp. Biochem. Physiol. A Comp. Physiol.* **65,** 375–380.

Diwan, A. D. (1975). Morpho-histology of the respiratory system of the freshwater crab, *Barytelphusa cunicularis* (Westwood, 1836). *Indian J. Zool. 16,* 11–20.

Diwan, A. D. and Nagabhushanam, R. (1976). Studies on heat tolerance in the freshwater crab, *Barytelphusa cunicularis* (Westwood, 1836). *Hydrobiologia* **50,** 65–70.

Dorgelo, J. (1977). Comparative ecophysiology of gammarids (Crustacea: Amphipoda) from marine, brackish, and fresh water habitats exposed to the influence of salinity-temperature combinations. IV. Blood sodium regulation. *Netherlands J. Sea Research.* **11,** 184–199.

Engelmann, M. D. (1966). Energetics, terrestrial field studies, and animal productivity. *Adv. Ecol. Res.* **3,** 73–115.

Epp, R. W., and Lewis, W. M., Jr. (1979). Metabolic responses to temperature change in a tropical freshwater copepod (*Mesocyclops brasilianus*) and their adaptive significance. *Oecologia* **42,** 123–189.

Farkas, T. (1979). Adaptation of fatty acid compositions to temperature—A study on planktonic crustaceans. *Comp. Biochem. Physiol.* **64B,** 71–76.

Fingerman, M., Oguro, C., Miyawaki, M., and McKinnell, R. G. (1964). The neurosecretory system in the head of the blind cave crayfish *Cambarus setosus* Faxon. *Amer. Midland Naturalist* **71,** 415–421.

Flint, R. W. (1977). Seasonal activity, migration and distribution of the crayfish, *Pacifastacus leniusculus,* in Lake Tahoe. *Am. Midland Naturalist* **97,** 280–292.

Fox, H. M., and Baldes, E. J. (1935). The vapour pressures of the blood of arthropoda from swift and still fresh waters. *J. Exp. Biol.* **12,** 174–178.

Fraser, J., Parkin, D. T., and Verspoor, E. (1978). Tolerance to lead in the freshwater isopod *Asellus aquaticus. Water Research.* **12,** 637–641.

Gerritsen, J. (1980). Sex and parthenogenesis in sparse populations. *Am. Naturalist* **115,** 718–742.

Gilles, R. (ed.). (1979). "Mechanisms of Osmoregulation in Animals; Maintenance of Cell Volume." 667 pp. John Wiley, New York.

Gophen, M. (1976). Temperature dependence of food intake, ammonia excretion and respiration in *Ceriodaphnia reticulata* (Jurine) (Lake Kinneret, Israel). Freshwater Biol. **6,** 451–455.

Gophen, M. (1977). Food and feeding habits of *Mesocyclops leuckarti* (Claus) in Lake Kinneret (Israel). *Freshwater Biol.* **7,** 513–518.

Green, J. D., and Chapman, M. A. (1977). Temperature effects on oxygen consumption by the copepod *Boeckella dilatata*. *N. Z. J. Mar. Freshwater Res.* **11,** 375–382.

Harris, J. E., and Mason, P. (1956). Vertical migration in eyeless *Daphnia*. *Proc. R. Soc. London, B* **145,** 280–290.

Harris, R. R., and Micallef, H. (1971). Osmotic and ionic regulation in *Potamon edulis,* a freshwater crab from Malta. *Comp. Biochem. Physiol.* **38A,** 769–776.

Hart, R. C., and Allanson, B. R. (1976). The distribution and diel vertical migration of *Pseudodiaptomus hessei* (Mrázek) (Calanoida: Copepoda) in a subtropical lake in southern Africa. *Freshwater Biol.* **6,** 183–198.

Hazel, J. Z., and Prosser, C. L. (1974). Molecular mechanism of temperature compensation in poikilotherms. *Physiol. Rev.* **54,** 620–677.

Hebert, P. D. N. (1978). Cyclomorphosis in natural populations of *Daphnia cephalata* King. *Freshwater Biol.* **8,** 79–90.

Holsinger, J. R., and Dickson, G. W. (1977). Burrowing as a means of survival in the troglobitic amphipod crustacean *Crangonyx antennatus* Packard (Crangonyctidae). *Hydrobiologia* **54,** 195–199.

Hutchinson, G. E. (1951). Copepodology for the ornithologist. *Ecology* **32,** 571–577.

Ivanova, M. B., and Klekowski, R. Z. (1972). Respiratory and filtration rates in *Simocephalus vetulus* (O. F. Muller) (Cladocera) at different pH. *Pol. Arch. Hydrobiol.* **19,** 303–3±9.

Kerfoot, W. C. (1977). Implications of copepod predation. *Limnol. Oceanog.* **22,** 316–325.

Khlebovich, V. V. (1969). Aspects of animal evolution related to critical salinity and internal state. *Mar. Biol.* **2,** 338–345.

Kivivuori, L. (1980). Effects of temperature and temperature acclimation on the motor and neural functions in the crayfish *Astacus astacus* L. *Comp. Biochem. physiol.* **65A,** 297–304.

Kotaiah, K., and Rajabai, B. S. (1972). Influence of thermal acclimation on oxygen consumption in the tropical freshwater crab, *Paratelphusa hydrodromus* (Herbst), with reference to size, sex, and sudden changes of temperature. *Indian J. Exp. Biol.* **10,** 375–378.

Kring, R. L., and O'Brien, W. J. (1976). Effect of varying oxygen concentrations on the filtering rate of *Daphnia pulex*. *Ecology* **57,** 808–814.

Lampert, W. (1978). A field study on the dependence of the fecundity of *Daphnia* spec. on food concentrration. *Oecologia* **36,** 363–369.

Larimer, J. L. (1966). A functional caudal photoreceptor in blind cavernicolous crayfish. *Nature (London)* **210,** 204–205.

Laybourn-Parry, J., and Strachan, I. M. (1980). Respiratory metabolism of *Cyclops bicuspidatus* (*sensu stricta*) (Claus) (Copepoda: Cyclopoida) from Esthwaite Water, Cumbria. *Oecologia* **46,** 386–390.

Lei, C. H., and Armitage, K. B. (1980). Ecological energetics of a *Daphnia ambigua* population. *Hydrobiologia* **70,** 133–143.

Maloeuf, N. S. R. (1937). Studies on the respiration and osmoregulation of animals. *Ztschr. Vgl. Physiol.* **25,** 1–28.

Mellon, D. F. (1977). Retention of oculomotor reflexes in blind cave-dwelling crayfish. *Brain Res.* **134,** 191–196.

Mellon, D. F., Jr., and Lnenicka, G. (1980). Structure and electrical properties of eye muscles in cave and surface-dwelling crayfishes. *J. Exp. Biol.* **84,** 187–199.

Mellon, D. F., and Lorton, E. D. (1977). Reflex actions of the functional divisions in the crayfish oculomotor system. *J. Comp. Physiol.* **121,** 367–380.

Mellon, D. F., Tufty, R. M., and Lorton, E. D. (1976). Analysis of spatial constancy of oculomotor neurons in the crayfish. *Brain Res.* **109,** 587–594.

Moore, W. G. (1967). Factors affecting egg hatching in *Streptocephalus seali* (Branchiopoda: Anostraca). *Proc. Symp. Crustacea,* 724–735.

Morris, J. E. (1971). Hydration, its reversibility, and the beginning of development in the brine shrimp *Artemia salina. Comp. Biochem. Physiol.* **39A,** 843–857.

Moshiri, G. A., Goldman, C. R., Godshalk, G. L., and Mull, D. R. (1970). The effect of variations in oxygen tension on certain aspects of respiratory metabolism in *Pacifastacus leniusculus* (Dana) (Crustacea: Decapoda). *Physiol. Zool.* **43,** 23–29.

Nagabhushanam, R., Hanumante, M. M., Mirajkar, M., and Kulkarni, G. K. (1979). Effects of thermal stress on the neurosecretory cells of the freshwater prawn, *Macrobrachium kistnensis* (Tiwari). *Hydrobiologia* **62,** 269–273.

Nelson, S. F., Armstrong, D. A., Knight, A. W., and Li, H. W. (1977). The effect of temperature and salinity on the metabolic rate of juvenile *Macrobrachium rosenbergii* (CrustaceaL Palaemonidae). *Comp. Biochem. Physiol.* **56A,** 533–537.

Nilssen, J. P. (1980).Acidification of a small watershed in southern Norway and some characteristics of acidic aquatic environments. *Int. Rev. Gesamten Hydrobiol.* **65,** 177–207.

O'Brien, W. J., and Vinyard, G. L. (1978). Polymorphism and predation: The effect of invertebrate predation on the distribution of two varieties of *Daphnia carinata* in South India ponds. *Limonol. Oceanogr.* **23,** 452–460.

Økland, K. A. (1978). Life history and growth of *Asellus aquaticus* (L.) in relation to environment in a Eutrophic lake in Norway. *Hydrobiologia* **59,** 243–259.

Ostapenya, A. P., Pechen, G. A., Babitskiy, V. A., and Pavlyutin, A. P. (1969). Metabolic rate of *Diaptomus graciloides* (Lill.) at low temperature. Hydrobiologia **5,** 88–91.

Petrusewicz, K., and Macfadyen, A. (1970). "Productivity of Terrestrial Animals: Principles and Methods." IBP Handbook No. 13. Blackwell Scientific Publ., Oxford.

Pillai, C. K. (1982). Studies on the breeding biology and egg development in a freshwater crab *Paratelphusa hydrodromous* (Herbst). Ph.D. Dissertation. Univ. Madras, India.

Ponomareva, S. A. (1975). Effect of salinity on the fresh-water shrimp *Dikerogammarus haemobaphes* (Eichwald) from the mouth of the Dnepr. *Hydrobiological J.* (U.S.S.R.) **11,** 67–68.

Potts, W. T. W., and Durning, C. T. (1980). Physiological evolution in the *Branchiopods. Comp. Biochem. Physiol.* **67B,** 475–484.

Rajyalakshmi, T. (1980). Comparative study of the biology of the fresh-water prawn *Macrobranchium malcolmsonii* of Godavari and Hoogty river system. *Proc. Indian Natl. Sci. Acad., Sec. B* **1,** 72–89.

Reid, G. K., and Wood, R. D. (1976). "Ecology of Inland Waters and Estuaries." 2nd Ed. Van Nostrand, New York.

Remane, A. (1934). Die Brackwasserfauna. *Zool. Anz.* **7,** 34–74.

Remane, A., and Schlieper, C. (1971). "Biology of Brackish Water." Revised, 2nd Ed. Wiley, New York.

Ricker, W. E. (1968). "Methods for Assessment of Fish Production in Fresh Waters." IBP Handbook No. 3. Blackwell Scientific Publications, Oxford.

Roux, C. (1975). Les variations de la courbe metabolisme/temperature de *Gammarus pulex* (L.) et *Gammarus fossarum* koch (Amphipodes) sous L'influence de divers facteurs ecologiques. *Crustaceana* **29,** 33–48.

Ryther, J. H. (1954). Inhibitory effects of phytoplankton upon the feeding of *Daphnia magna* with reference to growth, reproduction and survival. *Ecology* **35,** 522–533.

Sam, S. T., and Krishnaswamy, S. (1979). Effect of osmomolarity of the medium upon hatching of undried eggs of *Streptocephalus dichotomus* Baird (Anostraca-Crustacea). Arch. Hydrobiol. **86,** 125–130.

Sarvala, J. (1979). Effect of temperature on the duration of egg, nauplius and copepodite development of some freshwater benthic Copepoda. *Freshwater Biol.* **9,** 515–534.

Sarviro, V. S. (1980). Temperature aspects of the ecology of the amphipod *Gammarus lacustris*. *Sov. J. Ecol.* **11,** 40–46.
Schlüter, U. (1979). Über die temperaturabhängigkeit des wachstums und des häutungszyklus von *Argulus foliaceus* (L.) (Branchiura). *Crustaceana* **37,** 100–106.
Sigmon, C. (1979). Oxygen consumption in *Daphnia pulex* exposed to 2,4-D or 2,4,5-T. *Bull. Environ. Contam. Toxicol.* **21,** 822–825.
Slobodkin, L. B. (1962). Energy in animal ecology. *Adv. Ecol. Res.* **1,** 69–101.
Smith, G. A., Fitzpatrick, L. C., and Pearson, W. D. (1978). Metabolic relations to temperatures in the copepods *Diaptomus dorsalis* and *Mesocyclops edax* from North-Central Texas. *Comp. Biochem. Physiol.* **59A,** 325–326.
Smith, R. L. (1966). "Ecology and Field Biology." Harper and Row, New York.
Smyly, W. J. P. (1980). Effect of constant and alternating temperatures on adult longevity of the freshwater cyclopoid copepod, *Acanthocyclops viridis* (Jurine). *Arch. Hydrobiol.* **89,** 353–362.
Sprague, J. B. (1963). Resistance of four freshwater crustaceans to lethal high temperature and low oxygen. *J. Fish. Res. Bd. Canada.* **20,** 387–415.
Strong, D. R. (1972). Life history variation among populations of an amphipod (*Hyalella azteca*). *Ecology* 53, 1103–1111.
Styron, C. E. (1968). Ecology of two populations of an aquatic isopod, *Lirceus fontinalis* Raf. *Ecology* **49,** 629–636.
Styron, C. E., and Burbanck, W. D. (1967). Ecology of an aquatic isopod, *Lirceus fontinalis* Raf., Emphasizing radiation effects. *Am. Midland Nat.* **78,** 389–415.
Tadini, G. V., and Valentino, F. (1969). Richerche sulla determinazione della stasi riproduttiva in varie razze geografiche *di'Asellus aquaticus'* (Crust. Isop.). *Atti Accad. Naz. Lincei Memone* **9,** 3–50.
Thompson, L. (1970). Osmoregulation of the fresh water crabs *Metopaulias depressus (Grapsidae)* and *Pseudothelphusa jouyi* (*Pseudothelphusidae*). Ph.D. Thesis, Univ. of California, Berkeley (Univ. Microfilms, Ann Arbor, Michigan).
Troiani, D. (1954). La consummation d'oxygene de quelques Gammaridae. *Compt. Rend.* **239,** 1540–1542.
Vandel, A. (1926). La reconnaissance sexuelle chez les Aselles. *Bull. Soc. Zool. Fr.* **51,** 163–172.
Vernberg, F. J. (1962). Comparative physiology: latitudinal effects on physiological properties of animal populations. *Ann. Rev. Physiol.* **24,** 517–546.
Vernberg, F. J. (1972). Dissolved gases: Animals. *In* "Marine Ecology" (O. Kinne, ed.), pp. 1491–1526. Wiley-Interscience, London.
Vernberg, F. J. and Vernberg, W. B. (1970). "The Animal and the Environment," pp. 397. Holt Rinehart and Winston, New York.
Waters, T. F. (1962). Diurnal periodicity in the drift of stream invertebrates. *Ecology* **43,** 316–320.
Welton, J. S. and Clarke, R. T. (1980). Laboratory studies on the reproduction and growth of the amphipod, *Gammarus pulex* (L.). *J. Anim. Ecol.* **49,** 581–592.
Wolsky, A. (1934). Sauerstoffverbrauch und körpergewicht beim steinkrebs (*Potamobius torrentium* [Schrank] Ortmann), nebst kritischen bemerkungen über die methoden der bestimmung des sauerstoffrerbrauches. *Arb. Ung. Biol. Forsch. Inst.* **7,** 116–128.
Wulff, F. W. (1980). Animal community structure and energy budget calculations of *Daphnia magna* (Straus) population in relation to the rock pool environment. *Ecol. Model* **11,** 179–225.

Systematic Index—Crustacea*

*Note: Names that have been superseded appear in brackets. Parentheses around name of author of scientific name indicate that currently assigned genus is not the original one.

†For current taxonomic status of *Artemia*, see Persoone, G., P. Sorgeloos, O. Roels, and E. Jaspers (Eds.). "The Brine Shrimp *Artemia*," Vol. 1, Morphology, Genetics, Radiobiology, Toxicology, 1980. Wetteren, Belgium, Universa Press.

‡Editor's note: The crab *Cancer hydrodromus* Herbst, 1794 has over the years been placed in several different genera, including *Parathelphusa,* which incorrectly is also spelled *Paratelphusa.* In 1968, Bott erected the genus *Spiralothelphusa,* with *Cancer hydrodromus* as

---

the type species. The name *Spiralothelphusa hydrodroma* (Herbst, 1794) is now accepted by most taxonomic workers as the valid name. Yet *S. hydrodroma,* about 30 mm across the carapace, occurs in brackish water and is rather uncommon. In contrast, a distinct species, *Oziotelphusa senex* (Fabricius, 1798), which is about 45 mm across the carapace and superficially resembles *S. hydrodroma,* occurs commonly in fresh water (see Bott, R., 1970, Arkiv för Zoologi *22:* 631). The specific identity of crabs referred to *Parathelphusa hydrodromus* remains uncertain; it is possible that, in some cases, they may represent *O. senex* rather than *S. hydrodroma.*

Systematic Index—Non-Crustacea

Subject Index